Tracing Early Agriculture in the Highlands of New Guinea

In this book, historical narratives chart how people created forms of agriculture in the highlands of New Guinea and how these practices were transformed through time. The intention is twofold: to clearly establish New Guinea as a region of early agricultural development and plant domestication; and, to develop a contingent, practice-based interpretation of early agriculture that has broader application to other regions of the world.

The multidisciplinary record from the highlands has the potential to challenge and change long-held assumptions regarding early agriculture globally, which are usually based on domestication. Early agriculture in the highlands is charted by an exposition of the practices of plant exploitation and cultivation. Practices are ontologically prior because they ultimately produce the phenotypic and genotypic changes in plant species characterised as domestication, as well as the social and environmental transformations associated with agriculture. They are also methodologically prior because they emplace plants in specific historico-geographic contexts.

This book is aimed at a new audience, one that is not necessarily familiar with the geographical and historical nuances of the island of New Guinea. It has relevance to archaeologists studying the transitions from foraging to farming, as well as to researchers and students across of range of disciplines (including agronomists, anthropologists, geneticists, geographers and linguists) working in the Indo-Pacific region.

Tim Denham is Reader/Associate Professor of Archaeology at the Australian National University. He has undertaken fieldwork in Papua New Guinea, mostly in the highland interior, since 1990. His primary research has focussed on plant exploitation and the emergence of agriculture in the highlands during the Holocene. He has also published on the Holocene histories of Island Southeast Asia and northern Australia. Over the last decade, his interests have diversified to include the domestication of vegetatively propagated crops, especially bananas; geoarchaeology and environmental change, mainly in the wet tropics; and the application of new technologies to archaeological questions.

UCL Institute of Archaeology Publications

General Editor: Ruth Whitehouse

Director of the Institute: Sue Hamilton
Founding Series Editor: Peter Ucko

The Institute of Archaeology of University College London is one of the oldest, largest and most prestigious archaeology research facilities in the world. Its extensive publications programme includes the best theory, research, pedagogy and reference materials in archaeology, cultural heritage and cognate disciplines, through publishing exemplary work of scholars worldwide. Through its publications, the Institute brings together key areas of theoretical and substantive knowledge, improves archaeological and heritage practice and brings archaeological findings to the general public, researchers and practitioners. It also publishes staff research projects, site and survey reports, ethnographic work and conference proceedings. The publications programme, formerly developed in-house or in conjunction firstly with UCL Press and then with Left Coast Press, is now produced in partnership with Routledge, a part of the Taylor & Francis group. Details of the Institute's 80-plus publications can be found at www.routledge.com/UCL-Institute-of-Archaeology-Publications/book-series/UCL .

The Institute's publications programme consists of two series: a General Series, reflecting the Institute's wide-ranging archaeological research; and a Critical Cultural Heritage Series, promoting research that differs radically from the existing canon of cultural heritage texts.

This volume is part of the Institute's General Series.

UCL Institute of Archaeology can be accessed on line at   www.ucl.ac.uk/archaeology.

Tracing Early Agriculture in the Highlands of New Guinea

Plot, Mound and Ditch

Tim Denham

Routledge
Taylor & Francis Group

LONDON AND NEW YORK

First published 2018 by Routledge

2 Park Square, Milton Park, Abingdon, Oxfordshire OX14 4RN
52 Vanderbilt Avenue, New York, NY 10017

Routledge is an imprint of the Taylor & Francis Group, an informa business

First issued in paperback 2020

British Library Cataloguing-in-Publication Data
A catalogue record for this book is available from the British Library

Library of Congress Cataloging-in-Publication Data
Names: Denham, Tim, author.
Title: Tracing early agriculture in the highlands of New Guinea : plot, mound and ditch /
   Tim Denham.
Description: Milton Park, Abingdon, Oxon ; New York, NY : Routledge, 2018. | Series:
   UCL Institute of Archaeology Publications | Includes bibliographical references and
   index.
Identifiers: LCCN 2017060352 (print) | LCCN 2018010497 (ebook) | ISBN
   9781351115308 (Master) | ISBN 9781351115285 (ePub) | ISBN 9781351115292
   (Web PDF) | ISBN 9781351115278 (Mobi/Kindle) | ISBN 9780815361817
   (hardback : alk. paper)
Subjects: LCSH: Agriculture—New Guinea—History.
Classification: LCC S479.3.N43 (ebook) | LCC S479.3.N43 D46 2018 (print) |
   DDC 338.10995—dc23
LC record available at https://lccn.loc.gov/2017060352

ISBN 13: 978−0−8153−6181−7 (hbk)
ISBN 13: 978−0−367−58907−3 (pbk)

Typeset in Bembo
by Apex CoVantage, LLC

To mum, who cultivated my mind
To dad, who fostered my interest in farming

Contents

Figures

Tables

Foreword

This book is aimed at a new audience, one that is not familiar with the geographical and historical nuances of the New Guinea region, especially the mountainous interior. It is intended to move the multidisciplinary record derived from the highlands to a more central position in global debates on early agriculture, as well as to challenge and change long-held assumptions. In this book, I build upon and extend previous publications on early agriculture in the highlands of Papua New Guinea. Unlike most previous work, which is heavily focussed on lines of evidence, this book presents new concepts and methods developed to investigate early agriculture in the highlands.

It is difficult to do justice to nearly twenty years of research on early agriculture in the highlands of Papua New Guinea. Here, I draw on a rich set of data to interpret the practices of past cultivation in new ways. In this book, I initially trace the intellectual lineage from which my own conceptual framework emerged (Chapter 2). I introduce the reader to an understanding of the highlands as a place through historical, archaeological and geographical overviews (Chapter 3). I then discuss at length the practice-based method applied to the multidisciplinary investigation of cultivation practices in the past (Chapter 4), as well as the key plants of highland subsistence (Chapter 5). Historical narratives chart plant exploitation in the Pleistocene (Chapter 6), ambiguous traces of potentially novel cultivation practices (Chapter 7) and inferences of shifting cultivation during the early Holocene (Chapter 8), to clear traces of mound cultivation during the mid-Holocene (Chapter 9), ditched drainage in the late Holocene (Chapter 10) and subsequent innovations (Chapter 11). The last two chapters assess the resilience of agricultural practices in the highlands, especially given the social and environmental challenges faced by people living there today (Chapter 12), and the broader significance of early agriculture in the New Guinea highlands to world archaeology (Chapter 13).

Necessarily, various lines of evidence and previously published ideas are drawn upon and recast here, including a range of book chapters and journal articles. For the first time in this book, these disparate publications are brought together into a coherent and fluent narrative. Several chapters draw on and sometimes incorporate previously published sections of text, as well as figures and tables from numerous sources. The following sections in this book are reproduced, or incorporate amended sections of text, from previously published sources: Chapter 2 'Low-level food production and the "middle ground"' and 'The conditions of growth: a post-processual turn?' (from Denham 2007b: 4–8 and 8–11, respectively), 'Social dependence and environmental transformation' (from Denham 2007a: 94–97) and 'Towards a contingent conception of early agriculture' (from Denham 2011: S379–S383; courtesy of Wenner Gren Foundation); Chapter 4 'Ambiguity of past practices: questions of archaeological visibility' (from Denham 2005c: 293–294); Chapter 5 'Staple crops' (from Denham 2017: 43–48); Chapter 6 '*Pandanus* species in the highlands', 'More than hunting and the seasonal exploitation of *Pandanus*', and 'Rethinking occupation of the interior during the Pleistocene' (from Denham 2007c: 42–43, 43–45 and 45–46, respectively; courtesy of *Archaeology in Oceania*); Chapter 11 '*Casuarina* tree fallowing' and 'Sweet potato, pigs and big men' (from Denham 2013b: 114 and 114–115, respectively); and Chapter 13 'Plant exploitation in the tropics is different' (from Denham 2016b: 412–413). Although these chapters may include previously published sections of text, these are incorporated into up-to-date interpretations that follow long periods of gestation.

Acknowledgements

Over the course of my intellectual development, I am indebted to several mentors who have patiently guided me, despite my often wayward tendencies. In geography these are Michael Raw (Bradford Grammar School), Derek Gregory (Cambridge University), Peter Gould (Pennsylvania State University) and Peter Kershaw (Monash University). In archaeology they are Jack Golson and Matthew Spriggs (Australian National University) and Robin Torrence (Australian Museum). I have also benefitted from the intellectual input over many years from a number of colleagues and collaborators, principally Huw Barton, Chris Ballard, Mike Bourke, Mark Donohue, Simon Haberle and Geoff Hope.

As an undergraduate in geography in the mid-1980s, Tim Bayliss-Smith (Cambridge University) first piqued my interest in the highlands of Papua New Guinea. At that time, I met John Muke, a then PhD student in archaeology at Cambridge, who was from Tumba in the Middle Wahgi Valley and who has become a long-term colleague and friend on various fieldwork projects in Papua New Guinea. However, it was not until 1997, when I came to the Australian National University to start a PhD, that my research on early agriculture in the highlands started. I was originally recruited to undertake the PhD by Matthew Spriggs, whom I met while working as a contract archaeologist in Hawai'i during the early 1990s.

Since starting the PhD, my greatest intellectual debt is to Jack Golson, who guided me through the labyrinthine archaeological records of early agriculture in the highlands, principally from Kuk Swamp. Jack started excavating wetland sites in the highlands over fifty years ago; my research is a continuation of his work. I also thank Jack for unfettered access to, and use of, the Kuk archive.

The research presented here was funded by an International Postgraduate Research Scholarship (for my PhD at the ANU), a Monash Research Fellowship and three Australian Research Council Grants (DP0666524, DP1093191 and FT150100420). It is not possible to individually thank all the people who contributed, in some way, to this research, as there are so many. I would like to acknowledge the cartographers at the ANU (Kay Dancey, Jenny Sheehan and Karina Pelling) and Monash (Phil Scamp and Kara Rasmanis) who drafted most of the images used in this book. I also thank Mike Bourke and Robin Hide for sharing images and their expertise on agricultural practices and plants in New Guinea. All previously published and unpublished images are used with permission of the authors or originator, including Chris Ballard, Tim Bayliss-Smith, Mike Bourke, Richard Fullagar, Jack Golson, Simon Haberle, Robin Hide, Carol Lentfer, Peter Matthews, Mary-Jane Mountain, Jocelyn Powell, Glenn Summerhayes and Peter White.

I thank David Harris for originally suggesting that I translate my PhD research into a book, as well as Marion Cutting and Ruth Whitehouse at UCL Press who have been extremely patient in seeing this manuscript through to publication. The input of two, not-so-anonymous reviewers has greatly improved the final product. I also acknowledge my wife, Louise, who has put up with me while I completed this book, together with my children – Violet, Gabriel, Audrey and Bernadette.

Lastly, I thank all the communities and people across mainland Papua New Guinea who looked after me on numerous field trips to the country over a twenty-year period, especially the Kawelka at Kuk.

Part I

Rethinking early agriculture

1 Early agriculture in the highlands

An unexpected story

In many ways, the story of early agriculture in the highlands of New Guinea is unexpected. At the time of 'discovery' in 1933, few would have anticipated that the highlands would yield some of the earliest evidence for plant cultivation in the world. Similarly, few considered the New Guinea region to be a centre of plant domestication, including for globally significant crops such as bananas (*Musa* spp.) and sugarcane *(Saccharum officinarum)*. In contrast to other centres of early agriculture, cultivation in the highlands is heavily based on vegetative propagation, rather than the planting of seed. Thus new concepts and methods have been developed to investigate the history of early vegetative cultivation in the highlands.

In this volume, historical narratives chart how people first developed agriculture in the highlands and how these practices have been transformed through time. The intention is twofold: to clearly establish New Guinea as a region of early agricultural development and plant domestication and to develop a contingent, practice-based interpretation of early agriculture that has broader application to other regions of the world.

Why is early agriculture in New Guinea contentious?

> Prejudices are not necessarily unjustified and erroneous, so that they inevitably distort the truth. In fact, the historicity of our existence entails that prejudices, in the literal sense of the word, constitute the initial directedness of our whole ability to experience. Prejudices are biases of our openness to the world. They are simply conditions whereby we experience something – whereby what we encounter says something to us.
>
> (Gadamer 1976: 9)

Why do some find it so difficult to accept that agriculture was an early, seemingly independent development in New Guinea? Are these problems to do with the evidence, or do they reflect deeper presuppositions about agriculture, its origins and social forms? Conceptions of early agriculture persistently draw on packages of accompanying traits that usually include cereals, domesticated animals, pottery, sedentism, burial practices, storage and the control of surplus, the development of socio-political hierarchies and eventually the emergence of 'civilisation' (from Childe 1936 to Bellwood 2005). As will be shown in this book, none of these traits has a bearing on understanding the emergence of agriculture in the highlands of New Guinea. Indeed, these proxies arguably reflect little about early agriculture and much about the cultural traditions, education and experiences of those who study it (Denham 2006).

How can a single definition or concept embrace the diversity of early agricultural practices in different regions in the past? Can a single definition lend equal significance to seed-based cultivation in semi-arid regions of Southwest Asia and to vegetative propagation in the wet tropical rainforests of New Guinea? Are cultivation practices in open, arable fields comparable to horticultural plots in tropical rainforest? Are the cumulative effects of prolonged cultivation on plants, or domesticatory processes, similar for sexually and asexually (clonally) reproduced plants? Are the social associations (such as sedentism, material culture and socio-political stratification) and environmental transformations (such as burning, environmental degradation and forest clearance) likely to be equivalent in different regions given diverse cultural and geographical contexts? Namely, should we use the same measures to compare multidisciplinary evidence for past cultivation practices and domesticatory relationships between New Guinea and, for example, Southwest Asia?

Several aspects of agricultural practices and historical representations of the island seem to have prejudiced opinion against acceptance of New Guinea as a locus of early agriculture and plant domestication. There are lingering perceptions of the island as an historical 'backwater', of 'Stone Age' people forgotten in time (e.g., Blackwood 1950; Souter 1963; Gardner and Heider 1969). These views are reinforced by the perceived absence of developed socio-political hierarchies, excepting the 'big man' social institution (Sahlins 1967). Such perspectives echo rungs on the ladder of social evolution that have reverberated since the nineteenth century (see critiques in Golson 1977a; Denham 2005b; David and Denham 2006).

Early farmers in the highlands were probably not sedentary (Denham and Barton 2006). The claimed ages and floor plans for Pleistocene-age house structures at two sites in the highlands – Wañelek (S. Bulmer 1977a, 1991) and NFX (Watson and Cole 1977) – are open to question (Denham and Ballard 2003; Denham 2016a). Barring these two unsubstantiated claims, the earliest house structures are reliably dated to around the last 4500–4000 years in the highlands (Denham 2014). The antiquity of larger, nucleated settlements in the interior, as opposed to dispersed homesteads, is not known.

In terms of material culture, highlanders traditionally employed relatively simple tools for agriculture, primarily wooden digging sticks and spades, as well as an assortment of stone axes, adzes and hoes (Golson 1977b). Despite minor elaborations, the form and inferred function of many wooden tools have been relatively constant through time (Powell 1974; Gorecki 1978; Steensberg 1980; Golson and Steensberg 1985). Furthermore, metal and, in most inter-montane valleys, pottery were absent until the twentieth century. Although generally considered aceramic, pottery was manufactured and used in the eastern highlands within the last 3000 years (Swadling 1973; Huff 2016) and has been reported from 3430–2840 cal BP contexts at Wañelek (S. Bulmer 1977a, 1991; Gaffney, Summerhayes et al. 2015). The simple tools, static design, relative lack of pottery and absence of metal have fed notions of relative technological impoverishment when compared with other regions of early agricultural development globally.

Uniquely for a global centre of early agriculture, animal domestication did not occur on New Guinea. The principal animal domesticates of highland subsistence – chicken *(Gallus gallus)*, dog *(Canis familiaris)* and pig *(Sus scrofa)* – are of ultimate Eurasian origin. None of these domesticates was common in the highlands until the last 2000 years, and much more recently in some valleys (Sutton et al. 2009). Hence, the investigation of early agriculture in the highlands focuses on plants.

Other characteristics of agricultural practices on New Guinea are unusual, though not unique; they are shared by other tropical locations (Harris 1969, 1972; Piperno and Pearsall 1998). For instance, plant cultivation on New Guinea is predominantly vegetative rather than seed based. A wide range of plant types are cultivated vegetatively, including trees, palms and pandans, bananas and other herbs, shrubs, vegetables, grasses and root crops (Powell 1976; Sillitoe 1983; French 1986; Kennedy and Clarke 2004). Seed-based propagation and the consumption of grass seeds are only minor components of cultivation practices today.

If we step back from the New Guinea case for a moment, the transition to agriculture in the past is always likely to be tinged with uncertainty and open to contestation. We may strive to say whether something is agricultural or not, whereas the most significant questions are really what people were doing in the past and how those practices resonate with those in the present. This temporal interplay, or resonation between the past and present, is probably a better measure of whether ancient practices can be inferred to signify agriculture or not. Thus, it is essential to assess the multidisciplinary evidence for any region of early agricultural development on its own terms (Harris 1990, 2007). Namely, present-day agricultural practices within the highlands serve as a heuristic guide for the interpretation of multidisciplinary evidence in the past (Denham 2007a).

In turning to the highlands of New Guinea, the history of early agriculture may be considered an unusual story, which in comparison with other regions of the world is only beginning to be unravelled. The emerging multidisciplinary record for early agriculture is not out of step with characterisations of recent agricultural practices in the highlands (Clarke 1971; Powell et al. 1975; Sillitoe 1983; Bourke and Harwood 2009). If the idea is confronting, then reflect, lay any prejudices to one side and adopt an open stance. Perhaps early agriculture in New Guinea is unexpected and strange, yet it contributes to the global diversity of human experience and serves greatly to enrich the tapestry of human history.

Shifting perspective: a focus on practices

Hodder criticised attempts to provide an inclusive, ahistoric definition of agriculture and to determine origins in general. He claims:

> In all . . . cases the claim for origin is part of a wider narrative which ends with ourselves. Indeed, it could be said that it is the end point which has led to the focus on a particular starting point.
>
> (Hodder 1999: 175)

Following this line of reasoning, any definition of agriculture becomes determinate for the timing and location of its origins. In place of attempts to find ultimate origins, Hodder states:

> Contextual emphases on diversity, meaning, agency and contingency undermine any notions of 'the origin'. In a multivalent world, past and present, it becomes difficult to separate out a fixed 'thing' for which an origin can be

found. Rather than focussing on major transformations, it is possible to use archaeological data to gain an understanding of the indeterminate relations between large-scale processes and individual lives.

(Hodder 1999: 175)

Taking Hodder's criticism into account, upon what basis can an understanding of agriculture in New Guinea be made? Indeterminate relations can be characterised by what people were doing in the past, either individually or communally, within a given locale or landscape. One way of investigating indeterminate relations is through past practices that effectively link large-scale processes to meaningful activity (after Bourdieu 1990). In archaeology, practices represent human actions in the past, including habitual modes of behaviour and dispositions, as well as individual idiosyncrasies (Barrett 1994; Pauketat 2001; Denham 2005c, 2009; Denham and Haberle 2008; Bruno 2009; Jussuret 2010). Practices become inscribed in landscapes through time providing sequential evidence of human-environment interactions in the past.

From one perspective, cultivation practices are at the nexus of human-environment relations because they reconcile established dichotomies between social life, on the one hand, and the biosphere and geosphere, on the other (Denham and Haberle 2008: 483): 'Practices are structured by the environment within which they occur, while simultaneously acting upon and changing that environment'. From another perspective, the concept of practice resolves more abstract structure-agency dichotomies:

> The degrees to which someone's actions in the past, as in the present, were a product of structural determination (whether social, economic or mental), dispositions or individual improvisation are uncertain; initially, all we know is that something happened. The ultimate causes why something happened often remain hidden for an event in the present, let alone something that occurred millennia in the past.

(Denham and Haberle 2008: 483)

In the study of early agriculture, the concept of practice focusses analysis on what happened in the past, rather than upon more speculative interpretations of why something happened.

In an everyday sense, practices are what people do, or did, and comprise the multilayered constituent practices of different forms of plant exploitation, including gathering, management and cultivation. The adoption of a practice-based approach is intended to unpack 'monolithic' categories such as foraging, swidden cultivation and intensive cultivation as they are applied to the history of plant exploitation in New Guinea (Denham 2005c, 2007a, 2008a, 2009, 2011; Denham and Haberle 2008) and neighbouring regions (Latinis 2000; Terrell et al. 2003; Denham 2008b; Denham, Fullagar et al. 2009).

A focus on practices enables the continuities and discontinuities between forms of plant exploitation to be traced in space and time, and illustrates the ways practices were 'bundled', or deployed in conjunction with one another, by people living in specific landscapes in the past (Denham 2007a, 2009, 2011; Denham and Haberle 2008). Essentially, a practice-based framework is intended to reflect the porous and fluid character of early cultivation practices as they are reconstructed in the multidisciplinary record for a given locale.

Archaeology lends itself to a practice-based interpretation because it examines the evidence of people's actions in the past. By adopting a practice-based framework to investigate the emergence of agriculture, the analytical focus shifts away from archaeobotanically grounded narratives that emphasise domestication traits (from Harris 1989 to Smith 2001 and onwards) towards an understanding of how plants are enmeshed in cultivation practices in the past. Arguably, practices of cultivation are more fundamental, or prior, than subsequent effects such as domestication traits in animals and plants or environmental change. Thus, a practice-based framework for the investigation of early agriculture is adopted here in preference to more widely applied plant-based and domestication-focussed frameworks.

In the first part of this book, *Rethinking Early Agriculture* (Chapters 1 and 2), the question of early agriculture in the highlands of New Guinea is raised as a concern. The intention of Chapter 1 is to ask the reader to put aside, or bracket, any preconceptions regarding early agriculture in New Guinea. The intellectual traditions that underpin the conception of early agriculture proposed in this book are presented in Chapter 2.

The second part of the book, *Places, Practices and Plants* (Chapters, 3, 4 and 5) presents background information on the highlands, cultivation practices and crops grown, respectively. The multidisciplinary evidence for early agriculture arises from a landscape or region – namely, a place with its own temporal and geographical circumstances (Chapter 3). The character of cultivation practices specific to the highlands of New Guinea today act as a guide to the identification of practices there in the past (Chapter 4). Subsequently, the crop plants of highland agriculture today are introduced, together with practices of exploitation and cultivation (Chapter 5).

An understanding of early agriculture in the highlands requires exposition of the practices of plant exploitation and cultivation before detailed consideration of domesticatory relationships between plants and people. Practices are ontologically prior because they ultimately produce the phenotypic and genotypic changes in plant species characterised as domestication, as well as the social and environmental transformations associated with agriculture. Practices are also methodologically prior because practices emplace plants in specific historico-geographic contexts.

In the third part of the book, *Practices in the Past* (Chapters 6–11), historical narratives chronicle plant exploitation practices from the Pleistocene up to recent agricultural innovations. The temporalising of practices is not intended to be teleological or unilinear. There is no historical necessity to the emergence of agriculture in the highlands; rather, it is born of contingency. Further, early agriculture does not seem to be characterised by major technological or social thresholds and sequential replacement of one form of plant exploitation by another; rather, early practices are often ambiguous and more suggestive of continuity and fluidity, with elaboration of an expanding repertoire through time.

In the last part of the book, *Taking a Broader View*, the wider implications of this work are considered in terms of the challenges facing agricultural populations in the highlands today (Chapter 12), as well as the significance of the New Guinea record for global debates concerning the emergence of agriculture (Chapter 13). Although the past is important in its own right, contemporary challenges spur consideration of the relevance of the past to present global concerns. Furthermore, overcoming prejudicial interpretations of the past is a small step towards overcoming prejudice in the present.

2 Defining early agriculture in New Guinea

> The published literature on 'agricultural origins' is characterized by a confusing multiplicity of terms for the conceptual categories that define our discourse. There is little agreement about what precisely is meant by such terms as agriculture, horticulture, cultivation, domestication and husbandry. This semantic confusion militates against clear thinking about the phenomena we investigate, leads to misunderstanding and can provoke unnecessary disputes over interpretation of the evidence.
>
> (Harris 1996a: 3)

Any discussion of early agriculture, whether in New Guinea or elsewhere, requires an explicit definition and consideration of a fundamental question: how do we define agriculture in the past and thereby differentiate 'agriculture' from 'non-agriculture'? Further, how do we identify early agriculture in the New Guinea context?

In this chapter, the conceptual framework used to define agriculture and to identify it in the multidisciplinary record of the highlands is outlined. Before outlining a new conceptual framework, sufficient consideration should be given to previous attempts to define early agriculture, because these provide the intellectual foundation upon which anything new rests. Foremost are David Harris's continuum of human-environment interactions, Bruce Smith's low-level food production, Matthew Spriggs's socio-environmental definition of dependence and Tim Ingold's post-processual turn (based on Denham 2006, 2007a, 2007b).

A continuum of human-environment interactions

In an attempt to overcome semantic confusion, Harris proposed terminology and a continuum of human subsistence strategies against which agriculture can be identified (Figure 2.1; Harris 1989, 1996a, 2007; Harris and Fuller 2014; see Asouti and Fuller 2013). His perspectives echo those of many researchers who consider there to have been a continuum between 'hunting-gathering-fishing' and 'agriculture' in both the present and the past (e.g., Ford 1985; Smith 1998b, 2001; cf. Bellwood 2005). To exemplify, there are great difficulties in trying to classify the lifestyles of many groups studied archaeologically and ethnographically as either 'hunting-gathering-fishing'/'forager' or 'agriculture'/'farmer', particularly in New Guinea and adjacent regions. A multiplicity of terminology has been generated to connote this 'middle ground' including 'domiculture' (Hynes and Chase 1982), 'incipient agriculture' (Ford 1985), 'complex hunter-gatherers' (Zvelebil 1986), 'transitional' and 'proto-agriculture' (Yen 1989) and 'hunter-horticulturalism' (Guddemi 1992) (see Roscoe 2002; Sillitoe 2002; Specht 2003). As a result, there have been recurrent attempts to clarify terminology and to define the middle ground of 'in-between' groups.

Two entwined processes underpin Harris's continuum: the manipulation of biotic resources eventually leading to domestication, and the transformation of natural ecosystems to agroecosystems (Harris 1989, 2007). Types of plant exploitation are differentiated on the basis of episodic intensification across developmental thresholds (Harris 1989). For example, a system in which wild plant production is dominant can be differentiated from agriculture because the latter is 'based largely or exclusively on the cultivation of domesticated plants' (Harris 1996a: 4).

For Harris, agriculture entails the propagation and cultivation of genotypic and phenotypic variants, that is, domesticated crops and animals, as well as the establishment of agro-ecosystems. Agriculture is associated with:

> such activities as soil preparation, the maintenance of soil fertility, weeding, seed selection and storage, and the exclusion of potential predators attracted by the enlarged food-storage organs of domesticated plants.
>
> (Harris 1989: 21–22)

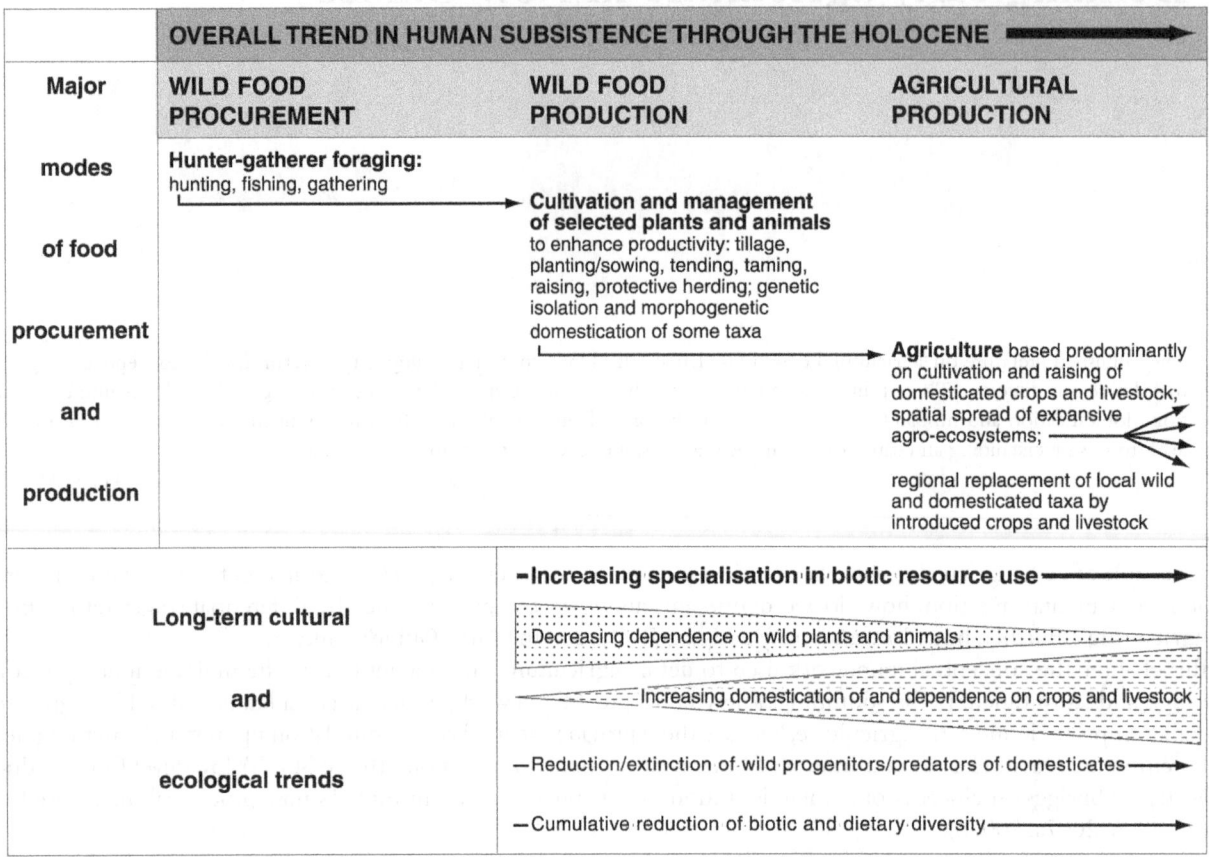

Figure 2.1 Harris's schematic representation of major transitions between modes of food procurement and production, together with long-term trends in human subsistence during the Holocene

Note: Harris discussed these transitions with respect to animals and plants, although only plants are relevant to the New Guinea context

Source: Harris 2007: Figure 2.1

In sequential formulations, his perspective became increasingly descriptive and non-directional (Harris 1989, 1996a, 2007), yet he retained correspondences between ecological effects, socio-economic trends and food-yielding systems through time.

Agriculture represents the most significant energetic threshold within Harris's schema. Domesticated species require greater degrees of sustained human intervention for development and to varying degrees 'have become dependent upon human assistance for their survival' (Harris 1989: 19). Human intervention produces desirable morphogenetic, or domestication, traits in plants and animals that increase productivity, ease of processing and so on. Tracking the emergence of these traits through time using macrofossil, microfossil and molecular techniques enables domestication and hence agriculture to be identified in the past.

Several factors call into question the value of using morphogenetic markers to differentiate subsistence practices in the Pacific context. For some plants, there may good correspondences between genetic markers and morphological transformations to determine domestication (for example, see papers in Zeder et al. 2006), whereas for other species these correspondences may be less clear.

Domesticated cultivars need not always be clearly demarcated from wild plants, either in terms of the archaeobotanical record or in terms of extant species. For example, the differentiation of wild and domesticated yams in the African context is difficult using morphological and genetic criteria (Chikwendu and Okezie 1989; Dumont and Vernier 2000; Mignouna and Dansi 2003). Many vegetatively propagated plants in the Pacific context, as well as plants in other parts of the world, have been characterised as 'semi-domesticated' (Yen 1985), or as having 'degrees of domestication' (Caballero 2004). On the basis of a consideration of vegetative cultivation in the humid tropics, many plants need not develop characteristic signatures similar to those documented for cereal grains (Harris 1996b: 568; Piperno and Pearsall 1998: 8; Smith 2001: 16–17; Yen 1985), or as rapidly (Fuller et al. 2014), although some do (Piperno 1998; Piperno et al. 2000).

From this perspective, domestication is not an all-or-nothing event. Rather, domestication can occur at different rates, be non-linear and discontinuous and should be considered an ongoing process for most plant and animals species. Indeed, genotypic and phenotypic changes may have been initiated in some plants and animals as soon as plants and animals were targeted by people (Hather 1996), and perhaps even incidentally.

Low-level food production and the 'middle ground'

Bruce Smith has drawn on previous classifications by Harris (1989, 1996a), as well as Ford (1985) and Zvelebil (1996), to reconceptualise the 'middle ground' between forager (food procurement) and farmer (agriculture) (this section reproduced in amended form from Denham 2007b: 4–8). Multidisciplinary research findings in different regions of the world, including Smith's research in the Americas (Smith 1992, 1998b), document diverse and numerous groups who actively and consciously engage(d) in various practices of food production – including cultivation of wild and domesticated plants – that fall 'in-between' foraging (hunting-gathering-fishing) and farming (agriculture). Smith (1998a, 2001) terms the 'in-between' conceptual territory as 'low-level food production'. He further differentiates this heterogenous category into 'low-level food production with domesticates' and 'low-level food production without domesticates'. Thus, Smith proposes a tripartite division: food procurement, low-level food production (with and without domesticates) and agriculture (Smith 2001: 27, 34).

Like Harris in his later formulations, Smith does not consider the mere presence of domesticated plants or animals in an archaeologically derived assemblage to necessarily signify agriculture in the past. Although domestication no longer defines the boundary between foraging and farming in Smith's schema, it does remain a central concept,

> because it represents such a significant level and form of intervention by humans in the life cycle of plant and animal species, and also because it is so clearly visible and recognizable across considerable spans of space and time.
>
> (Smith 2001: 17)

Of most relevance, Smith views the transition from 'low-level food production with domesticates' to 'agriculture' to lie along a cline from 30% to 50% caloric dependence upon the contribution of domesticates to annual caloric budgets. The cline accounts for the various societies who produce(d) food using a combination of wild, minimally managed and domesticated species in a variety of cultural and environmental settings.

The conceptual differentiation of 'food procurement' and 'low-level food production without domesticates' is complex and difficult. For Smith, previous schemas differentiated food-procurers from food-producers based on the nature of their practices. Rather than there being a sharp boundary between the two lifeways, there are clines that attenuated in either realm:

> As a result, any efforts to determine where exactly to place societies in this complex boundary zone of intensive food procurement and low-level food production is not simply a matter of ascertaining the presence or absence of certain forms of life-cycle-intervention activities on the part of humans, but rather should include consideration of the intensity, intentionality, species focus, and total range of such activities that are present in a group's economic repertoire.
>
> (Smith 2001: 29)

Smith acknowledges that people in different regions followed diverse pathways of food production and, as a critique of teleological and deterministic arguments, he states that all types of food procurement and food production should be seen as 'stable solutions, as end points and destinations worthy of study in and of themselves' (Smith 2001: 24). The middle ground is not 'transitional'; it does not imply that people were inevitably developing from food procurement to low-level food production (and from 'without domesticates' to 'with domesticates'), and on to agriculture. Rather the middle ground represents a diverse array of 'successful long-term socioeconomic solutions, fine-tuned to a wide range of local cultural and environmental contexts' (Smith 2001: 34).

Social dependence and environmental transformation

The continued reliance on domestication to define agriculture in the past, which is evident in Harris's and Smith's schemas, has arguably given too much significance to one epiphenomenon of food production practices (this section reproduced in amended form from Denham 2007a: 94–97). A domestication focus has sometimes shifted the emphasis from studying 'what people were doing in the past' to a fixation on defining these practices solely in terms of their effects on plant and animal morphogenetics. Effects can be used to trace more fundamental causes, and morphogenetic transformations in animal and plant species under human selection are significant processes. However, morphogenetic transformations do not necessarily provide a reliable guide to the degree of involvement or dependence of people on different modes of subsistence, and are even more inappropriate in the tropics where biological signatures of vegetative cultivation may be different to those of seed-based practices.

Spriggs (1996: 525–526) modified Harris's schema for application to the Pacific context, a modification accepted in part by Harris (1995: 849–850). Drawing on Yen's (1985) observation that many Pacific cultigens are not fully domesticated today, Spriggs abandons Harris's reliance on domestication as a threshold demarcating cultivation and agriculture because it 'no longer seems relevant' (Spriggs 1996: 525). Spriggs replaces 'domestication' with,

> when dependence on agriculture began, defined here in terms of the creation of agro-ecosystems that limit subsistence choice because of environmental transformation or labour demands.
>
> (Spriggs 1996: 525)

Spriggs is refocussing the investigation of agriculture away from plant and animal morphogenetics onto 'human behaviour and organization' (Spriggs 1996: 525). This shift recognises that the investigation of agriculture in the past is primarily a study of people's subsistence in the past. It reflects an increasingly pervasive awareness that agriculture needs to be decoupled from domestication and to be considered in more social terms (Hather 1996: 548; Ingold 1996: 21; Thomas 1999: 7–33); namely, as 'a strategy that people have come to rely on for subsistence' (Hather 1996: 548).

According to Harris's and Smith's schemas, swidden cultivation would be considered agriculture only when accompanied by a reliance on domesticated plants. From that perspective, and given the reported incomplete domestication of some traditional cultivars of taro *(Colocasia esculenta)*, some yam species *(Dioscorea* spp.) and some bananas *(Musa* cvs.) in New Guinea (after Yen 1985), many contemporary cultivation practices in New Guinea would not be considered agricultural even though they patently are (Bourke and Harwood 2009). Swidden cultivation accompanied by extensive clearance and systematic tillage, with only a weak-to-moderate reliance on domesticated plants, would be considered a system of wild food plant production (after Harris 1996a: 4).

A fundamental problem with Harris's and Smith's schemas is the focus on plant domestication to the near exclusion of other spheres of human involvement, dependence and impacts on the environment. Such distinctions are not necessarily relevant to all historical contexts. A reliance on wild food resources characterised early forms of cultivation in Southwest Asia (after Hillman 1981: 189), the lowland neotropics (Piperno and Pearsall 1998: 8), lowland Britain (Thomas 1999: 24–25) and China (Fuller et al. 2009); yet these same societies are usually accepted as being agricultural.

For Spriggs, large-scale swidden cultivation is 'agriculture' if these activities were depended upon for subsistence and limited alternative subsistence choices due to 'environmental transformation or labour demands' (Spriggs 1996: 525). These criteria apply to many forms of swidden cultivation irrespective of the relative composition of wild and domesticated resources. Spriggs's scenario better accords with depictions of traditional cultivation practices in New Guinea and the Pacific, as well as the multidisciplinary evidence for them in the past.

At present, most of the social aspects accompanying subsistence practices in the early to mid-Holocene in the highlands are invisible archaeologically and, therefore, unknown (after Harris 1995: 849–850; Yen 1998: 164; Denham 2006). Consequently, palaeoecological signals of cultivated landscapes (Kennedy and Clarke 2004), domesticated environments (Yen 1989), domesticated landscapes (Terrell et al. 2003), social landscapes (Bayliss-Smith and Golson 1999) and agricultural landscapes (Haberle 2003) have been used to provide a guide to the nature and extent of different subsistence practices in the past and the likely levels of human dependence upon them (Pearsall 2007). As people increasingly transformed their environment through disturbance, degradation and clearance of forests, they limited the availability of wild plant and animal resources, thereby increasing their dependence upon cultivated food.

As with using morphogenetic markers of animals and plants, landscape transformation is just another effect, epiphenomenon or proxy record (Jones and Colledge 2001: 395) used to track a more fundamental cause, that is, agriculture. Problems of archaeological visibility and the identification of agricultural thresholds remain (Harris 1995: 850; Yen 1998: 164). As is the case with morphogenetic markers, there is no necessary correspondence between landscape transformation and agriculture. Just as non-agricultural practices exert selective pressures that may account for morphogenetic transformations in some tree species, for example, *Canarium* sp. (Yen 1996) and tubers and corms (Hather 1996), so too extensive alteration of vegetation communities in Australia is not associated with agricultural activities, for example, the Atherton Tablelands (Kershaw et al. 1997). Conversely, practices considered to represent early agriculture in Southwest Asia (Hillman 1996), the lowland neotropics (Piperno and Pearsall 1998) and lowland Britain (Thomas 1999: 7–33) did not produce major, synchronous palaeoecological signals of forest clearance. Hodder has drawn on asynchronies in plant domestication and environmental transformation to claim: 'It is increasingly difficult to identify any point within a 4000 year period at which agriculture "began" in the Near East' (1999: 175). Thus agriculture need not produce large-scale environmental change, and the latter need not indicate the former. Given these problems of inferring agriculture from its morphogenetic and palaeoecological signals, on what basis can a framework for the interpretation of early agriculture in New Guinea be made?

On the basis of these qualifications, it is possible to revise Harris's and Smith's schemas for application to the New Guinea context. A reliance on domesticated plants is replaced with a 'dependence on cultivation' as the differentiating criterion for agriculture. Certainly, plants in more intensive forms of vegetative cultivation within degraded inter-montane environments have undergone greater directed selection and genetic isolation than in less intensive practices, even though the precise degree of domestication may be unclear. The 'middle ground' becomes more defined – and includes various forms of indeterminate strategy, such as resource intensification, transplantation and planting – both in the present and the past. Some forms of cultivation lie in the middle ground, but only as incidental activities – whether based on wild plants or domesticates – that do not require a reorientation of social life focussed on cultivated plots and the rhythms of crop production. It is the reorientation of social life resulting from the dependence upon the cultivation of plants in plots that gives agriculture its distinctive signature.

The conditions of growth: a post-processual turn?

Until relatively recently, debates about the nature of agriculture, its origins and its diffusion remained isolated from broader conceptual developments within the discipline. Most participants in debates on early agriculture are situated squarely within processual traditions (e.g., Harris 1996a, 1996b, 2007; Piperno and Pearsall 1998; Smith 2001; Bellwood and Renfrew 2002; Bellwood 2005). Over the last decade, more post-processual perspectives on how to conceive and interpret agriculture, particularly early agricultural practices, have come to the fore (this section extracted and amended from Denham 2007b: 8–11).

As discussed, definitions of agriculture are increasingly being decoupled from the concepts of domestication (Hather 1996; Spriggs 1996) and sedentism, as well as other material cultural traits, for example, pottery (Thomas 1996b, 1999: 7–33). Several advocate more socially oriented interpretations in which agriculture, domestication and production are not divorced from broader realms of social life (e.g., Hastorf 1998, 2016; cf. Bender 1978). From this perspective, morphological and genetic evidence of domestication is not so significant for determining the presence of agriculture in the past. One such position has been eloquently argued by Ingold (1996: 21, 2000: 86) who defines agriculture in terms of 'the relative scope of human involvement in establishing the conditions for growth'.

In a thought-provoking essay, Ingold (1996, revised as 2000: 77–88) unravels, challenges and rejects several historically received ideas that underlie orthodox, domestication-based conceptions of agriculture. For Ingold, these ideas are 'embedded in a grand narrative of the human transcendence of nature, in which the domestication of plants and animals figures as the counterpart of the self-domestication of humanity in the process of civilisation' (2000: 77). Prominent among these ideas are a series of dichotomies that historically have permeated the distinction between hunting-gathering and agriculture (Figure 2.2).

SUBSISTENCE	PRIMARY ACTIVITY	WORLD-VIEW	EXPLICIT RATIONALE	RESOURCES	LOCALE	INTERPRETATIVE EMPHASIS
Hunting-gathering-fishing	Collection	Food is ready-made in nature	Plants grow themselves	Wild/Feral	Landscape	More natural
↕	↕	↕	↕	↕	↕	↕
Agriculture	Production	Food is made by people	Plants are grown by people	Cultivated/Domesticated	Plot/Garden	More social and historical

Figure 2.2 Homologues illustrating conceptual tensions between hunting-gathering-fishing practices and agriculture, as relevant to the highlands of New Guinea

Note: Although often portrayed as dichotomies, both tendencies co-occur in plant exploitation practices in the highlands

Source: Updated version of Denham 2007b: Figure 1.1 drawing upon ideas in Ingold 2000

Ingold's starting point is the division between collecting and production. A 'modern emphasis on production' attaches 'special significance . . . to the so-called "artificial selection" of plants and animals as the key criterion for distinguishing food-production from food-collection, and hence for determining the point of transition from hunting and gathering to agriculture and pastoralism' (Ingold 2000: 85). For Ingold, the idea of domestication implies that people 'make' an animal or plant in terms of some preconceived end, that is, it is teleological and implies some deliberate design. The relative duration and interlocking of the life cycles of people with animals and plants would seem to undermine the putative deliberateness of domestication except for the fastest-growing species, for example, annual and biennial plants and relatively short-lived animals. By contrast, given the lifespan of most species of tree is much greater than that of humans, the idea of planned intervention in the generation of domesticates is untenable. Ingold effectively reorients the analytic focus from domestication, that is, the production of genotypic and phenotypic variants, namely domesticates, to the 'conditions of growth', that is, the practices people engaged in to grow plants and rear animals.

He argues that the dichotomy between collection and production is the fundamental cause of problems in attempts to define agriculture and distinguish it from other forms of obtaining food:

> In terms of this dichotomy, human beings must *either* find their food ready-made in nature *or* make it themselves. Yet ask any farmer and he or she will say, with good cause, that the produce of the farm is no more made than it is found ready-made. It is *grown* . . . [yet] *what do we mean by growing things?* On the answer to this question must hinge the distinctions between gathering and cultivation, and between hunting and animal husbandry.
>
> (Ingold 2000: 85; emphasis in original)

As his argument develops, Ingold shifts the emphasis from 'making' to 'growing', and in doing so, he seeks to undermine the most basic of all Western dichotomies, that between nature and culture (see Lévi-Strauss 1969 and critique by Derrida 1978). Ingold does not envisage a distinction between social and natural worlds; he does not consider that people 'make', or inscribe themselves on nature either through the construction of landscape or through plant and animal domestication. Rather, humans and the environment are mutually constitutive for the transformation of the world; they grow together in 'a single, continuous field of relationships' (Ingold 2000: 87).

If Ingold's perspective is adopted and applied to studies of early agriculture, the analytical focus moves from documenting genotypic and phenotypic changes in plants and animals to studying the degrees to which people establish 'the conditions for growth' as a means of distinguishing 'regimes of plant and animal husbandry' (Ingold 2000: 86). Problems of application remain. In the example given by Ingold (2000: 85–86), garden cultivation clearly displays a greater degree of intervention than gathering plant foods, but there is still a lack of clarity in how the varying degrees of 'human intervention' are to be evaluated in more complex scenarios. The categorical demarcation between 'agriculture' and 'non-agriculture' still has to be made; namely, the line has to be drawn somewhere – whether based on plant or animal domestication or 'conditions of growth'.

One clear way in which Ingold's ideas are relevant to the New Guinea context is the distinction between plants that grow themselves and plants that are grown by people. Many people in New Guinea distinguish foods that are planted in plots or gardens, even if left to go feral, from those that are gathered in the landscape (Sillitoe 1983). People may be aware if the plant is wild or cultivated based on phenotype, even though the distinction between feral plants in old gardens and truly 'wild-types' is sometimes blurred and dependent on collective memory.

In Ingold's schema, agriculture is associated with plants grown by people in cultivated plots. Clearly this involves a greater intervention in the conditions of growth than occurs through gathering, management of wild plants and resource intensification in the landscape. Ambiguities remain in terms of measuring degrees of intervention and using them as a basis of categorical distinction. Nevertheless, Ingold's work marks a dramatic shift in the way agriculture is conceived. As yet, few studies have seriously attempted to directly apply his thinking, though his ideas have tangentially informed practice-centred, contingent and more ambiguous interpretations of early agriculture and arboriculture (e.g., Denham 2004b, 2009, 2011; Fairbairn 2005).

Towards a contingent conception of early agriculture

Animal and plant exploitation, including pastoralism and agriculture, are one of the most important subsets of human-environment relations, both in terms of social transformation (Bellwood 2005; Barker 2006) and environmental change (Ruddiman et al. 2015). As seen earlier, divergent perspectives have been proposed for the interpretation of agriculture in the past. How can this divergence be accommodated? Here, a broader conceptual framework is proposed that can be applied in different social and historical contexts. This framework is not atheoretical (contra Gremillion et al. 2014);

rather, it is rooted in an engagement with a broad range of ideas from the biological, environmental and social sciences (this section reproduced in amended form from Denham 2011: S379–S383).

The interpretation of agriculture in the past requires us to consider the multiplicity of factors that converge in any given human-environmental interaction, including those of the biophysical realm – such as climate, environment and the biology of cultivated plants and tended livestock – as well as the social realm – namely people and various facets of their practices, including cultures, societies and technologies. Inherently, agriculture has multiple socio-environmental dimensions that are mutually transformative; namely, they are historical and each acts upon and changes the other through time (e.g., from $Time_x$ to $Time_{x+1}$ in Figure 2.3). Understanding and encapsulating the totality of these dimensions is essential to characterise the emergence of agriculture in any given historico-geographical setting.

An inclusive conception of agriculture as a subset of human-environment interactions sheds a critical light on the nature of attempts to elicit a singular definition, or an ultimate 'cause', of agricultural development in the past. Any singular definition of agriculture will ultimately founder because it prioritises one epiphenomenon of human-environment interactions over others. For example, definitions focussed on the identification of domesticated animals or plants in the archaeobotanical record, or on inferences regarding the dependence of people on domesticates, fail to fully acknowledge that the morphogenetic transformation of species are not uniform in the past or the present. Numerous factors affect the relative propensity of a species to accumulate anthropically selected traits through time, some are biological (e.g., Ladizinsky 1998; Larson et al. 2014), whereas others are environmental (e.g., Pearsall 2007; Smith 2012) and social, technical or practical (e.g., Marshall et al. 2014). Multiple dimensions of domesticatory relationships influence not only the degree to which plants and animals exhibit domestication traits, but also their archaeological visibility.

A priori, the recursivity of human-environment interactions hinders the interpretation of a singular, or ultimate, cause of early agriculture. The positing of any biophysical or social phenomenon as a primary cause is arbitrary, because any explanation inevitably folds back into the duality of human-environment relations (following Giddens 1984), or into Ingold's (2000: 87) 'continuous field of relationships'. It is effectively impossible to determine ultimate causation within a recursive spiral or continuous field.

To exemplify, if climatic amelioration and stabilisation at the beginning of the Holocene are posited as the ultimate cause for the emergence of agriculture across the globe (e.g., Richerson et al. 2001), this explanation fails to account for the restricted number of locations in which this actually occurred: why did agriculture emerge in some places and not in others, and why is there so much temporal variation? Any account of ultimate causation is soon beset by qualifiers, such as resource availability and species' susceptibility to domestication. Geographical variations in the availability and susceptibility of animal and plant resources to domestication do not account for why agriculture emerged in some places and not in others during the early Holocene. For instance, indigenous plants were evidently domesticated in other regions much later, for example, India (Fuller et al. 2004), parts of Africa (Kahlheber and Neumann 2007) and North America (Asch and Hart 2004). Any climatic explanation soon slips sideways from the biophysical realm to the social realm in order to account for the observed spatiotemporal variations. The social realm necessitates a much broader consideration of how people in different locales engaged with their environments in order to understand why some developed agriculture and others did not. In logical terms, what may initially be characterised as a cause – whether climatic amelioration, environmental degradation or social transformation – soon becomes a relatively widespread precondition, which in turn becomes a relatively benign context.

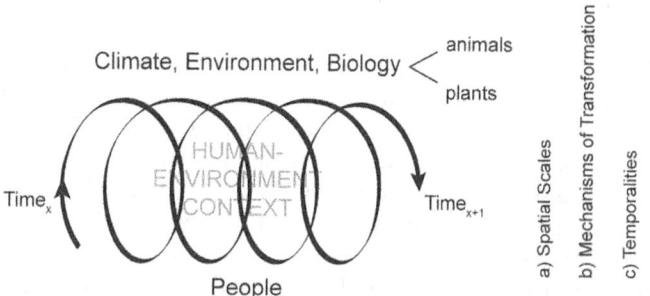

Figure 2.3 Schematic representation of the multiple dimensions of early agriculture, as a subset of human-environment interactions

Note: The dimensions (a to c) are characterised more fully in Figure 2.4

Source: Denham 2011: Figure 1

Three cross-articulating dimensions of human-environment interaction are relevant to the characterisation of early agriculture in any historico-geographical context:

1   spatial scales of analysis, type of method and lines of evidence;
2   transformative mechanisms and archaeological expressions of agriculture; and,
3   temporalities of associated phenomena.

Each dimension is briefly discussed here.

Articulating space and place

There are considerable variations in the spatial scale of analysis through which early agriculture can be inferred. The scale of analysis adopted has implications for the ways in which evidence is used – whether conflated, low resolution or particular, high resolution – and tends to be associated with a specific methodology – either comparative or contextual, respectively (Figure 2.4a). Characterisations of early agriculture at continental and subcontinental scales may conflate data from different locales and of slightly different ages to draw a general, comparative picture of agricultural development (Renfrew 2002; Bellwood 2005). Others draw on locally generated, heavily contextualised evidence to characterise early agricultural development within given landscapes or places (Pearsall 2007; Denham and Haberle 2008) and emphasise intra-regional variability (Barker 2006).

There are conceptual tensions between comparative and contextual approaches (cf. Renfrew 1973, Thomas 1996b, Renfrew 2002), but the two perspectives are not necessarily incompatible. For those seeking to understand early agriculture, reconstructions of highly specific and contextual information at the local level can be cautiously situated within broader historical and geographical processes at the regional, inter-regional and continental scales. The converse is more problematic due to the lack of resolution and specificity in conflated datasets.

Domains, transformative mechanisms and archaeological expression

Agriculture is predicated on varying degrees of human intervention in the life cycle of plants (focus here), which in turn yield intended and unintended consequences in terms of plant biology and the utility of managed plants for people. The nature of the mutually transformative relationship between plants and people is differentially expressed in given historico-geographical contexts; it is not restricted to the biological domain, even though issues of biological domestication have been the focus of most debate. Rather, four domains have a bearing on the character of early agriculture, namely how agriculture is expressed archaeologically, in any historico-geographical context (Figure 2.4b; cf. Sayer 1984): biological, social, environmental and technological.

In the biological domain, phenotypic and genotypic changes in a plant can result from the advertent and inadvertent influences of people on selection and genetic isolation. In essence, domestication represents human intervention in the evolution of an organism (Larson et al. 2014). The 'degrees' to which plants become domesticated vary greatly in terms of rates, fixation of traits and correspondence between acquired phenotypic characteristics and genetic markers. Environmental controls on gene expression and phenotypic plasticity may complicate the use of morphologically derived criteria for discriminating wild and cultivated plants in archaeobotanical assemblages. These issues are further complicated in New Guinea where vegetative propagation is the dominant mode of plant reproduction and where the ecology, phenology and genetic history of most wild and cultivated plants are under-studied and poorly known. To date, there is limited archaeobotanical evidence for the processing, consumption and discard of cultivated plants at occupation sites in the highlands (Christensen 1975a; Donoghue 1989; Summerhayes et al. 2010; Lewis et al. 2016); these records shed little light on the degree of dependence communities had upon plant cultivation through time.

Various social characteristics and practices have been associated with agricultural communities, for instance, major demographic growth leads to demic expansion and the development of socio-political hierarchies. Farmers are oft-associated with 'Neolithic' cultural traits, such as stone axe-adzes, sedentism and pottery. Needless to say, these are primarily assumptions drawn on experience in specific regions, usually Eurasian, and need not apply to New Guinea (Golson 1977a; Denham 2005b; Denham et al. 2007). Agricultural communities are by definition reliant on cultivated food for their subsistence (Hather 1996: 548), yet relatively few studies have sought to focus on the cosmological shifts accompanying the emergence or adoption of agriculture in a given locale – notable exceptions include Cauvin (2000), Hodder (2010) and Deitrich et al. (2012). To date, the ways in which the material cultures and social worlds of New Guinea highlanders changed with the advent of agriculture are poorly known and under-theorised.

In terms of the environmental domain, the palaeoecology of the New Guinea highlands from the Pleistocene to the present is reasonably well understood (Hope and Haberle 2005; Haberle 2007). Most palaeoecological records are based on

a) Spatial Scales

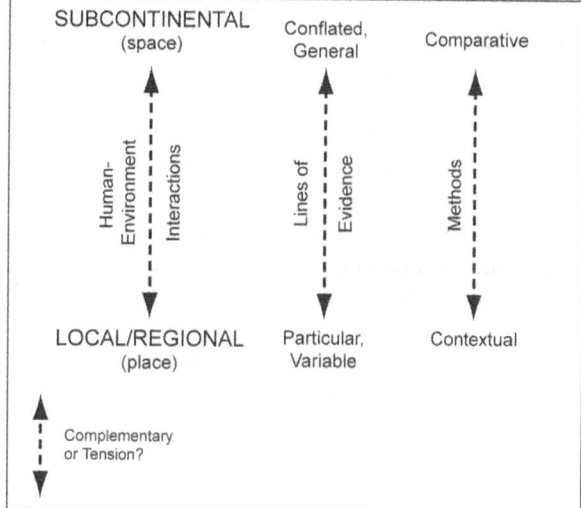

b) Mechanisms of Transformation

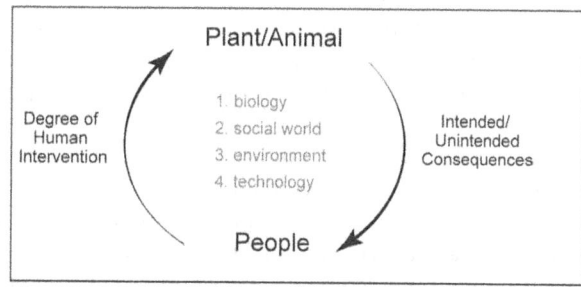

c) Temporalities

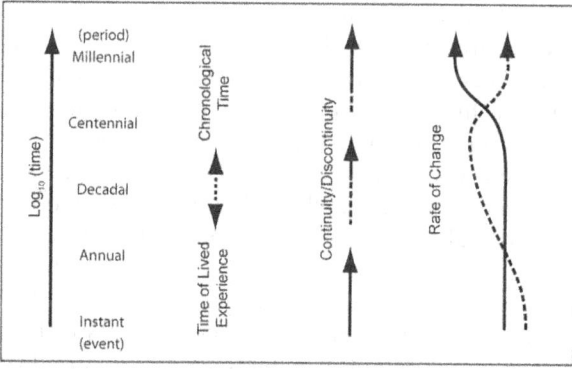

Figure 2.4 Three dimensions of early agriculture

Note: (a) correspondences between spatial scale, lines of evidence and methods; (b) four domains for the mutually transformative nature of domesticatory relationships between animals/plants and people; (c) temporalities of associated phenomena

Source: Denham 2011: Figure 2

pollen and microcharcoal records from wetlands; these require the distillation of climatic, volcanic and anthropic components at local and regional scales. Particular focus has recently been applied to investigating rates of change and stepped changes within pre-existing historic ranges of variability in order to infer anthropic, especially agricultural, influences against a changing backdrop of climatic shifts and episodic tectonic events (see Haberle 1994, 2007). The climatic sensitising of landscapes to anthropic transformation is being recognised; these synergistic effects are most clear in widespread burning of tropical rainforest during El Niño years. Similar climatic sensitising of environments to anthropic influences almost certainly occurred in the past. Thus, shifts in palaeoecological indicators – such as key plant taxa and charcoal frequencies – need not always be associated with dramatic shifts in what people were doing; they may represent continuities of practice in climatically sensitised environments that are highly susceptible to change.

Technology provides the means through which people exploit plants and transform their environment. In the highlands context, technologies are associated with multifarious practices of cultivation, such as vegetative propagation, translocation and transplantation, digging, staking, mounding, ditch digging and so on. These practices are visible either through direct archaeological evidence of features and the tools used, or through indirect or proxy lines of evidence. In order for people to practise certain kinds of cultivation, they need to have certain kinds of technologies. In terms of the four transformative domains associated with the emergence of agriculture, archaeological remains of past practices is the strongest line of evidence for the New Guinea highlands.

Different considerations of early agriculture draw variably on different domains and lines of evidence to construct arguments for or against early agriculture in different parts of the globe (Bellwood 2005; Barker 2006; Denham et al. 2007). In part, these positions reflect the definitions of agriculture adopted, which are usually either inherited from research in Eurasia or developed to suit the available evidence within a region. These domains are variably expressed, articulated and aligned in different instances of early agriculture. Particular factors seem to be important and correspond in some regions, whereas in others they do not; however, they are all relevant.

Temporalities of associated phenomena

Debates concerning early agriculture tend to project unilinear or multilinear trajectories from the past to the present; namely, they are teleological. Time is viewed as continuous; processes are viewed as cumulative (e.g., Richerson et al. 2001), as if they lead somewhere significant other than solely towards the present. Three aspects of time and the temporality of things (namely, the temporal extension of something, or its being in time; Thomas 1996b) are significant. Although time and temporality are implicit to any discussion of early agriculture and subsequent transformations, they are rarely made explicit (Figure 2.4c).

First, the temporality of things is usually assumed to be continuous, or semi-continuous. In part, this reflects the punctuated nature of archaeological finds and the need to place fragmentary finds into chronological-geographic sequences (e.g., Scarre 1988). Interpretations adopt various lines of reasoning – from uniformitarian, to Ockham's Razor, to historical materialist, to post-processualist – to place archaeological finds in time, namely to temporalise them, to infer their chronological extension and to position them in a sequence. Discontinuities can be recognised in the deep past, such as the abandonment of a crop or technology; however, they may be hard to determine with confidence due to the absence of evidence, which does not suffice as evidence of absence.

Second, rates of change are rarely considered, which in part can be a function of the records and partly a function of perspective. As discussed earlier, domestication is a process which operates at variable rates for different species and subspecies in different historico-geographic contexts (Fuller et al. 2014). Some species – perhaps those with annual life cycles subject to intensive human selection and a high degree of genetic isolation – would be anticipated to accumulate traits resulting from human management at a relatively rapid rate. Conversely, these same species not subject to the same degrees of human selection and genetic isolation, as well as other species with longer life cycles (such as trees and some animals), would be anticipated to accumulate traits at a slower rate. The rate of accumulation of domestication traits within an organism is a function of the human-environment context, namely a function of the multiple domains associated with that domesticatory relationship. Thus experimental farming can yield high rates of change in cereals within decades (Hillman and Davies 1990, 1999), perhaps due to the high degrees of human selection and genetic isolation, whereas archaeobotany suggests the accumulation of these traits actually occurred over thousands of years in Southwest Asia (Tanno and Willcox 2006), perhaps due to variable management practices through time and gene flow between wild and cultivated stock (see Jones and Brown 2007). Following this type of reasoning, the accumulation of domestication traits in trees could feasibly occur over centuries or millennia (Yen 1996).

Genotypic and phenotypic changes are a continuum of change, along which measures of domesticity, as opposed to wildness, are determined. Variations in the rate of domestication, whether measured genetically or phenotypically (as is customary in archaeobotany), can be anticipated to vary depending upon whether a plant is propagated sexually or clonally; the accumulation of selected somatic mutations in the latter is a qualitatively different type of process to the accumulation of selected mutations through sexual reproduction (Yen 2003; Fuller et al. 2014). Additionally, genotypic and phenotypic traits resulting from human management should not be anticipated to accumulate at the same rate within a species, or between species, especially given latent issues of gene expression and phenotypic plasticity for some plants under cultivation (Gremillion and Piperno 2009; Gremillion et al. 2014). The generation of phenotypic varieties in some plants, such as bananas and yams, need not correspond to genotypic change; they may represent a phenotypic response, or the differential expression of a gene, due to the environment of cultivation and growth.

Third, the archaeology of early agriculture has tended to view time in the abstract, namely chronologically, and not from the perspective of lived experience, or experiential time. Issues of plant domestication are generally considered from the perspective of how many years before traits *x* and *y* become apparent in the archaeobotanical record; they are rarely considered from the perspective of how these traits accumulated through the day-to-day activities of people and were passed on from generation to generation. The time of lived experience is a precondition for the constitution of chronological time (Heidegger 1962) and processes thereby inferred, yet it is largely absent from discussions of early agriculture.

Existential aspects of time are glossed, or avoided, because they can be considered to be attempts to get inside the minds of people in the past. Although partly true, this is always the case, for discussion of domestication invariably considers whether traits were intentionally or unintentionally accumulated, namely to understand the intent, or mindset, of the people involved. The character of the domesticatory process requires a consideration of experiential time, whether to understand the deliberate selection of a taro corm or cereal grain, or the qualitatively different temporal perspective of planting a tree. The former yields within a year, whereas the latter may take decades before yielding and requires an inter-generational perspective (Ingold 2000; Terrell 2002).

Context, specificity and visibility

Following Hather (1996), Ingold (1996) and Spriggs (1996), agriculture needs to be defined in social terms of dependence and involvement:

> Given the acknowledged problems in measuring such a social basis in the distant past (Harris 1998: 88; Yen 1998: 164), early agriculture in New Guinea has been differentiated from other forms of plant exploitation using an evidential triumvirate through which its effects are potentially visible [Figure 2.5]. The effects of agriculture are evident in plant cultivation and use (which may or may not lead to domestication), environmental transformation and in the archaeological remains of former cultivation practices. The nature and diversity of contemporary practices serve as heuristic guides [or modes of calibration] for translating these lines of evidence into specific and contingent interpretations of early agriculture: contingent because different conditions and varieties of practice produce different signals, and specific because any interpretation needs to be grounded in the physical evidence of those past practices, [that is] archaeological remains of former cultivation.
>
> (Denham 2007a: 99)

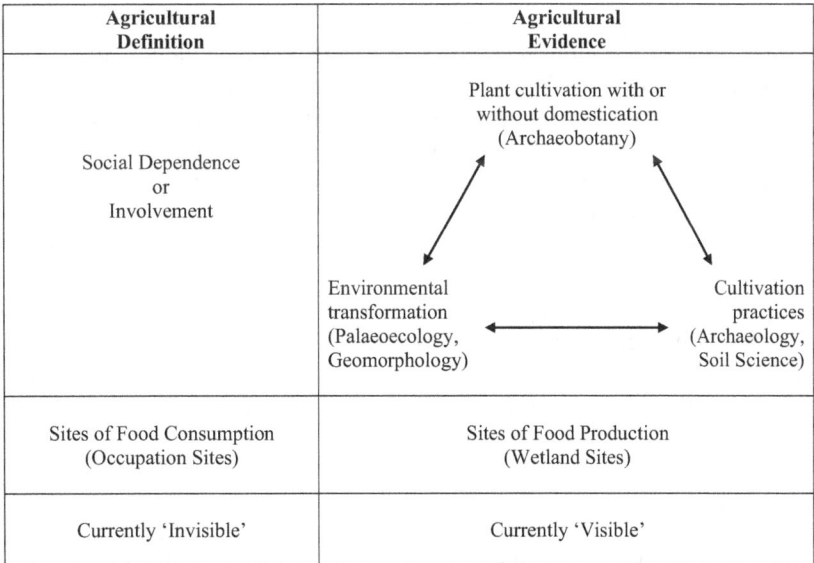

Figure 2.5 Schematic representation for the evidential 'effects' through which early agriculture has been identified in the highlands of New Guinea

Note: Currently, the degree of social dependence on cultivation cannot be determined from archaeobotanical assemblages at occupation sites; rather, social dependence or involvement is inferred on the basis of the reconstruction of cultivation practices using a triumvirate of evidence (principally, archaeobotany, archaeology and palaeoecology) at wetland sites of agricultural production

Source: Updated version of Denham 2007a: Figure 5.4

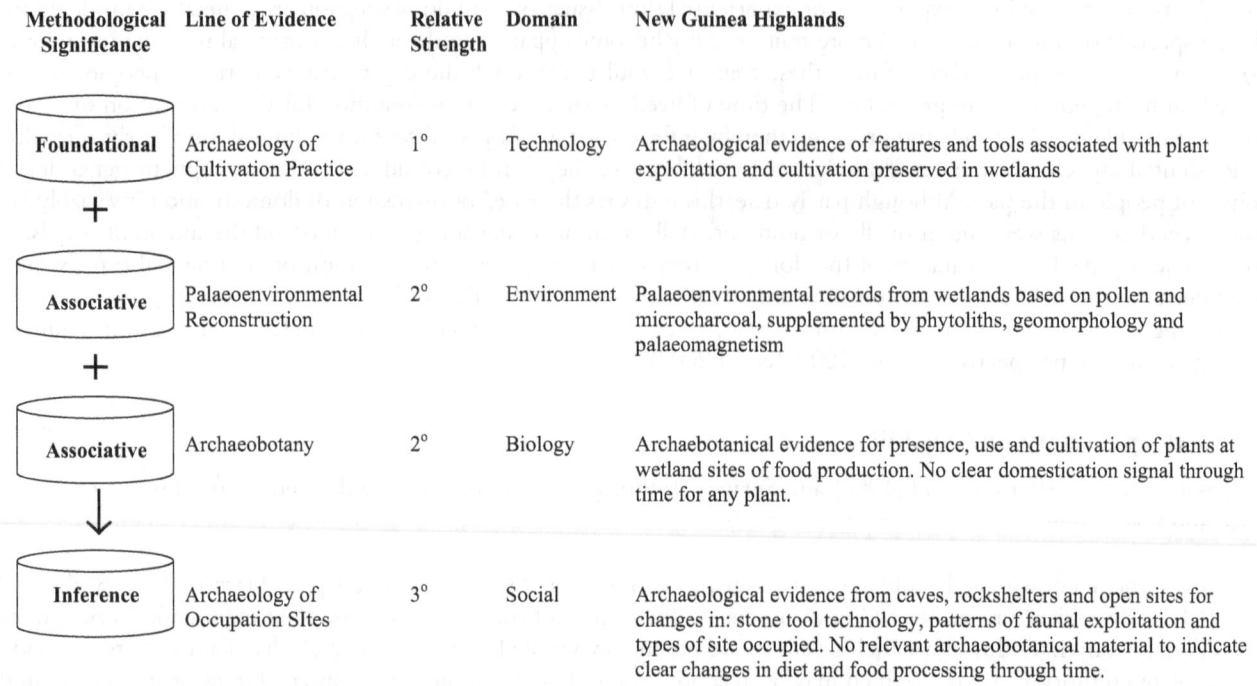

Methodological Significance	Line of Evidence	Relative Strength	Domain	New Guinea Highlands
Foundational	Archaeology of Cultivation Practice	1°	Technology	Archaeological evidence of features and tools associated with plant exploitation and cultivation preserved in wetlands
Associative	Palaeoenvironmental Reconstruction	2°	Environment	Palaeoenvironmental records from wetlands based on pollen and microcharcoal, supplemented by phytoliths, geomorphology and palaeomagnetism
Associative	Archaeobotany	2°	Biology	Archaebotanical evidence for presence, use and cultivation of plants at wetland sites of food production. No clear domestication signal through time for any plant.
Inference	Archaeology of Occupation SItes	3°	Social	Archaeological evidence from caves, rockshelters and open sites for changes in: stone tool technology, patterns of faunal exploitation and types of site occupied. No relevant archaeobotanical material to indicate clear changes in diet and food processing through time.

Figure 2.6 'Combination-lock' motif for the determination of early agriculture in the past

Note: Different domains have precedence in different debates about early agriculture in different parts of the world: archaeological evidence of cultivation practices takes precedence in New Guinea (depicted here)

Ethnographic and historical accounts are guides to infer early agriculture in any given locale. The use of ethnographic analogues is not rigid, because novel or non-analogue forms of agriculture may have existed in the past. However, they provide contingent, contextual and relevant starting points for historical interpretation.

In the highlands of New Guinea, archaeological features derived from past subsistence practices are preserved in the wetlands, and this evidence grounds the interpretation of associated archaeobotanical and palaeoecological information. Direct, physical remains of past practices, such as features and associated artefacts and materials, enable multiple lines of contextual evidence to be woven into robust and specific interpretations. Archaeological remains enable specificity in interpreting what people were doing in the past, their use of plants and the cumulative effects of these practices on the environment. Such specificity is necessarily contingent on given biological, environmental and social contexts.

As an analogy, different lines of evidence should be considered in terms of a combination (Figure 2.6). The combination or alignment of multiple lines of evidence enables early agriculture to be 'unlocked' or determined from sometimes faint multidisciplinary traces. The precise combinations of early agriculture vary from place to place and through time; hence the model is not prescriptive even though the combinations always comprise these different elements. For comparison with New Guinea, the identification of early agriculture in Southwest Asia is usually predicated on archaeobotanical evidence of domesticated plants (foundational), in conjunction with archaeological evidence from settlements (associative) and palaeoecological evidence of environmental transformations (associative), whereas the character of early cultivation practices is largely invisible (inferential).

Clarification of terminology

Debates about early agriculture can become embroiled in terminological and substantive disputes about what does and what does not constitute evidence of agriculture in the past. Here, a flexible and contextual understanding of early agriculture relevant for different regions of the world is advocated. A contingent conceptual framework has been proposed in which multiple lines of evidence are compared against ethnographic and historical accounts of different forms of plant exploitation for a given region.

The term 'plant exploitation' is a generic term to connote the use of plants by people, with an emphasis on obtaining food, whether as part of extensive foraging practices or intensive forms of systematic agriculture. Plant exploitation has traditionally been examined through two entwined themes: the manipulation of plant resources and the transformation of natural to artificial ecosystems. In this book, a similar dual approach is adopted: a focus on the spatial and temporal bundling of practices that are constitutive for different forms of plant exploitation (Chapter 4), and the effects of these practices, and the forms of plant exploitation they comprise, upon plant resources in the past (Chapter 5).

A working definition of agriculture is proposed here for the highlands of New Guinea. Agriculture applies when communities are reliant upon the cultivation of plants for food. Cultivation connotes the planting of crops in cleared and specially prepared plots; planting can be affected by vegetative propagation or by seed. In the New Guinea context, agriculture is effectively synonymous with horticulture. The domestication status of the plants can range from vegetatively propagated 'wild' plants to sterile cultivars. The plots can range from ephemeral swidden plots in the rainforest to more permanent raised bed cultivation in grasslands and wetlands. These practices are agricultural when they require people to reorient their social life to the cultivation of plants for food.

Arboriculture refers to the intensification of tree, palm and pandanus resources in the landscape through a variety of practices. Arboriculture includes the burning, clearing and tending of self-reproducing stands and groves; the vegetative propagation of suckers or other plant parts; and the transplantation of germinated seedlings. On New Guinea, arboricultural management is intended for fruit, nut and sago-production, the production of construction materials and other uses, as well as fallowing practices.

Domestication is a multi-tiered concept that encapsulates the bringing of plants, animals and landscapes into the sphere of human engagement (social domestication) to the creation of genotypic and genotypic variants through the advertent or inadvertent intervention in plant life cycle (biological domestication). The former meaning is important for understanding social space, the social construction of landscape and the creation of place. The latter meaning is more specific and restricted to the human-mediated creation of new phenotypic and genotypic variants.

Part II

Places, practices and plants

3   The importance of place

Ages of discovery

On 8 March 1933, a charted Junkers aeroplane flew Australian-based gold prospectors and colonial administrators over the Wahgi Valley, one of the largest inter-montane valleys on the island of New Guinea. They encountered the following:

> Spread out for a hundred miles between the high mountains was the green, sunlit immensity of the Wahgi. A big river fed by innumerable streams ran through its centre, and oblong houses in homestead groups of four or five dotted across a continuous patchwork of neat, square gardens. The Wahgi, enormous, fertile and heavily populated, greatest of New Guinea's highland valleys, had been 'discovered'.
>
> (Connolly and Anderson 1987: 79)

At that time, the international community was amazed to find the floor of the valley so densely populated and cultivated. The mountain ranges of the Central Highlands had previously been considered uninhabited, and they soon became known in the popular imagination as the *Land that Time Forgot* (Leahy 1936; Leahy and Crain 1937). No-one suspected that this region would provide the longest history of cultivation on the island, including evidence for some of the earliest agricultural practices in the world.

At 'First Contact', the antiquity of occupation and agriculture in the highlands, or those areas of New Guinea above 1200 m altitude, could only be guessed. From the late 1950s onwards, archaeological and palaeoecological investigations sought to determine how long people had been living and cultivating in the main highland valleys (Bulmer and Bulmer 1964).

The earliest archaeological excavations in the highlands were conducted by Sue Bulmer in 1959–1960 (S. Bulmer 1966; Bulmer and Bulmer 1964; Gaffney, Ford et al. 2015; Denham 2016a). On the basis of her excavations at two rockshelters, Kiowa and Yuku, she surmised that agriculture was at least 4000 years old in the highlands (S. Bulmer 1966: 156a). She qualified her conclusions:

> Direct proof of agriculture must depend on pollen analysis, the future fortunes of archaeology in obtaining organic remains, and the analysis and dating of ditches and drains of agricultural derivation.
>
> (S. Bulmer 1966: 152)

These lines of evidence – palaeoecology, archaeobotany, archaeology and radiocarbon dating – have underpinned the investigation of early agriculture in the highlands of New Guinea ever since.

Only a handful of archaeological excavations at occupation sites over the last sixty years shed light on the long antiquity of cultural change in the main valleys along the highland spine of New Guinea (Figure 3.1; Table 3.1; S. Bulmer 1966, 1977a; White et al. 1970; White 1972; Christensen 1975a; Watson and Cole 1977; Swadling 1983; Mountain 1991a; Summerhayes et al. 2010). These investigations have demonstrated changing stone tool technologies (S. Bulmer 1977b, 2005; Evans and Mountain 2005; Golson 2005; Gaffney, Ford et al. 2015), megafaunal extinctions (Mountain 1991a) and patterns of hunting (Sutton et al. 2009), and the antiquity of localised networks of exchange in the interior (Hughes 1977). Archaeological excavations of occupation sites have yielded comparatively little reliable evidence of the plants exploited by people (cf. Christensen 1975a and Summerhayes et al. 2010), or direct evidence for the emergence of agriculture in the highlands.

By 1970, a provisional antiquity of agriculture in the highlands had been established from the results of archaeological and palaeoecological investigations at several wetlands in the Wahgi Valley (Figures 3.2–3.3; Table 3.2). These wetlands were primarily sites of agricultural production that preserved archaeological remains of former cultivation features and

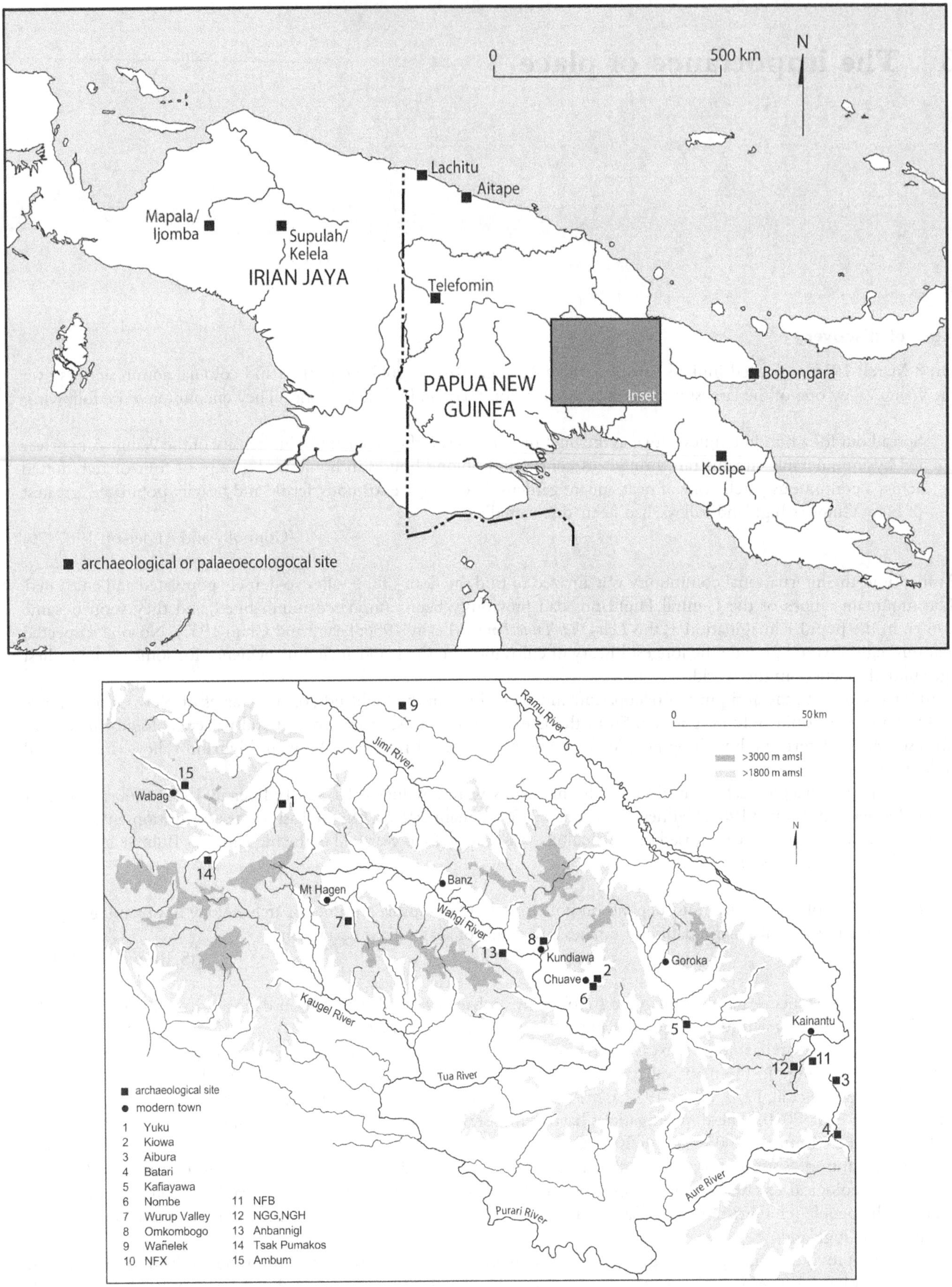

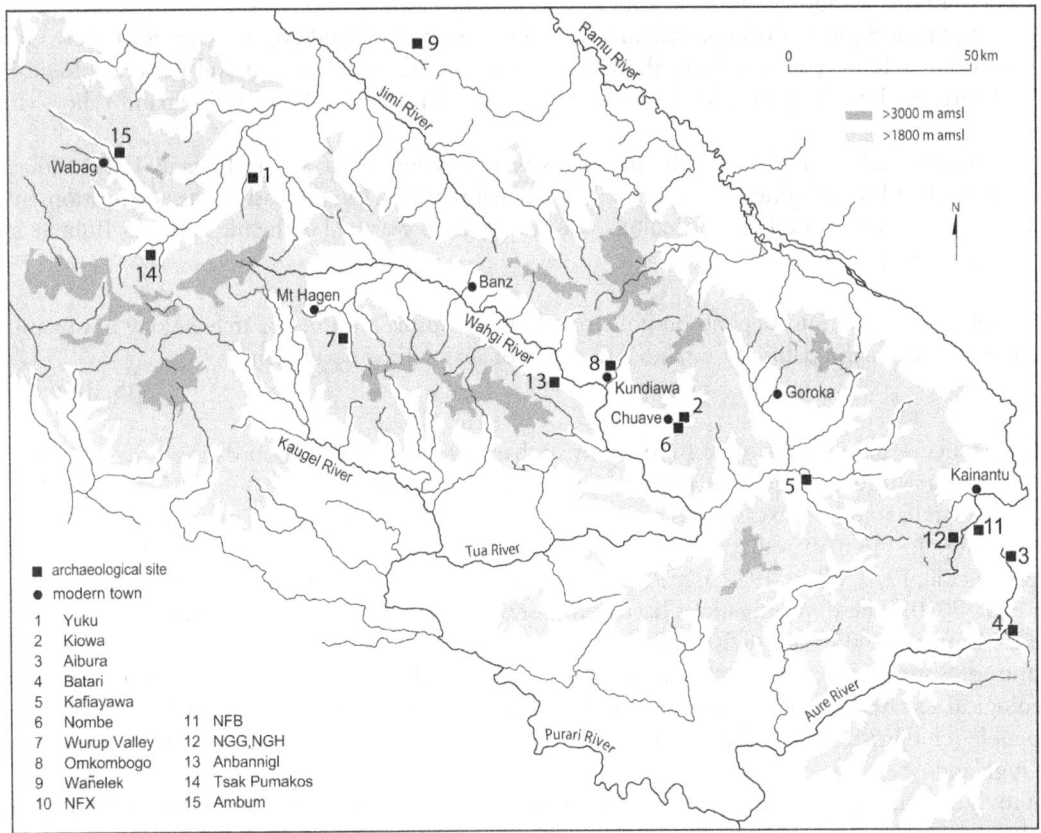

Figure 3.1 Map of key occupation sites in the highlands of New Guinea

Note: Lower inset shows archaeological sites in the central highlands of Papua New Guinea

Source: Sutton et al. 2009: Figure 1

Table 3.1 Earliest occupation of pre-2000 cal BP cave, rockshelters and open sites in the highlands of New Guinea (excluding wetlands)

Site Name	Altitude (m AMSL)	Earliest Occupation (cal BP)	Primary References
Mapala	3996	6500–5900	Hope and Hope 1976a
Kamapuk	2050	5300–4800	Christensen 1975a; Aplin 1981
Kosipe/Ivane Valley	2000	49,000–43,000	White et al. 1970; Fairbairn et al. 2006; Summerhayes et al. 2010
Manim 2	1770	10,800–10,200	Christensen 1975a; Mangi 1984[1]
Wañelek	1710	19,400–17,300? 4100–2800	S. Bulmer 1973, 1977a[2], 1991; Gaffney, Summerhayes et al. 2015; Denham 2016a
NGH	1700	4400–3800	Watson and Cole 1977
NGG	1680	3800–3300	Watson and Cole 1977
Nombe	1660	25,500–19,600	Mountain 1991a; Denham and Mountain 2016
NFB	1650	4800–4000	Watson and Cole 1977
Aibura	1640	4500–3900	White 1972
NFX	1550	23,600–20,000?	Watson and Cole 1977[3]
Kiowa	1530	12,600–11,600	S. Bulmer 1964, 1966, 1975; Gaffney, Ford et al. 2015, 2016; Denham 2016a
Kafiavana	1350	13,300–11,600	White 1972
Batari	1300	9500–8700	White 1972[4]
Yuku	1280	15,300–13,300	S. Bulmer 1964, 1966, 1975[5]; Denham 2016a

Notes: Two sites (Lemouru and Uweka) excavated by Mountain were not reported in detail; the earliest reported date for Lemouru is 1800 ± 80 BP (ANU 2575, Gillieson et al. 1986: 322) and both sites probably post-date 2000 cal BP. No dates were reported for three sites (Anbannigl, Tsak Pumakos A and B) excavated by Kobayashi and Hayakawa (n.d.), and one site (Omkombongo) excavated by White (1967: 33–39). The depth of cultural material suggests considerable antiquity at Tsak Pumakos B (to at least 2 m below current ground surface), Omkombongo (to at least 2.5 m) and, to a lesser degree, Anbannigl (over 1 m deep).

1 Mangi (1984: 23–25) provides the earliest reliable date for Manim 2 (ANU-1467; the range for ANU-1468 is 9870 ± 610 BP and is too broad to facilitate interpretation).

2 The earliest reported occupation at Wañelek is based on a single date (GX-3331, see S. Bulmer 1977a) and the context and associations of this date are uncertain. More reliable evidence indicates multiple occupations during the period 4100–2800 cal BP (Denham 2016a).

3 Watson and Cole (1977: 194) report the earliest dated charcoal at NFX to be 'mineralized'. This sample could be contaminated and may thus be unreliable for interpretation. Three other dates at the site could be more reliable indicators of its earliest occupation and cluster later in the Pleistocene: UW-262 (11,510 ± 140 BP), I-7284 (12,620 ± 280 BP) and I-7284C (13,210 ± 270 BP).

4 An earlier date has been obtained from Batari (ANU-40, 16,850 ± 700 BP, charcoal), although White (1972: 16) does not consider it to be associated with human activity.

5 S. Bulmer (1975: 30) considers archaeological remains at Yuku to pre-date GX-3112B; however, there is extensive disturbance at the site that confounds ready interpretation (Denham 2016a).

ditches, as well as palynological evidence of vegetation history. Archaeological investigations occurred at Warrawau Plantation in 1966, including the Mantons' site (Golson et al. 1967; Lampert 1967; Powell 1970a); Minjigina in 1967 (Lampert 1970); Kindeng in 1968 (unpublished); and Kuk in 1969 (Allen 1970). Palaeoecological records of forest disturbance and clearance at Warrawau and Draepi-Minjigina (Powell 1970a) suggested agriculture was of greater antiquity than 2300 years-old antiquity of a wooden digging stick at Warrawau.

Following initial reconnaissance in 1969, Kuk was chosen to be the focus of large-scale multidisciplinary investigations of early agriculture. Taking place from 1972 to 1977, these were designed to:

> confirm and extend . . . the antiquity of agriculture in the New Guinea highlands; the age and circumstances of the move to cultivation into the swamplands; the nature and organisation of agricultural activities there and associated settlement; and the date, causes and extent of their abandonment.
>
> (Golson 1976a: 209)

The significance of the findings at Kuk changed in step with the increasing antiquity of archaeological evidence of ancient cultivation, as well as with new botanical research indicating the importance of the New Guinea region to understanding the origins of Pacific agriculture (Barrau 1963; Yen 1973, 1982, 1985, 1991; Powell 1976). Powell (1976) provided an indication of the range of economically useful plants and Yen (1982) undertook a phytogeographic study of important edible plants, such as taro *(Colocasia esculenta)*. Early indications suggested Australimusa (now Callimusa) bananas *(Musa* spp.), sugarcane *(Saccharum officinarum)* and a host of minor crop plants were domesticated in New Guinea (Yen 1973). These unfolding studies informed the interpretation of archaeological and palaeoecological research at Kuk and other wetland sites in the highlands (Golson 1989: 678–683).

Initially the Kuk project was intended to establish the antiquity of pre-Ipomoean agriculture in the highlands (Golson 1976b, 1977c), namely, the history of agriculture before the introduction of the sweet potato *(Ipomoea batatas)*, a South

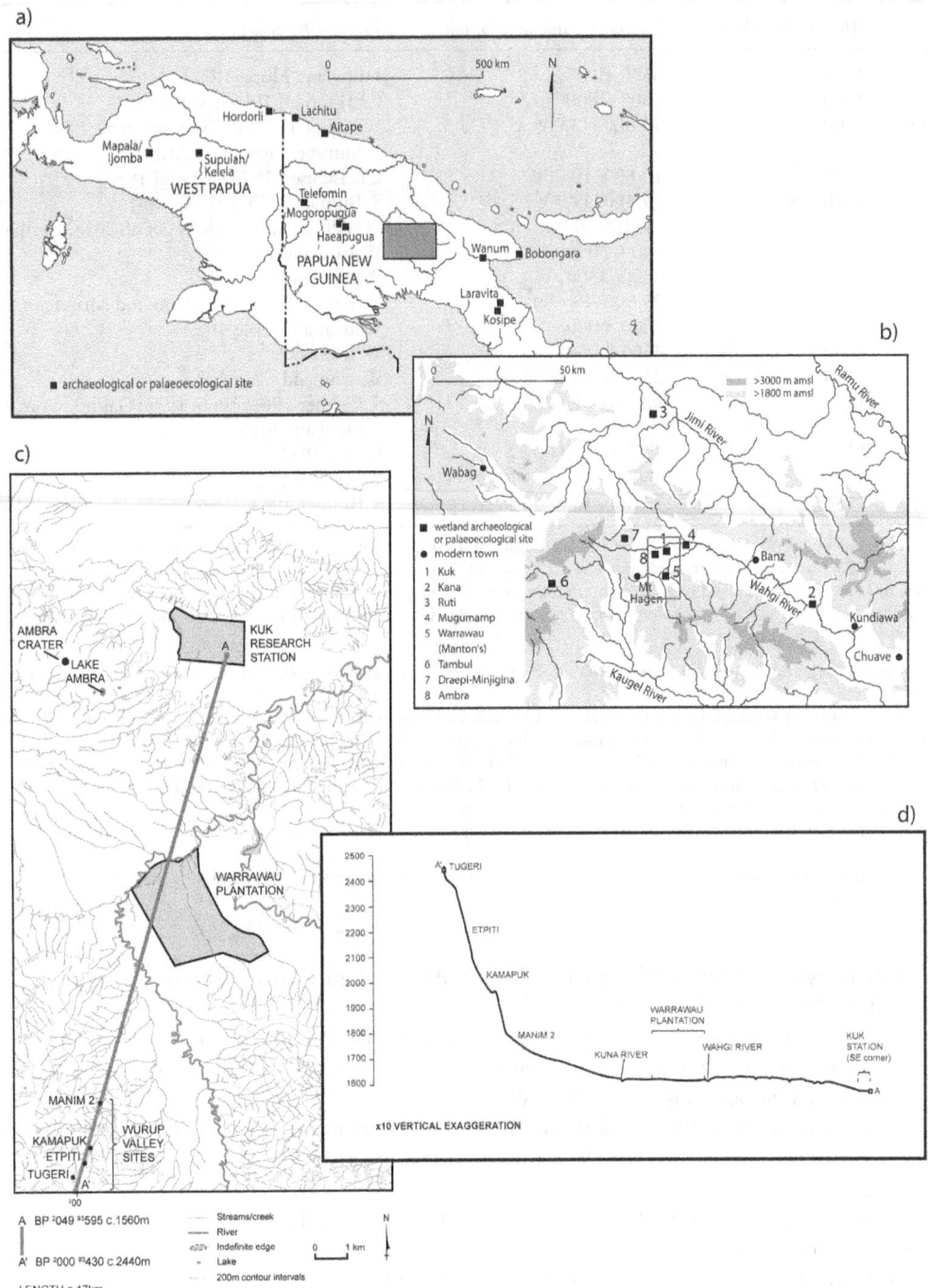

Figure 3.2 Depictions of significant archaeological and palaeoecological sites at wetlands on the island of New Guinea

Note: Maps of (a) the island of New Guinea, (b) the central highlands of Papua New Guinea (location indicated by shaded box on panel a), and (c) the Upper Wahgi Valley (location indicated by shaded box on panel b). The altitudinal transect across the Upper Wahgi Valley (d) shows the respective altitudes of key wetland and occupation sites (transect location indicated by the bold line on panel c)

Source: Denham and Haberle 2008: Figure 3

American domesticate, which has become the dominant staple crop in large parts of the highlands over the last few hundred years. During the 1970s, the primary domesticates of New Guinea agriculture, as well as associated agricultural technologies, were thought to be of ultimate Southeast Asian origin (Golson 1976a, 1985; Golson and Hughes 1980). During the 1980s, the interpretative emphasis shifted to ideas of early and independent agricultural development and

Figure 3.3 Photographs of the Upper Wahgi Valley landscape

Note: Banana plot growing at Kuk Swamp (Upper, Source: Tim Denham, 1999); Tibi Tea Plantation (Middle; Source: Tim Denham, 1998); view from Ep Ridge southwest across the floor of the Upper Wahgi Valley (Lower; broadly following bold line in Figure 3.2c; Source: Tim Denham, 1998)

Table 3.2 The earliest archaeological evidence for cultivation at wetland sites in the highlands

Site Name[1]	Altitude (m)	Location	Main Field Seasons	Earliest Evidence (cal BP)[2]	Key Publications
Tambul	2170	U. Kaugel Valley	1976	4600–4100	Golson 1997a
Mogoropugua	1890	Tari Basin	1980	700–300	Ballard 1995
Minjigina	1890	U. Wahgi Valley	1967	c. 2000–1000	Lampert 1970; Powell 1970a; Golson 1982a
Ambra Crater	1760	U. Wahgi Valley	1999	560–500	Sniderman et al. 2009
Haeapugua	1650	Tari Basin	1991–1992	3000–2000	Ballard 1995, 2001
Kindeng	1600	U. Wahgi Valley	1968	n/a[3]	Unpublished
Warrawau (Manton's)	1590	U. Wahgi Valley	1966, 1977	c. 6000–5000	Golson et al. 1967; Lampert 1967; Powell 1970a; Golson 2002
Kuk	1560	U. Wahgi Valley	1972–1977, 1998–1999	10,000[4] 7000–6400	Golson 1977c; Denham 2003a, 2005a, 2007a; Denham et al. 2003; Denham, Haberle et al. 2004; Denham, Golson et al. 2004; Golson et al. 2017
Mugumamp	1560	U. Wahgi Valley	1977	c. 4000	Harris and Hughes 1978
Kana	1480	M. Wahgi Valley	1993–1994	3000–2000	Muke and Mandui 2003

Notes

1 Other wetland sites were inspected by archaeologists, although none was investigated in detail. For example, the site at Kotna (1580 m) in the Upper Wahgi Valley was village land under drainage for coffee. The site was visited by Jack Golson and John Muke in 1988, at which time they sought permission to record features exposed in drain walls. Permission was refused, but while waiting they were able to look at some stretches of drain wall, in which ditches comparable to those of Phase 5 at Kuk were exposed (Jack Golson pers. comm. 2002).

2 The earliest dates designated with c. are based on inter-site cross-correlations with the Kuk sequence using ditch morphology and tephrochronology, rather than direct radiocarbon dating of associated materials (see Golson 1982a; Denham 2003b, 2005a; Coulter et al. 2009).

3 The archaeological finds at Kindeng have not been cross-correlated to those at other wetland sites (Jack Golson pers. comm. 2001).

4 Claims for c. 10,000 cal BP cultivation at Kuk are uncertain.

plant domestication in the New Guinea region (Golson 1989, 1990, 1991a, 1991b). By the 1990s, claims for early and independent agriculture on New Guinea were not universally accepted (Bayliss-Smith 1996; Spriggs 1996).

Foremost, and despite decades of research, there was limited publication of the primary archaeological evidence associated with early agriculture in the highlands. Without substantial publication of the excavation results, the claims for early agriculture could not be independently assessed. In the limited publication of the archaeological evidence, the form and function of feature types associated with the earliest practices were uncertain. The shifting interpretations contributed to uncertainty regarding the character of several key features; were they anthropic and did they represent cultivation?

Particularly limiting was archaeobotanical evidence for the presence, use and cultivation of plants. The strongest lines of evidence were macrobotanical, namely, seeds and wood preserved in wetland sediments (Powell 1970a, 1970b, 1982a). Although macrobotany indicated the potential range of plants available for human use at various times in the past (Powell 1982a: 32), most species were part of the highland flora and associations with periods of wetland cultivation were not clear. Significantly, an early phytolith study yielded evidence for bananas (*Musa* spp.), which were associated with periods of wetland cultivation (Wilson 1985; see Bowdery 1999). At that time, Wilson was unable to securely discriminate between sections, species and subspecies of bananas nor to discriminate between cultivars and wild-types. The paucity of clear archaeobotanical evidence suggestive of cultivation was particularly problematic: how can there be early agriculture if there is limited archaeobotanical evidence of the plants people were growing?

In addition, there was a lack of palaeoecological evidence for forest clearance potentially associated with the earliest agricultural practices between c. 10,000 to 5000 years ago. Powell (1970a, 1980, 1981, 1982a) documented disturbance and clearance of lower montane forests at the base of pollen diagrams at several sites in the Upper Wahgi Valley dating from 5500 to 5000 years ago. These vegetation histories missed the key period of likely agricultural development during which forests were cleared before 5000 years ago. As a result, there was no broader palaeoecological context to understand the timing or extent of early agricultural practices documented in archaeological excavations at Kuk and other sites.

In order to address evidential deficiencies with previous work and to construct a more robust foundation for claims of early agriculture in the highlands, a new phase of multidisciplinary research was initiated at Kuk, including excavations in 1998 and 1999 (Figures 3.4–3.5; Denham 2003a; Denham et al. 2003; Golson et al. 2017). A suite of relatively new techniques from the archaeological sciences was applied to the investigation of periods of wetland drainage and use pre-dating 2000 years ago, including plant microfossil analyses (primarily phytolith and starch grain), mixed-method stratigraphic

Figure 3.4 Map showing the extent of archaeological excavations at Kuk Swamp during the two major periods of excavation

Note: 1972–1977 (directed by Jack Golson and Philip Hughes) and 1998–1999 (directed by Tim Denham and Jack Golson)

analyses (X-radiography and soil micromorphology) and Accelerator Mass Spectrometry (AMS) dating of very small charcoal fragments. These techniques had been developed, or had become more reliable, since the initial investigations at Kuk in the 1970s (Denham et al. 2003, 2004b; Denham, Fullagar et al. 2009; Denham, Haberle et al. 2009; Fullagar et al. 2006; Denham and Grono 2017).

Primarily due to the sequential, multidisciplinary investigations at Kuk, the antiquity of agriculture in the highlands is now established to be at least 7000–6400 years old (Denham et al. 2003), and possibly 10,000 years old (Denham, Golson et al. 2004; Golson 2007; Golson et al. 2017; Table 3.3). Kuk is the 'type-site' for the investigation of early agriculture on New Guinea. It is the most intensively investigated site and preserves the oldest and most continuous sequence of wetland manipulation for plant exploitation and cultivation on the island. No comparable investigations have yet occurred in the lowlands of New Guinea (Fairbairn 2005) or Island Melanesia (Lentfer and Torrence 2007; Lentfer et al. 2010). Although now generally accepted by much of the archaeological community (i.e., Neumann 2003; Barker 2006; Larson et al. 2014), the New Guinea evidence has been qualified by some (Bellwood 2005).

Archaeological frames of reference

People first colonised New Guinea at least 55,000–50,000 years ago (Groube et al. 1986; O'Connell and Allen 2007; Summerhayes et al. 2010; Clarkson et al. 2017). At that time, New Guinea was joined to Australia in a contiguous land mass, called Sahul. Since the period of human colonisation, New Guinea has been an island for approximately only

Figure 3.5 Photographs showing the character of the archaeological excavations at Kuk Swamp

Note: Jack Golson is present wearing a white shirt in both photos. Herman Mandui (former Head of Archaeology at the PNG National Museum and Art Gallery) is wearing a striped shirt in middle of lower image and excavating together with local Kawelka men

Source: Tim Denham, 1998

8000 years, following post-glacial sea level rise that flooded the land bridge to create the Torres Strait (Lambeck and Chappell 2001; Chappell 2005).

The long-term history of New Guinea is generally portrayed in terms of two major colonisation events: initial settlement by modern humans by c. 55,000–50,000 years ago, and the later arrival of Austronesian farmer-voyagers from Island Southeast Asia at c. 3500–3000 years ago (see Summerhayes et al. 2009 for a recent formulation of this position). From this perspective, New Guinea was largely isolated from areas to the west between initial colonisation and the arrival

Table 3.3 The antiquities and lines of evidence for archaeological phases at Kuk Swamp

Phase	Age (cal BP)	Description	Artificial Palaeochannels[1]	Palaeosrufaces[2]	Ditches	Stone Artefacts	Wooden Artefacts[3]	House Sites[4]
1	c. 10,000	amorphous palaeosurface		X		X		
2[5]	7000–6400	mounded palaeosurface		X		X		
3	4350–4000	early sub-phase:	X	?	X	X		
	2700–2400	rectilinear ditch networks late sub-phase: rectilinear/dendritic ditch networks	X	?	X	X		
4	2000–1230/970	grid-like field systems	X		X	X	X	?
5	700–c. 290	grid-like field systems	X		X	X	X	X
6	250–50	grid-like field systems	X		X	X	X	X

Notes: Amended version of Denham 2007a: Table 2 and Golson 2017a: Table 1.2. The phases developed for Kuk Swamp by Golson and colleagues are not adopted here; in this book, the transformative character of plant exploitation practices through time is emphasised.

1 Artificial palaeochannels are differentiated from ditches at Kuk on the basis of scale and function, although the mode of formation of some palaeochannels is uncertain (see discussion in Denham, Golson et al. 2004). Palaeochannels function as major water disposal conduits for ditch networks.
2 Occasional features interpreted to represent 'within plot' cultivation have been recorded for late Phase 3.
3 No wooden artefacts were collected from Phase 1–3 contexts (Powell 1982a: Table 2).
4 Ed Harris in 1977 noted unexcavated house remains at a multi-occupation site that potentially pre-date Phase 5.
5 A late Phase 2 sub-phase possibly pre-dates Kim (R) tephra deposition at c. 3980–3630 cal BP, although it is not well characterised.

of Austronesian language speakers. Any agricultural activities on New Guinea before c. 3500 years ago would thus be inferred to represent the indigenous, or independent, emergence of agriculture on the island.

The two-wave colonisation model for New Guinea and adjacent islands is highly contested. The idea that New Guinea was relatively isolated up to c. 3500 years ago following colonisation by modern humans seems unlikely because of documented maritime interactions in the circum-New Guinea region. These include interaction of New Guinea with the islands of Wallacea and the Bismarck Archipelago during the Terminal Pleistocene (post-Last Glacial Maximum) and during the early and mid-Holocene (e.g., Denham 2004b; Szabó and O'Connor 2004; Terrell 2004; Bulbeck 2008; Donohue and Denham 2010; Wright et al. 2013; Specht et al. 2014). These maritime interactions were plausibly of low frequency, yet they were sufficient to introduce viable populations of marsupials from New Guinea to islands in Wallacea during the early Holocene and to the Bismarck Archipelago during the Pleistocene (Heinsohn 2010). Maritime interactions during the mid-Holocene are evidenced through the movement of artefacts, including obsidian and stone mortars, pestles and figurines (Summerhayes and Allen 1993; Summerhayes 2003; Torrence and Swadling 2008), as well as exchanges within Island Southeast Asia (Bulbeck 2008). Rather than being viewed as isolated, New Guinea was periodically incorporated in spheres of interaction that in cumulative form extended west to mainland Southeast Asia; the degrees of interaction increased during the late Holocene (Donohue and Denham 2010).

Given that New Guinea was plausibly enmeshed in spheres of interaction that extended into Island Southeast Asia before 3500 years ago, how is early agriculture to be conceived? Is it an independent innovation or derivative from Eurasia? As will be discussed at length in this book, early agricultural practices in the highlands most likely emerged from preexisting foraging practices there (Denham and Barton 2006). This agricultural efflorescence did not necessarily mark a major disjuncture from pre-existing practices, rather it is better characterised in terms of continuity. Given highly variable social and environmental contexts across New Guinea, as well as across Southeast Asia and possibly northern Australia, there were mosaics of plant exploitation strategies – primarily differentiated in terms of major food plants and the nature of exploitation practices – across these regions during the early and mid-Holocene. Even though the plants and practices varied, there were considerable commonalities shared by people living within these tropical rainforests (e.g., Denham 2008b; Denham, Fullagar et al. 2009; Denham, Donohue et al. 2009; Barton and Denham 2011, 2017).

Commonalities reflect a shared orientation of modern humans to their world, albeit differentially expressed in specific historical and geographic settings. Common practices within rainforest environments include disturbance of the forest, primarily through burning and the deliberate, local modification and management of species composition; exploitation of fauna through gathering, hunting and scavenging; a focus upon oil, protein and starch-rich plants, especially arboreal resources and tuberous plants; and mobility (Barton et al. 2012: 2).

During the Holocene, different emphases accumulated within these plant exploitation mosaics to create the diversity of forms seen in the recent past, ranging from extensive hunting-gathering-fishing to farming and a host of intermediate strategies. These historical transformations did not require the crossing of major energetic thresholds, as characterised for Eurasian agriculture (Harris 1996a), rather they resulted from slight shifts in emphasis that accumulated to yield marked differences in kind (Denham, Fullagar et al. 2009). At present, a long-term history for the emergence and transformation of agriculture has been reconstructed only for the Upper Wahgi Valley landscape; comparable regional records have not yet been reconstructed for places in lowland New Guinea, Island Melanesia and Island Southeast Asia.

From this perspective, agriculture in the highlands is 'early' on a global scale and dates to at least 7000–6400 cal BP. But is it 'independent'? The idea of independence implies without direct or indirect interference from outside, which seems unlikely for most communities on the globe even at such an early time. Alternatively, independent can be considered to refer to the way a novel form of plant exploitation emerged in a particular place from pre-existing orientations and practices, while acknowledging that these communities were engaged in proximal networks of interaction along which ideas, plants and things diffused. Thus, independent does not presuppose isolation; it encompasses the ebbs and flows of social life through time.

An introduction to highland environments

A mountainous 'spine' runs east-west along the island of New Guinea. This highland region contains a string of intermontane valleys and basins with floors at 1200–2000 m altitude. At the time of European exploration of the interior in the 1930s, these valleys and basins were densely settled and cultivated by upwards of a million people. The floors of many basins and valleys contain extensive wetlands, which are highly productive for agriculture and which subsequently have been foci for human activity in the distant and recent pasts, especially given endemic malaria below c. 1000 m that has limited population densities in the lowlands.

The primary controls on cultivation in highland valleys are temperature, soil moisture availability and soil fertility (following Brookfield 1964; Allen and Bourke 2009; Bourke 2017). Climate is a major factor affecting plant growth and the suitability of different places for growing specific crop plants. Rainfall in the highlands is predominantly orographic, with relatively high levels of cloud cover and decreasing temperature with altitude (Allen and Bourke 2009). Climates are increasingly seasonal and dry seasons are longer further eastward along the highland spine (Brookfield 1964; McAlpine et al. 1983). Variations in rainfall seasonality influence the timing of crop production and masting in trees (Bourke et al. 2004) and have been implicated in the development of regionally diverse agricultural systems in the highlands (Feil 1987). Only in the eastern part of the highlands are soil moisture deficits sufficient to limit crop growth during the dry season.

Throughout most of the highlands, temperature is the primary control on crop growth and the crop base decreases with altitude. Temperature sets the upper limit to intensive agriculture and permanent settlement in the highlands at c. 2200–2400 m (Brookfield 1964), with some cultivation extending up to 2800 m (Allen and Bourke 2009). For a small number of highland domesticates – such as karuka (*Pandanus* spp.) – temperature sets a lower altitudinal limit to crop growth; namely, these highland plants do not grow well, if at all, in the lowlands. Within the 1200–1800 m altitudinal band, climates are relatively temperate and frost becomes an issue for cultivation only above 2200 m. Indeed, diurnal temperature range on the floor of the main highland valleys is considered ideal for plant productivity (Allen and Bourke 2009). Yields are greater or comparable to lowland locales for many crops, even though crop maturation is usually longer with altitude.

Soils in the highlands can be divided into three dominant groups (after Allen and Bourke 2009: 81–86):

1   Inceptisols are moderately weathered and well-drained soils, often formed on volcanic ash (andisols);
2   Entisols are young and weakly developed soils formed on recently deposited alluvial or on steep slopes subject to erosion; and,
3   Histosols are organic-rich soils formed in wetlands.

Nutrient availability in highland soils under cultivation has not been well studied (exceptions include Latham 1980; Bleeker 1983; Wood 1987). Some soils are thought to have undergone nutrient enrichment due to periodic volcanic ash falls during the Holocene. The soil nutrient benefits of these distal tephra falls derived from eruptions hundreds of kilometres away are, though, largely assumed (Blong 2017).

In general, soils on valley slopes and floors are of low-to-moderate quality, whereas wetlands provide environments suitable for prolonged plant growth and cultivation. Not only are wetlands relatively protected from seasonal, inter-annual or longer-term climatic fluctuations, especially with respect to rainfall, the admixture of inwashed (allochthanous), alluvial sediments and *in situ* (autogenic), organic-rich deposits creates soils with prolonged and high fertility. Thus, wetlands provide high quality environments for cultivation within low-to-moderate quality landscapes. As a consequence, all major wetlands in highland valleys between 1200–2200 m have been used for cultivation by people in the recent or distant pasts (Ballard et al. 2013; Ballard 2017).

The suitability of highland environments for crop growth and cultivation, particularly in the western highlands, is clearly demonstrable for the Upper Wahgi Valley. This valley is particularly significant, because the history of agriculture in the highlands has largely been reconstructed using multidisciplinary data derived from a single landscape within it (Figures 3.2–3.3; Golson 1982a; Denham and Haberle 2008).

The Wahgi Valley is one of the largest inter-montane valleys in the highlands of New Guinea. The valley contains some of the most extensive wetlands in the highlands, totalling c. 250 km^2 (Haantjens et al. 1970: 64–65; Hughes et al. 2017). The Upper Waghi Valley landscape extends from the valley floor at c. 1500 m to the upper valley walls above 2400 m. The landscape can be classified into four main classes, each with associated soil types: wetlands (histosols); the Wahgi River and its drainage network (entisols); the slightly irregular topography of the valley floor, which consists of weathered tephra-mantled lahar deposits (andisols); and, relatively steep and unstable valley walls (entisols). The landscape sustains diverse wetland, riparian and intra-montane habitats.

The soils on the valley floor formed on lahars created by a massive debris avalanche that occurred at least 100,000 to 80,000 years ago and probably more than 400,000 years ago (Blong 1986; Pain et al. 1987: 275). The lahars are mantled by a sequence of major tephras, the youngest of which, Tomba Tephra, is at least 40,000 years old and thought to originate from an eruption of a vent on Mount Hagen (Pain and Blong 1976). The cover of Tomba Tephra on these hills is variable and 'on hills is commonly less than 1 m but it is thicker, perhaps several metres, on their flanks' (Hughes et al. 1991: 231).

Fluvial processes have modified this landscape and led to the formation of extensive alluvial fans and wetlands in low-lying and poorly drained areas on the valley floor. Over the last 40,000 years or more, these wetlands have accumulated autogenic (peat) and inwashed (silt and clay) sediments. Numerous more recent distal tephras were deposited across the Upper Wahgi Valley landscape during the Terminal Pleistocene and Holocene (Pain and Blong 1976). These are poorly differentiated on the valley floor and walls due to erosion and pedogenesis (Chartres and Pain 1984), whereas they are relatively well preserved as distinct, contiguous or semi-contiguous bands in wetlands. A tephrochronology has been established for Pleistocene and Holocene tephras in the highlands, which has been key for the relative dating of archaeological features preserved in wetlands (Table 3.4; Pain and Blong 1976; Coulter et al. 2009; Blong et al. 2017).

The Upper Wahgi Valley has a lower montane humid climate with an average annual temperature of 19°C and annual rainfall of c. 2700 mm (Hughes et al. 1991: 229). Climate in the Upper Wahgi Valley is moderately aseasonal and like much of the highlands is dominated by local orographic effects (Powell et al. 1975: 2). Seasonal variations in mean monthly rainfall, temperature and humidity are moderate, slight and slight, respectively (McAlpine et al. 1983: 70, 93 and 104–106, respectively), with very low variability in annual rainfall (McAlpine et al. 1983: 73). A slight dry season occurs in the Upper Wahgi Valley between May and June, though soil water content does not usually limit plant growth (McAlpine et al. 1983: 74, 137). Extreme climatic conditions associated with El Niño Southern Oscillation (ENSO) events caused

Table 3.4 Tephrochronology of major Holocene tephras at wetland archaeological sites in the Upper Wahgi Valley

Tephra Name/Code	Blong et al. (2017) (cal BP)	Bayesian Model (Coulter et al. 2009) (cal BP)[1]	Comparable Data (Coulter et al. 2009) (cal BP)
Tibito (Z)	304–282[2]	–	305–270[3]
Kenta (S1)	680/310–930/690	–	c. 500[4]
Olgaboli (Q)	1190–970[3]	1110 ± 171	1190–970[3]
Kuning (S2)	1690/1420–1120/930[1]	1296 ± 260	c. 1500[4]
Baglaga (Y)	2700/2340–2110/1820[1]	2190 ± 210	2650–1950[5]
Mun (NP)	2730–2120[5]	–	c. 2500[4]
Kim (R)	4140/3730–3980/3690[1]	3917 ± 49	3980–3630[5]
Komun (R+W)[6]	6440–5990[5]	–	–

Notes: Using most likely date ranges in Blong et al. (2017), Bayesian model date ranges in Coulter et al. (2009) and comparable data presented in Coulter et al. (2009).

1 Based on dates for Ambra Crater in Sniderman et al. (2009)
2 Tibito (Z) tephra most likely fell in AD 1660s (Blong 2017)
3 Following Haberle (1998)
4 Estimates based on archaeological, chronological and stratigraphic data at Kuk Swamp (Denham 2003a: Table 2.1)
5 Based on Denham et al. (2003: Table S2)
6 Denham (2003a) does not consider this to be a tephra, but a deposit primarily comprised of phytoliths (Denham and Grono 2017)

mild drought conditions in the Wahgi Valley in 1997, which were less severe in the wetlands and in comparison with other valleys in the highlands (Allen 2000).

As Bourke (2017: 64) has stated:

> The upper Wahgi Valley has an ideal environment for agricultural production, with an adequate but not excessive rainfall; mild temperatures in which crops mature more slowly and have time to accumulate more carbohydrate; less cloud cover to limit plant growth; and very fertile soils, particularly when swamps are drained of excessive water. Plant productivity is high in the valley. . . . Perhaps it is not surprising that the upper Wahgi Valley is the site of some of the earliest documented agriculture in the world.

4 Cultivation practices in the highlands

Agriculture in the past is defined in terms of the degree of involvement and dependence of communities upon the cultivation of plants for food, whether those plants are phenotypically and genotypically domesticated or not. Consequently, early agriculture in New Guinea is hereby differentiated from other forms of plant exploitation using an evidential triumvirate through which its effects are potentially visible: plant cultivation and/or use (with or without domestication), environmental transformations associated with persistent cultivation, and archaeological remains of former cultivation practices. As will be shown, the nature and diversity of contemporary practices in the highlands serve as heuristic guides for translating these lines of multidisciplinary evidence into specific and contingent interpretations of agriculture in the past. These interpretations are contingent because different conditions and varieties of practice produce different signals; they are specific because any interpretation is grounded in the archaeological evidence of past cultivation practices (Denham 2007a: 99). As a result, agricultural chronology in the highlands is heavily reliant on the multidisciplinary record from wetland sites of food production.

Traditional agricultural practices in the highlands have been extensively documented by agronomists, anthropologists and geographers (e.g., Brookfield and Brown 1963; Brookfield and Hart 1971; Clarke 1971; Powell et al. 1975; Steensberg 1980; Powell and Harrison 1982; Sillitoe 1983, 1996; Allen and Ballard 2001; Bourke and Harwood 2009). In this chapter, several key characteristics of cultivation practices are reviewed, insofar as they are relevant to the investigation of agricultural practices in the past. This chapter essentially provides a road map for the historical narratives based on detailed treatments of archaeological and multidisciplinary evidence in Part III (Chapters 6–11). This chapter lays out a novel approach to agricultural origins. Rather than starting with culture history or evidence of domestication, cultivation practices are effectively mapped spatially in the present and then projected temporally back in time.

A vegetative disposition

> Most perennial plants possess two modes of regeneration: sexual reproduction through seed and clonal reproduction through some form of vegetative propagation. . . . The relative importance of sexual vs. clonal recruitment may vary widely among plant species as well as among populations within species. For example, there are many anecdotal reports in the literature of species having abandoned sexual reproduction for some form of clonal reproduction, at least in some habitats or parts of their geographic range.
>
> (Eckert 2002: 279)

A vegetative disposition, or the use of plants through an awareness of their vegetative reproductive capacity, is a fundamental characteristic of plant exploitation practices, especially cultivation, on New Guinea. Vegetative propagation in cultivation entails the removal and planting of a reproductively viable plant part – such as a subterranean storage organ, aerial root fragment, stem cutting, sucker or vine slip (e.g., Sauer 1952; Hather 1996; Barton and Denham 2017); the resultant offspring is a clone of its parent. People also exploit feral or wild plants vegetatively, for example, by gathering tubers of yams or corms of taro while leaving sufficient in the ground to enable the plant to continue growing and produce another crop.

Vegetative propagation is often characterised as a technique used for the reproduction of root crops, such as gingers (*Zingiber* spp.), taro (*Colocasia esculenta*) and yams (*Dioscorea* spp.) (Hather 1996). In New Guinea, it predominates as the primary mode of reproduction for a wide range of crops and plant types, including: fruiting herbaceous plants (*Musa* spp.) and leafy herbaceous plants (*Abelmoschus manhiot, Rungia klossii*); grasses (*Saccharum officinarum, Setaria palmifolia*), palms (*Metroxylon sagu*), pandans (*Pandanus conoideus*), as well as root crops. These plants are cultivated for various edible plant parts, including buds, efflorescences/flowers, fruits, leaves, nuts, stem pith and a variety of subterranean storage organs (roots, rhizomes, tubers, corms) (Powell et al. 1975; Powell 1976, 1982b; French 1986).

Unusually in global terms, several grasses were domesticated in the New Guinea region under vegetative propagation for the consumption of sugar-rich pith, young shoots and stems, or inflorescences (see Sillitoe 1983; French 1986;

Bayliss-Smith 1988). Elsewhere in the world, grasses have been cultivated from seed, with domestication traits resulting from human selection for increased seed yields (glume size, grain numbers) and ease of harvesting and processing of seeds (non-shattering rachis, apical dominance) and so on (see Zohary et al. 2012). Vegetative propagation of grasses is effected by the planting of a stem-segment, stalk-segment, top of the stalk or by division of the root mass. The sugar-rich stem of *Saccharum robustum* is generally considered to have undergone initial domestication in New Guinea, with subsequent hybridisation with *S. spontaneum* most probably in ISEA or potentially in New Guinea, to generate sugarcane *(S. officinarum)* (Daniels and Daniels 1993; Grivet et al. 2004). Other vegetatively propagated grass domesticates include *Saccharum edule* (for unopened flowers) and *Setaria palmifolia* (for edible shoots and stem hearts), which are considered to have been domesticated in the lowlands and highlands, respectively.

Some more recently introduced crop plants, including the South American domesticates sweet potato *(Ipomoea batatas)* and manioc *(Manihot esculenta)*, have been readily incorporated into New Guinea cultivation practices, probably in part, because they are vegetatively propagated. However, New Guinea agriculture is not, and may never have been, exclusively vegetative. For example, two introduced plants – bottle gourd *(Lagenaria siceraria)* and cultivated wax gourd *(Benincasa hispida)* – are usually planted from seed (Sillitoe 1983 and French 1986, respectively), as are some traditional plants – amaranth *(Amaranthus tricolor)*. Similarly, other crops reproduced from seed – such as maize *(Zea mays)* and even rice *(Oryza* sp.) – are increasingly being adopted and inter-cropped in mixed gardens.

In the highlands, *Casuarina oligodon* is usually transplanted as a system of tree fallowing using seedlings rather than from seed. The seedlings develop from self-sown seed. Seedling transplantation is a cognate vegetative practice because it entails the removal, movement and replanting of the seedling rather than reproduction from the planting of seed. Similarly, grafting is another cognate vegetative practice, because it entails cultivation without sexual reproduction.

Even though people know that many vegetatively propagated plants produce viable seed and can reproduce sexually, they preferentially employ vegetative methods of propagation and transplantation, a preference that has been documented elsewhere in the world (e.g., for *Ensete ventricosum* in Ethiopia; Hildebrand 2007). Vegetative propagation is likely preferred because it offers greater control over the gene pool through genetic isolation of the curated stock and the best opportunity to reproduce desired phenotypic characteristics, such as colour, shape, size, taste, toxicity and so on. On occasions when vegetatively cultivated crops adventitiously reproduce from seed, as noted for some traditional varieties of taro, cultivators may adopt the resultant lineage as a new variety within their vegetatively propagated stock (Kennedy and Clarke 2004; see Clement et al. 2010).

Cultivation practices in the highlands have probably always been predominantly vegetative. Vegetative propagation is also characteristic of plant cultivation in permanently wet equatorial rainforests across the globe, including the Americas, Africa and Southeast Asia (Harris 1972, 1973; see papers in Denham et al. 2007 and Barton et al. 2012). The earliest forms of cultivation in the lowland neotropics (Piperno and Pearsall 1998), Southeast Asia (Barker 2014), parts of East Asia (Zhao 2011; Yang et al. 2013) and New Guinea (Yen 1985), as well as elsewhere (Weiss et al. 2006), are thought to include vegetative forms of propagation. Although seed-based and vegetative forms of cultivation are often contrasted (e.g., Harris 1969; Hildebrand 2007), they are rarely mutually exclusive. For example, early seed-based agriculture in Southwest Asia is thought to include a vegetative component (Weiss et al. 2006) and several minor crops in the New Guinea highlands are planted from seed.

The term 'vegeculture' is applied to forms of cultivation that are dependent upon vegetative propagation, as opposed to reproduction from seed (Yen 1973; Hather 1996; Shuji and Matthews 2002; Denham and Barton 2014). Vegeculture can be more broadly viewed as a common orientation to the vegetative reproductive capacity of plant resources. This orientation can be shared by foragers as well as cultivators, and be associated with distinctive social practices (Barton and Denham 2011, 2017).

Diversity of plant exploitation in New Guinea

Highly variable subsistence practices have been documented across the island of New Guinea in terms of plants, cultivation practices and people's reliance on cultivated food. In general terms, these include hunting, gathering of invertebrates, fishing (largely confined to the lowlands), rearing of domesticated animals and the exploitation of plants. The exploitation of plants for food includes variable dependence on the gathering of wild plants, the collection of tree crops and the cultivation of mostly vegetatively propagated root crops and vegetables. Staples, cultivation practices and intensity of cultivation – which is often measured in terms of technology, frequency of use and labour inputs per unit area of land – vary greatly from place to place (Brookfield and Hart 1971: 94–124; Powell 1976; Bourke 2001; Bourke and Harwood 2009).

In a recent survey of agricultural practices across Papua New Guinea, intensity of land use was measured in terms of duration of cultivation relative to the length of the cultivation cycle (active cultivation and fallow) (Allen and Bourke 2009: 39–40). Degree of intensification usually corresponds to land availability within highland altitudinal zones, rather than population density *per se*. Less intensive practices usually occur where sufficient land is available to support extensive forms of cultivation or plant exploitation, that is, shifting cultivation and arboriculture, respectively. Where social pressures on land use are higher, more intensive forms of cultivation are practised. Social pressures on land are not just demographic, namely to feed relatively large populations; they include the generation of surpluses for ceremonial use, exchange and redistribution (Modjeska 1982) and arise through socio-spatial constraints on territorial expansion in densely populated highland valleys (Brookfield and Brown 1963).

Wetlands in the main highland valleys are unusual, as they appear to have long-been foci of intensive cultivation (Golson 1977c; Gorecki 1986; Ballard 1995). People may have semi-continuously cultivated them for their prolonged fertility, resistance to drought and extended periods of crop yields. Despite overall variations in land use intensity, most communities in New Guinea employ multiple plant production and procurement strategies of variable intensity across the landscapes they inhabit or periodically visit.

Plant exploitation practices in the highlands and lowlands of New Guinea are often characterised as vegeculture in the highlands and arboriculture in the lowlands. To a degree, such a distinction has merit: there is a greater number and wider diversity of pandanus, palm and tree crops in the lowlands of New Guinea that are exploited for fruits, nuts and sago (Bourke 1996); whereas the highlands are relatively depauperate in tree crops, excepting *karuka* Pandanus (mostly *Pandanus brosimos/iwen/julianettii* with minor use of *P. antaresensis*; Stone 1982a, 1984; Denham 2007c) and *Castanopsis* nuts (R. Bulmer 1964). From a different perspective, people utilise a common suite of practices to exploit plants in the lowlands and highlands; the emphases vary because of different resource bases.

In the highlands and highland fringes, people are reliant to a greater degree on the cultivation of starch-rich staples and vegetables; their diets are supplemented by the periodic or seasonal exploitation of arboreal resources, primarily *karuka* pandanus (*Pandanus brosimos/iwen/julianettii* complex) where available (Denham 2005c). Cultivation practices vary greatly in type and intensity, including drainage of wetlands for cultivation (e.g., Ballard 2001), semi-permanent, mound and raised bed cultivation on valley slopes (e.g., Waddell 1972) and shifting cultivation in rainforests (e.g., Clarke 1971). Even for intensive agricultural practices in some of the large inter-montane valleys (Brookfield and Brown 1963; Powell et al. 1975), people maintain a repertoire of strategies, including hunting and gathering, as well as extensive and intensive modes of cultivation in plots across a range of environments (e.g., Bowers 1968; Waddell 1972; Ballard 1995). For instance, people in the highlands today may intensively cultivate drained wetlands, while maintaining house plots and dispersed mixed plots on valley slopes, claims over fruit- and nut-bearing trees and bird of paradise display trees, and rights to hunt and gather in tracts of forest.

Many highland groups have access to lower-lying and higher-altitude areas where they maintain rights and the practical knowledge to exploit tree crops, including sago stands and karuka groves, respectively. Thus, communities are not usually restricted to a particular altitudinal zone, even though they may predominantly live and cultivate within inter-montane valleys. The degrees of mobility and access to resources in different altitudinal ranges were probably greater in the past (Denham and Barton 2006). Consequently, the lowland-highland dichotomy is not a particularly useful way of organising our thinking about people living in the interior of New Guinea in the past. There are altitudinal variations in climate, animal and plant distributions, resource availability, cultures and languages, yet the lowland-highland division is an artificial way of thinking about the past. The highlands were designated as an apparently coherent entity only following 'discovery' of dense populations living there in the 1930s; the category has subsequently pervaded thinking on a whole range of other issues (Denham 2007c: 46).

In lower altitudes, there is generally a greater emphasis on tree crops and cultivation supplemented by hunting, gathering and fishing. Groups who are heavily reliant on sago *(Metroxylon sagu)* have been classified as hunter-gatherers (Roscoe 2002), even though many sago stands are anthropic – whether tended or planted. Most lowland groups are dependent to varying degrees on plant cultivation (Serpenti 1965) whereas some are difficult to classify in terms of standard 'forager-farmer' terminology (Dwyer and Minnegal 1991; Terrell 2002; Specht 2003).

Practices of cultivation

Against this backdrop of diversity, several constituent practices and technologies are common to different forms of plant exploitation in the highlands, as well as many lowland locales. Not all of them are ubiquitous; many have restricted geographic distributions. Different practices are associated with plot clearance, planting and harvesting, ground preparation and earthworks, and fallowing and nutrient cycling. These constituent practices are briefly discussed with respect to shifting cultivation, semi-permanent cultivation and different types of plots (Bourke and Harwood 2009).

Plot clearance

Plots within rainforest were traditionally cleared of undergrowth and small trees using stone and wooden tools (Figure 4.1). The resultant debris was either burned at the end of the dry season, or mulched in wetter and aseasonal areas.

Figure 4.1 Swidden plots cleared in rainforest, southern flanks of Bismarck Mountain Range

Note: Cleared plot with dead vegetation and prior to burning (left), newly planted plots (middle and right)

Source: Tim Denham, 1990

Figure 4.2 Swidden plot cleared in rainforest with retention of large tree trunks

Note: Notice tree trunks pollarded and retained within the plot in the foreground, southern flanks of Bismarck Mountain Range

Source: Tim Denham, 1990

These 'slash and burn' or 'slash and mulch' practices are designed to release nutrients into the surface of the soil, making them available for crop growth. Burning of debris within cleared plots usually occurs at the end of the dry season that releases nutrients rapidly into the soil. Given the high rainfall experienced in most areas, these nutrients are rapidly eroded or leached from the soil profile rather than bound within the soil. In contrast, mulching enables a slower release of nutrients from decaying organic matter left on the surface of a prepared plot.

Large trees are often left in prepared plots after being trimmed of low-lying branches, pollarded or ring-barked, and in some cases fires are set in holes within the trunk to kill the tree (Figure 4.2). Pollarding refers to cutting off the upper part of a tree while retaining the lower portion of the trunk after it has been trimmed of branches. Ring-barking is a technique whereby a section of bark is removed all the way around the trunk of the tree to kill it; the technique is usually applied to trees that are too large to chop down. It is often impractical to fell massive trees and root systems assist with binding the soil and preventing soil creep and landslides in exposed plots. Additionally, some trees are economically important, associated with ancestors or bird of paradise display trees; these types of significant trees are often retained within plots.

Extensive grasslands mark the location of formerly cultivated land in the highlands (Figure 4.3; Powell 1982b). In these areas, recurrent cultivation and burning hinders the regeneration of forest. Plots in grasslands are traditionally considered more difficult to prepare for cultivation than those in rainforest. The dense root mats in grasslands are hard to till, especially without the aid of pigs. This type of cultivation entails burning off the grass and complete tillage of the soil, and is often accompanied by types of ground preparation to prolong use of the plot (Figure 4.4).

Shifting cultivation plots within the rainforest are usually surrounded by substantial wooden fences to keep out feral pigs, and domestic pigs are released to roam within and around dispersed settlements. By contrast, nucleated settlements and semi-permanent plots are often entirely surrounded by wooden fences, wooden palisades or large ditches to keep out pigs and, formerly, people. Pigs are usually tethered or kept in houses within the settlement. Prior to the introduction of pigs to New Guinea within the last few thousand years (Sutton et al. 2009), people would not have needed to construct fences around plots.

Ground preparation and earthworks

Under shifting cultivation within rainforest, minimal preparation of the soil occurs prior to planting by dibbling. By contrast, extensive tillage of the topsoil may occur under grasslands. Various types of ground preparation and earthworking may occur to prolong cultivation within a plot and minimise erosion.

Figure 4.3 Forest-grassland mosaics

Note: (upper) southern flanks of Bismarck Mountain Range (Source: Tim Denham, 1990); (lower) Simbai Valley, Bismarck Mountain Range (Source: Tim Denham, 2007)

Figure 4.4 Settlement and cultivation of grassed clearing within rainforest, Lower Tagali Valley

Source: Tim Denham, 2009

Mounds

On the floors of the main highland valleys, namely at altitudes from c. 1200–2000 m, prolonged cultivation occurs using earthen mounds (of different sizes) or raised beds (of different shape) (Figures 4.5 and 4.6). These earthworks are often supplemented by the deposition of ash and composting within the mound or raised bed, as well as by *Casuarina* tree fallowing or leguminous crop rotation (Powell et al. 1975; Bourke 2009).

Sub-circular or circular mounds form part of cultivation practices across New Guinea (Waddell 1972; Bourke and Allen 2009; Hitchcock 2010). Mound dimensions vary in terms of diameter and height, with different sizes serving different functions and having discrete geographical patterns (following Bourke and Allen 2009: 251–254). Small mounds (measuring 10–40 cm high, 40–100 cm in diameter) are designed to increase the depth of topsoil and to create a drier growth environment. Today, small mounds are relatively ubiquitous across the island and are associated with sweet potato cultivation. Medium (40–70 cm high, 100–250 cm diameter) and large (> 70 cm high, > 250 cm diameter) mounds are primarily restricted to highland valleys, with compost and burnt waste incorporated into the mound prior to its formation and planting. The decomposition of organic compost within the mound increases the available nutrients for plant growth, especially on depleted volcanic soils in some highland valleys, as well as raising the soil temperature within the interior of the mound at higher altitudes. Furthermore, the microtopography of the mound shape facilitates cold air drainage around the mound and away from the plants growing within it. Mounds are an excellent innovation for they prolong cultivation, shorten fallow periods, and extend cultivation practices to higher altitudes, onto wetter soils and onto poorer soils.

The design and layout of the mounds within plots vary. Some differences are no doubt associated with specific groups and regions, whereas others are associated with the ways in which mounds are opened up during harvesting and then reformed for planting. To illustrate, two types of mound cultivation and short fallow are practised by Meldpa-speaking populations at altitudes above c. 1830 m in the Mount Hagen region: *kongderemen* (shifting-mounds) and *wenderemen* (opening-out) (Figure 4.7; Powell et al. 1975: 11).

In *kongderemen*, the mound is left to fallow for 3 to 4 months after the tubers are harvested. Prior to planting, any weeds or grasses that grow in the plot are pulled out and heaped between the mounds. Then part of the soil from the pre-existing mound is thrown on to the compost heap on each side. An individual mound shifts location between each cultivation cycle, namely, each time the plot is reworked for planting.

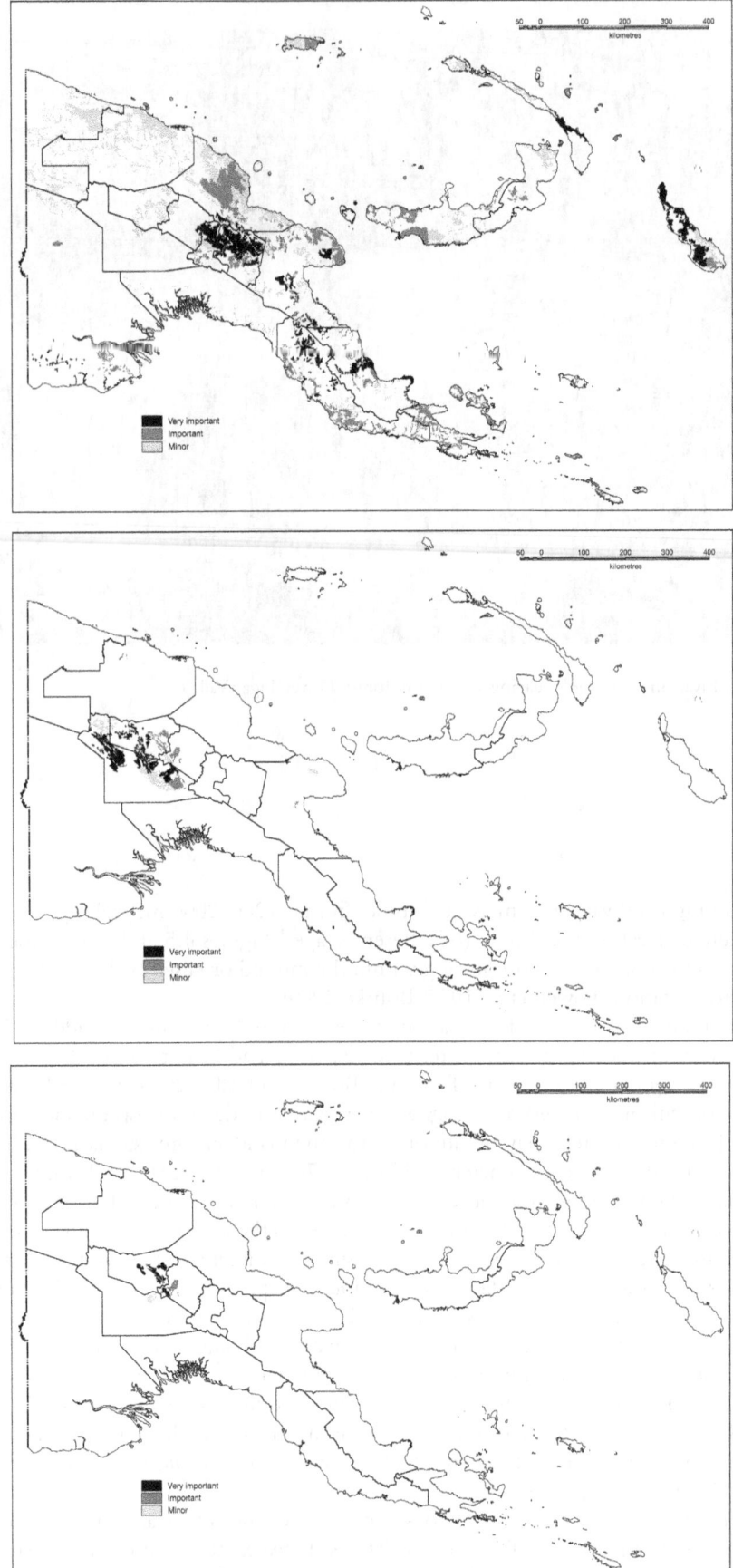

Figure 4.5 Maps showing the distribution and relative importance of mounds

Note: (upper) small mounds; (middle) – medium mounds; (lower) – large mounds

Source: Bourke and Allen 2009: Figures 3.11.1, 3.11.2 and 3.11.3, respectively

Figure 4.6a Large composted mounds under construction for planting with sweet potato *(Ipomoea batatas)*, Sirunki
Source: Robin Hide, 1990

Figure 4.6b Large mounds, newly planted with sweet potato *(Ipomoea batatas)*, Sirunki
Source: Robin Hide, 1990

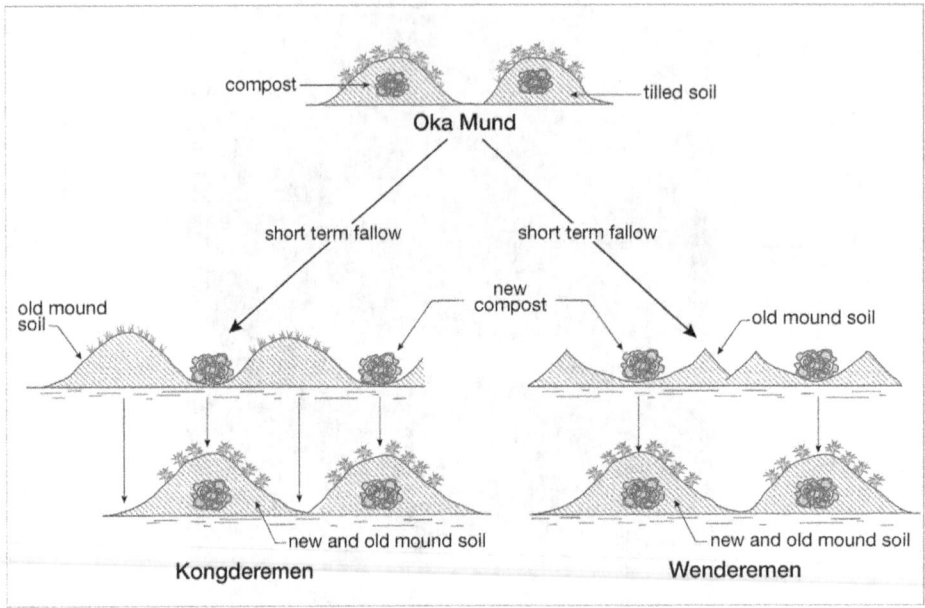

Figure 4.7 Different types of mound composting and building in the Mount Hagen area

Note: Under *kongdremen* the mound location moves during each successive planting, whereas under *wenderemen* the mound location stays the same during successive plantings

Source: Redrafted version of Powell et al. 1975: Figure 5

In *wenderemen*, the mound is opened out and left to fallow for 3 to 4 months after the tubers are harvested. Any waste is thrown into the depression created in the centre of the former mound. Upon completion of the short-fallow cycle and prior to planting, any weeds and grasses are pulled up and thrown onto the compost within the mound crater. The surrounding soil is then reworked and piled up over the central compost to form a mound, which is approximately in the same location of the former mound.

Crops are planted into the freshly tilled soil of the newly constructed mound surface. According to Powell et al. (1975: 11), yields are the same from both types of mound cultivation. The use of mounds enables the cultivation of plots for extended periods without a long fallow; 'a 30–40 year fallow is considered necessary, however, once the ground is left' (Powell et al. 1975: 11).

Raised beds

Bourke and Allen (2009: 254–255) describe the construction of raised beds (Figures 4.8 and 3.1):

> Beds are constructed by digging a grid of shallow drains at regular intervals across the garden site and throwing spoil from the drains onto the surface between the drains. Grasses, weeds and other vegetation are usually left to dry and decompose on the surface and are buried by the spoil from the drains. Additional green material may be added, in which case the beds are composted. After the harvest and a short fallow, the beds are reconstructed by digging a grid of new drains through the middle of the previous beds, filling in the old drains and throwing the spoil onto the surface of what are the new beds.

Raised bed cultivation predominantly occurs in the highlands and is locally important in some lowland locales (Serpenti 1965). Raised beds tend to be square on level and gently sloping ground, with long beds occurring on steeper sloping ground. Geographically, the distributions of the two raised beds are different within the highlands. Powell et al. (1975: 6) provide some additional information on raised bed preparation and associated composting practices in the Mount Hagen area:

> If the plots are to be used for mixed crops the ground is levelled and the soil is beaten further by the women using a short (30 cm) heavy club. . . . Any roots, or small stumps are pulled out, heaped in the centre of each [square bed] and burned. Stones and strong roots are thrown outside the garden. The ash from the burnt debris is spread over the centre of the plot.

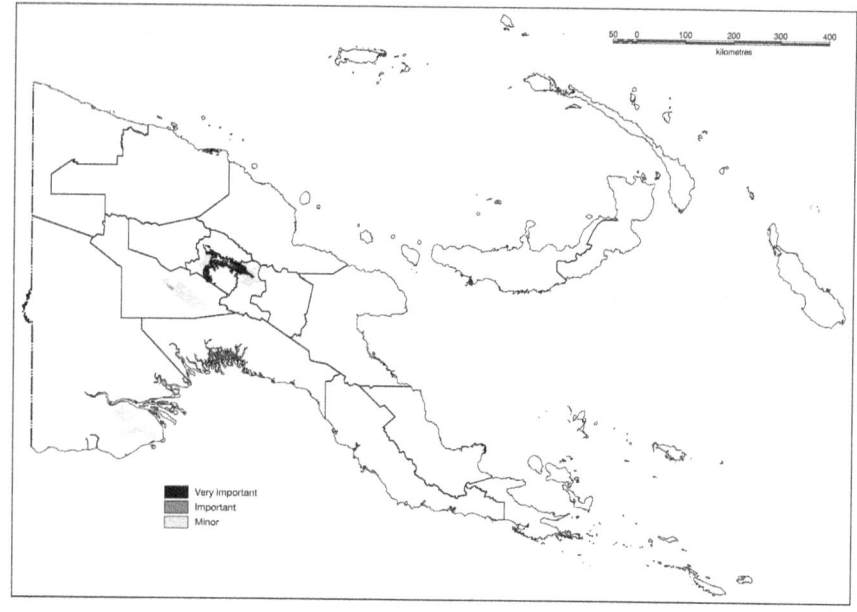

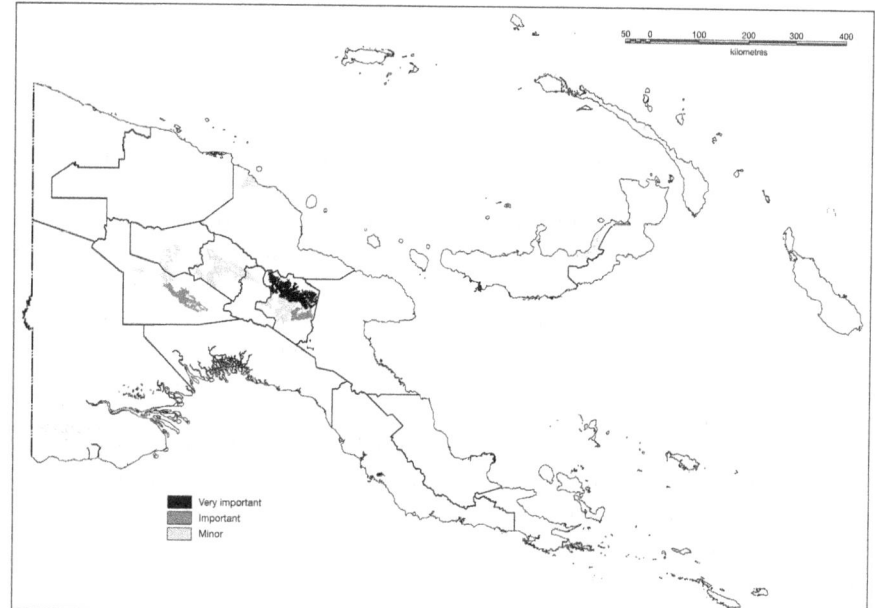

Figure 4.8 Map showing distribution and relative importance of raised bed cultivation

Note: (upper) square beds; (lower) long beds

Source: Bourke and Allen 2009: Figures 3.11.4 and 3.11.5, respectively

Both types of raised bed increase the depth of topsoil and rooting zone, prolong fertility and cultivation of a plot and enable soil water management. With altitude, raised bed cultivation gives way to mound cultivation; indeed, in some cases, raised beds are used initially to 'soften' the soil prior to mound construction (Powell et al. 1975: 10).

Ditches

The primary agronomic functions of ditches in New Guinea are to drain wetlands for cultivation and to manage water flow on valley slopes, rather than for irrigation (Figures 4.9 and 4.10). Other uses include the delimitation of boundaries, to manage pigs around settlements, for defence and so on (Ballard 2017). Ditches are differentiated here from plot drains. Plot drains are relative shallow features within or around cultivated plots used to manage surface water flow; they are relatively ubiquitous in cultivated plots across New Guinea. Ditches are larger and deeper features

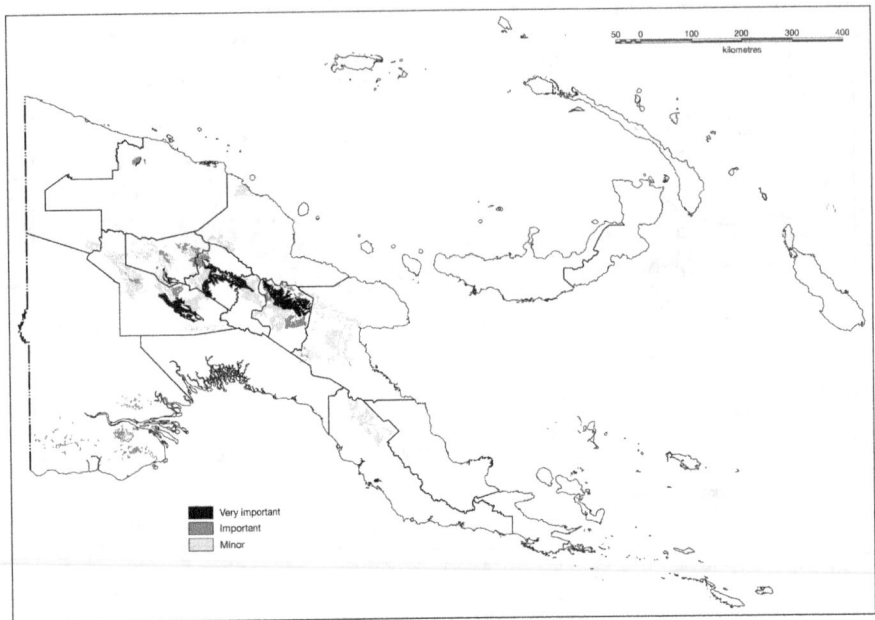

Figure 4.9 Map showing distribution and relative significance of plot drains and ditches

Source: Bourke and Allen 2009: Figure 3.12.1

used in wetlands to lower the water table sufficiently to enable cultivation on adjacent land. There can be a functional overlap between drains and ditches, for instance, relatively shallow drains may lower water tables within some plots permanently or after periods of prolonged rainfall. Additionally, drainage networks may include articulated plot drains that then flow into larger ditches. Thus, the distinction between ditches and drains proposed here is heuristic rather than precise.

Ditches have been used to drain wetlands for cultivation across the highlands of New Guinea, as well as in some low-land locales. Most suitable wetlands in the highlands have, at one time or another, been artificially drained (Ballard et al. 2013; Ballard 2017). In general, a drainage hierarchy exists within ditch networks that enables the flow of water away from cultivated areas and into a river or stream.

Even though some of the drainage networks constructed across New Guinea are extensive, they were all made by local, communal labour (Steensberg 1980; Ballard 1995; Bayliss-Smith 2007). These networks were made using relatively simple tools, such as wooden digging sticks, wooden spades, wooden rakes and hands. Following the layout and digging of ditches, cultivation of wetlands occurs using mounds and/or raised beds within the drained plots. Like grassland plots on the valley floors, some wetland plots are cultivated for extended periods, and even on a semi-permanent basis, before being left to long fallow.

An ability to grow plants in wetlands confers numerous agronomic advantages. Foremost, wetland soils are nutrient rich, because they represent the accumulation of fine alluvial sediments and nutrients ultimately derived from the top-soils and subsoils within the drainage basin. In the highlands, wetland soils often contain autogenic, or autochthonous, peats that accumulated *in situ*, as well as distal tephras from periodic volcanic eruptions. The resultant complexing of predominantly silts, clays and organics within wetlands forms durable soils with prolonged fertility and water retention (Denham and Grono 2017; Hughes et al. 2017). During the construction and routine maintenance of drainage networks, organic and nutrient rich mud is dug up and spread across the cultivated surface, thereby acting as an episodic fertiliser.

Wetland soils are not subject to the same rates of weathering and erosion as soils on valley slopes. Most highland soils on valley slopes have deep weathering profiles and are subject to various processes of erosion and leaching; they are relatively nutrient poor (Wood 1987). By contrast, wetland soils accumulate through time and, rather than being subject to chemical weathering and leaching, they are periodically replenished with dissolved and suspended sediments; they are relatively nutrient rich.

Cultivation in wetlands is not subject to the same types of climatic vagaries as plots on valley slopes. High magnitude rainfall events can rapidly erode cultivated plots through surface wash, gullying and landslides, whereas periods of drought can lead to reductions in yield or total loss of crops. As has been documented, the effects of drought are minimised in wetlands where water tables and yields are ordinarily maintained (Ballard 2000).

Figure 4.10 Photographs of field drains and ditches

Note: Former gridded field systems visible as vegetation marks on the wetland margin at Kuk (Upper; Source: Courtesy of Jack Golson, originally taken by Jim Bowler, 1972); ditched drainage of wetland for cultivation, Grand Valley Baliem (Middle; Source: Chris Ballard, 1987); cleaning of a major water disposal channel at Haeapugua (Lower; Source: Chris Ballard, 1990)

Terraces

The use of terracing is less prevalent in New Guinea than many areas of Island Southeast Asia and the Pacific (Hawai'i, Aneiytum and so on). Contour-terracing is widely used in swidden and more established plots to retard soil erosion and for soil retention. It usually takes the form of laying tree trunks along contours to capture sediment entrained and washed downslope in surface flow (see Figure 4.1). Earthen benching occurs in some lowland locales, but in the highlands is associated with the preparation of house sites. The construction of more robust terracing systems has not been reported in contemporary practices, although people have modified river terraces for cultivation.

Terraces on dryland slopes have been reported for the Arona, Goroka and Kainantu Valleys (Sullivan and Hughes 1986; Sullivan et al. 1986, 1987). The terraces were originally thought to be alluvial terraces, before a human origin was proposed (Sullivan et al. 1986: 33–35, 1987: 202). Terraces displayed 'horizontal to near horizontal platforms with steep backwalls' (Sullivan et al. 1987: 203) and varied greatly in terms of height (most 3–4 m), length (20–200 m) and width (most 10–15 m). There was considerable variability in terms of the archaeological features and sediments; for example, a drain-like feature ran along the centre of one terrace parallel to the backwall (Sullivan et al. 1986: 36, 1987: 205). From existing reports, the reported anthropogenic construction of the terraces is unlikely (Golson and Gardner 1990: 410) and the drain and other features most likely represent modification of an alluvial terrace surface for cultivation.

Planting, weeding and harvesting

Planting can comprise minimal tillage directly into ash-strewn, mulched or tilled soil, as well as into unprepared ground. A dibbling stick is used to make a hole into which the plant part, or more rarely seed, is planted. The soil is then backfilled by hand. A range of plants may be staked within plots, whether to encourage the growth of yam vines, to support stands of sugarcane or to support fruiting bananas. Weeding occurs periodically and, like harvesting, is done by hand or using a wooden digging stick to dislodge soil.

Fallowing and nutrient cycling

Rainforest soils in New Guinea are usually nutrient poor. Slash-and-burn and slash-and-mulch techniques release nutrients stored in the vegetation into the soil as fertiliser prior to planting. However, the speed of nutrient cycling within the soils of rainforests, especially at lower altitudes, results in rapid nutrient leaching, turnover and depletion. The impacts of these processes are exacerbated in newly cleared plots because exposure of the ground surface can lead to rapid erosion of topsoils and subsoils, as well as nutrient depletion. Shifting cultivation plots in lower altitude rainforests are usually cultivated only for one to three plantings before abandonment to fallow for at least fifteen years (Figures 4.11 and 4.12).

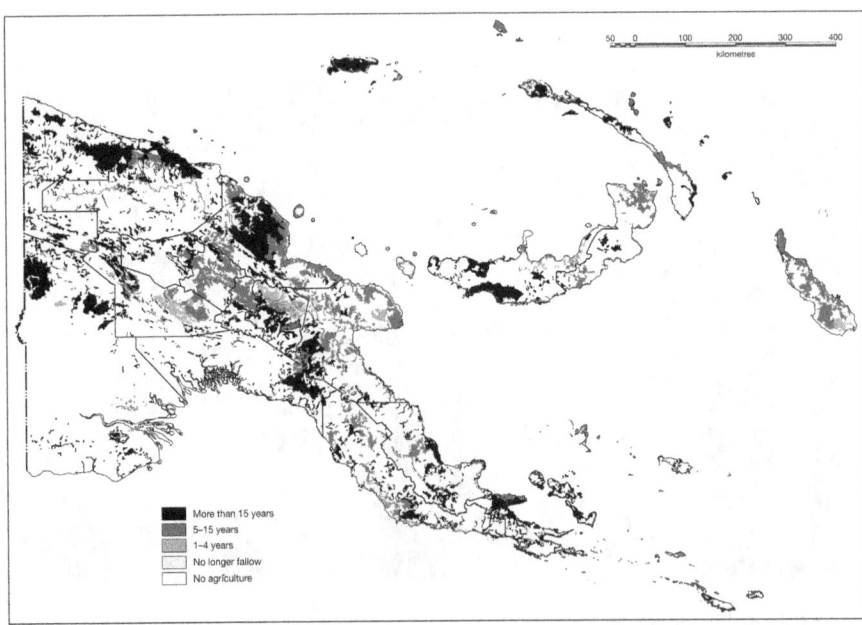

Figure 4.11 Map showing the distribution and duration of fallow period following cultivation

Source: Bourke and Allen 2009: Figure 3.8.3

Figure 4.12 Former gardens left to fallow

Note: (upper) abandoned mounds covered in grass, Lower Tagali Valley vicinity (Source: Tim Denham, 2009); (lower) saplings growing within abandoned shifting cultivation plot in rainforest, southern flanks of Bismarck Mountain Range (Source: Tim Denham, 1990)

The slightly slower nutrient cycling in montane rainforests, largely a function of decreased temperature with altitude, results in slightly more prolonged use – perhaps of two to five cycles – before abandonment to fallow for generally fewer than fifteen years. Abandoned shifting cultivation plots are still visited to obtain cuttings for propagation and to harvest perennials, including tree crops.

Under extensive swiddening practices with long fallow periods, secondary forest species will become established and eventually give way to mature forest species. For lower montane forests in the highlands today, secondary forest species commonly include *Trema*, *Dodonaea* and *Macaranga*, whereas mature species include *Castanopsis*, *Lithocarpus* and *Nothofagus* (Powell 1970a, 1982b; Haberle 2003). A mosaic of forest patches in successive stages of regrowth can occur across the landscape. If fallow periods are shortened, or if the area is subject to repeated burning, then secondary forest species will not become re-established. Rather grasses, primarily *Miscanthus* and *Imperata* species, come to predominate in abandoned plots. Through time – with continued cultivation, shortening of fallow and anthropic burning – the montane forests give way to grasslands.

Several practices have been devised to prolong fertility and cultivation of mound, raised bed and wetland plots. Usually, these involve some form of composting with ash and organic matter, often within the mound or raised bed (see Figure 3.7). Additionally, some groups within the 1400–2100 m altitudinal range practise *Casuarina* tree fallowing, whereby the seedlings of *Casuarina oligodon* (*yar* in pidgin) are transplanted in newly abandoned plots (Figures 4.13 – upper and 4.14; Powell et al. 1975; Bourke and Allen 2009: 245–247). After 8–12 years of growth, the trees are cut down and used for construction, firewood and tool-making. Although the processes are not fully understood, *Casuarina* fix nitrogen and carbon in the soil at a faster rate than areas otherwise left to fallow under secondary regrowth or grass.

In other areas, people employ crop rotations using leguminous crops (Figure 4.13 – lower). Traditionally winged bean *(Psophocarpus tetragonolobus)* and, more commonly from the 1930s–1950s, peanuts *(Arachis hypogaea)* are planted following harvesting in sweet potato plots (Bourke and Allen 2009: 248–249). These complementary plants are also inter-cropped to enhance nitrogen levels in the soil to enhance the growth of other crops, especially staples and, more recently, coffee.

Types of plot

Three types of cultivated plots are readily identified in New Guinea: house garden, mixed-crop plot and single-crop plot. House gardens are ubiquitous across Papua New Guinea. Traditionally house gardens were planted with sugarcane, green

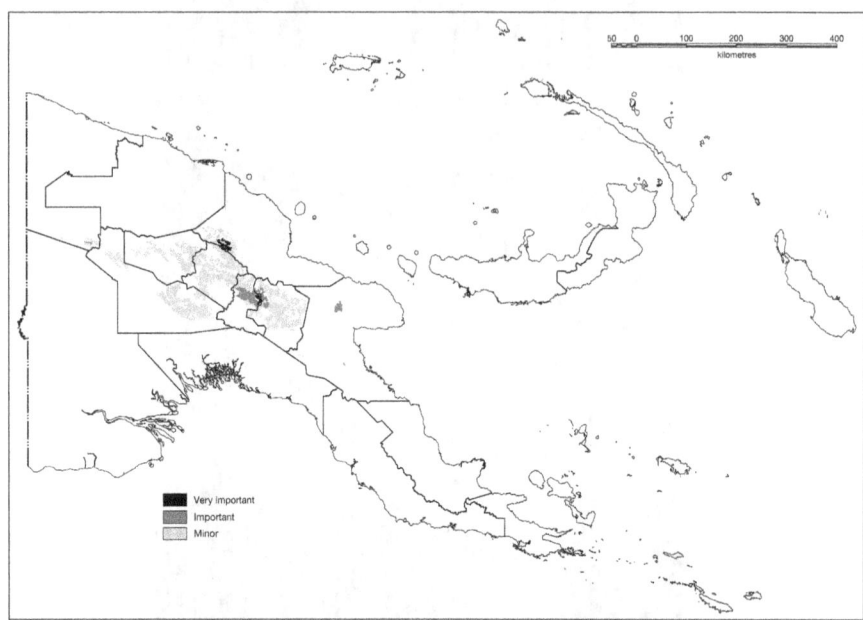

Figure 4.13 Maps depicting distribution and relative importance of tree fallowing (upper; primarily *Casuarina oligodon*) and legume rotations (lower).

Source: Bourke and Allen 2009: Figures. 3.11.6, 3.10.1 and 3.10.3, respectively

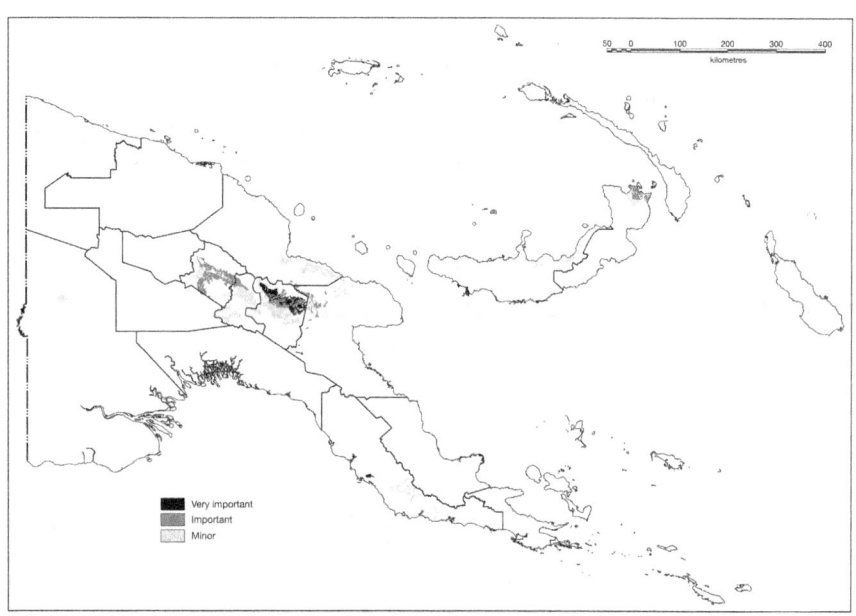

Figure 4.13 (Continued)

Figure 4.14 Photograph of *Casuarina oligodon* planted in a former cultivation plot (background) and dead *Casuarina* trees standing in a newly planted plot (foreground), Simbai Valley

Source: Tim Denham, 2007

leafy vegetables, fruit trees and ornamentals. These gardens are adjacent to houses, fertilised with domestic waste, including ash and sweepings, and continuously cultivated for extended periods.

Plots in either forest or grassland can be planted with a single crop (monoculture) or mixed crops (polyculture). A range of staples are cultivated in monocultural plots (Figure 4.15) – including bananas (*Musa* spp.), sweet potato *(Ipomoea batatas)*, taro *(Colocasia esculenta)* and yams *(Dioscorea* spp.) – and these same plants can be incorporated into polycultural plots. Mixed-crop plots are planted with a variety of inter-cropped vegetables (Figures 4.16 and 4.17). The advantage of polyculture, or mixed-crop cultivation, is complementarity in nutrient usage and potential

Figure 4.15 Photographs of cultivation plots

Note: (upper) planted with sweet potato *(Ipomoea batatas)*, with clumps of taro *(Colocasia esculenta)* and stands of sugarcane *(Saccharum officinarum)*, Simbai Valley (Source: Tim Denham, 2007); (lower) newly established taro growing in foreground, with stakes supporting winged bean *(Psophocarpus tetragonolobus)*, Karimui (Source: Tim Denham, 2008)

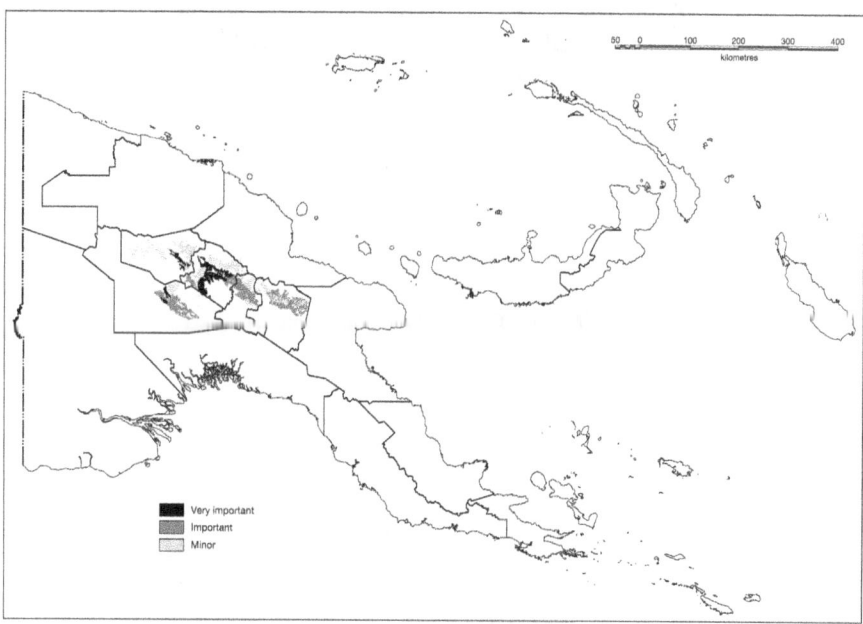

Figure 4.16 Map depicting distribution and relative importance of mixed gardens
Source: Bourke and Allen 2009: Figure 3.12.2

nutrient-fixation by different plants, as well as replication of ground cover and understorey levels of vegetation structure. Several authors have studied the interplay between mixed and single-crop cultivation systems among different cultural groups in the highlands (Bowers 1968; Ballard 1995). Often, the different types of plot form separate indigenous categories, each with their own practical and cultural associations.

Tools of cultivation

Agriculture in the highlands was traditionally undertaken using relatively simple tools (Golson 1977c). Stone adzes and axes would enable trees and understorey vegetation to be cut, perhaps aided by the use of stone hoes to clear ground cover and turn the soil (S. Bulmer 2005). Working of the soil in dryland contexts – whether forest or grassland – could be done using wooden digging sticks. In wetlands, wooden spades and rakes were used to dig and maintain ditches and associated infrastructure, whereas digging sticks were used for ground preparation, planting and harvesting. The forms of these stone and wooden tools have remained relatively constant for thousands of years (Powell 1974; Gorecki 1978).

Today, steel machetes and steel axes are the primary tools of forest clearance and steel spades are used for digging ditches. Some steel implements were traded into the highlands along localised exchange networks decades, and perhaps several decades, before direct contact with Europeans (Hughes 1977). Even so, wooden digging and dibbling sticks are still widely used for tilling soil, planting and harvesting.

Ambiguity of past practices: questions of archaeological visibility

Groube (1989) proposed that people were reliant on plants and were intentionally modifying environments to increase productivity following initial colonisation of New Guinea by at least 40,000 years ago. A landscape-based, practice-oriented approach provokes us to envisage how people engaged in a range of exploitative practices of varying intensity in diverse habitats across the highlands, and to envisage how these practices changed through time (this section reproduced in amended form from Denham 2005c: 293–294). Here, the intensity of a specific practice refers in a general way to the frequency and degree of disturbance to the biota and soils. The archaeological and palaeoecological visibility of different plant exploitation strategies varies greatly across space and through time.

Figure 4.17 Photographs of mixed cultivation plots

Note: (upper) swidden plot comprising taro *(Colocasia esculenta)* with banana *(Musa* cv.) and sugarcane *(Saccharum officinarum)* growing in the foreground and mixed taro, sweet potato *(Ipomoea batatas)* and greens in newly planted plot in the background, Bismarck Mountain Range (Source: Tim Denham, 1990); (lower) bound stands of sugarcane, with bananas in foreground, Kuk Swamp (Source: Tim Denham, 1998)

Lower-intensity practices

The exploitation of forest resources, such as hunting, foraging, modification of habitats by ring-barking, tending of favoured plants, localised burning and limited vegetation clearance and soil preparation, may leave no diagnostic trace in archaeological and palaeoecological records. Perhaps only weak signals can be inferred when the cumulative effects of these practices over millennia become apparent. For example, gradual increases in the relative proportions of useful plants within palaeoecological assemblages may be indicative of prolonged low-intensity intervention in habitats to promote the growth of favoured species, such as groves of pandans (*Pandanus* spp.) in the highlands and stands of sago *(Metroxylon sagu)* in the lowlands. Even then, the causes of these increases are not always readily identifiable. For example, elevated *Pandanus* spp. pollen frequencies in early Holocene contexts at Kuk, as well as in the pollen records of other wetlands in the highlands (Haberle 1998), may represent the successional development of swamp forest in a human-modified environment rather than anthropic resource intensification or the deliberate planting of groves (Denham, Haberle et al. 2004; Haberle et al. 2012).

The archaeobotanical visibility of low-intensity resource exploitation associated with the tending, deliberate targeting and consumption of food plant resources is sometimes claimed from the movement of species outside their natural geographic range. Such arguments have been proposed to account for the presence of a range of starch-rich crop plants in the highlands, which are thought to be of ultimate lowland origin (e.g.,Yen 1995).Vegetative propagation became the favoured means of reproduction for these plants within inter-montane valleys because they were beyond the natural range for sexual reproduction. Such a scenario is highly speculative. Only limited research has been undertaken on the natural distribution of these plants in New Guinea and on the effects of altitude and other environmental factors on their mode of reproduction.

The palaeoecological visibility of lower-intensity practices is open to question because they are less fixed in space and time; namely, they occur across a landscape rather than being focussed on a defined plot. The reading of anthropic indicators in pollen diagrams in New Guinea requires consideration of climatic and volcanic factors that may differentially sensitise environments to human-induced disturbance (e.g., Haberle 1994, 2003). The environmental impacts of similar practices may vary greatly across space and through time in response to climatic and volcanic perturbations, as witnessed during the 1997 El Niño-induced drought in the highlands of New Guinea (Allen 2000). Due to problems of differentiating human and climatic-induced contributions, caution is needed when eliciting and distinguishing the nature of lower-intensity practices solely from palaeoecological records for the Pleistocene or early Holocene.

Higher-intensity practices

In the highlands, practices with higher degrees of intervention in the management of biological and soil resources include the construction and cultivation of plots. Whether part of intensive wetland drainage systems with raised bed cultivation or a swidden horticultural plot used for only a year, agricultural practices entail the clearance of vegetation, delineation of a plot, soil preparation and planting (see Powell et al. 1975; Powell 1976). These practices are more archaeologically visible because they are more fixed in time and place.

The clearest means to identify early agriculture in the highlands is to unearth direct evidence of former cultivation practices. Early agriculture can be identified from archaeological remains of soil preparation within former plots, cultivation and planting features (e.g., stakeholes and postholes, mounds, pits), drainage features (e.g., microtopographic drainage such as runnels and large-scale ditch networks), agricultural and plant-processing artefacts (e.g., mostly wooden digging tools and some stone artefacts) and associated plant remains. Indeed, archaeological finds enable often ambiguous archaeobotanical and palaeoecological records to be anchored and directly related to cultivation practices in the past (Denham 2007a).

A practice-based method for the investigation of early agriculture

Practice refers to the habitual activities of people, whether they are a product of structuring influences, dispositions or individual whim (Bourdieu 1990). Practices are what people do. As such, practices emplace people, plants and technologies.

A practice-centred method has been developed to understand the long-term history of agriculture in New Guinea (Denham 2005c, 2009, 2011; Denham and Haberle 2008). This method is intended to create a common conceptual basis and language to discuss all forms of plant exploitation in the past or present (following Latinis 2000; Bourke 2001; Terrell et al. 2003). The intention is to free debate from the seemingly ever-present problems of differentiating foragers or hunter-gatherers from farmers, as well as attempts to clarify the transitional 'middle ground' between them. In doing so, there is recognition that forms of plant exploitation in different parts of the world need contingent, or more malleable, conceptual frameworks for classification; each region needs to be viewed on its own terms (Harris 1990, 2007). Within each region, different combinations of evidence will be significant for identifying and understanding practices that are constitutive for different forms of plant exploitation. Within tropical forest locations, the conceptual demarcations for traditional forms of plant domestication, agriculture and horticulture are likely to be especially unclear and porous.

A chronology of constituent practices

Contemporary cultivation practices in the highlands of New Guinea act as a heuristic guide for interpreting practices in the past. From the ethnographic literature, discrete practices can be identified that are constitutive for multiple forms of plant exploitation (Figure 4.18). These are effectively generic practices – such as burning, digging, exploiting tubers, exploiting nuts and so on – that are common to diverse forms of plant exploitation, ranging from foraging to farming. Other constituent practices – such as making mounds, digging ditches and constructing fences – are associated with more specific forms of cultivation.

A chronology of constituent practices can be reconstructed for a given locale using multidisciplinary evidence. For instance, in the Upper Wahgi Valley, complementary lines of evidence have been brought together from multiple sites across a defined landscape, including (Figure 4.19; see Figures. 3.3–3.4; Denham and Haberle 2008):

- archaeological excavations in wetlands reveal evidence of topographical manipulation, mounding, ditched drainage and associated practices for the last 10,000 years;
- palaeoecological reconstructions of wetlands at Kuk, Warrawau, Lake Ambra and Ambra Crater, as well as at Manim 2 rockshelter, collectively show environmental changes over the last 30,000 years;

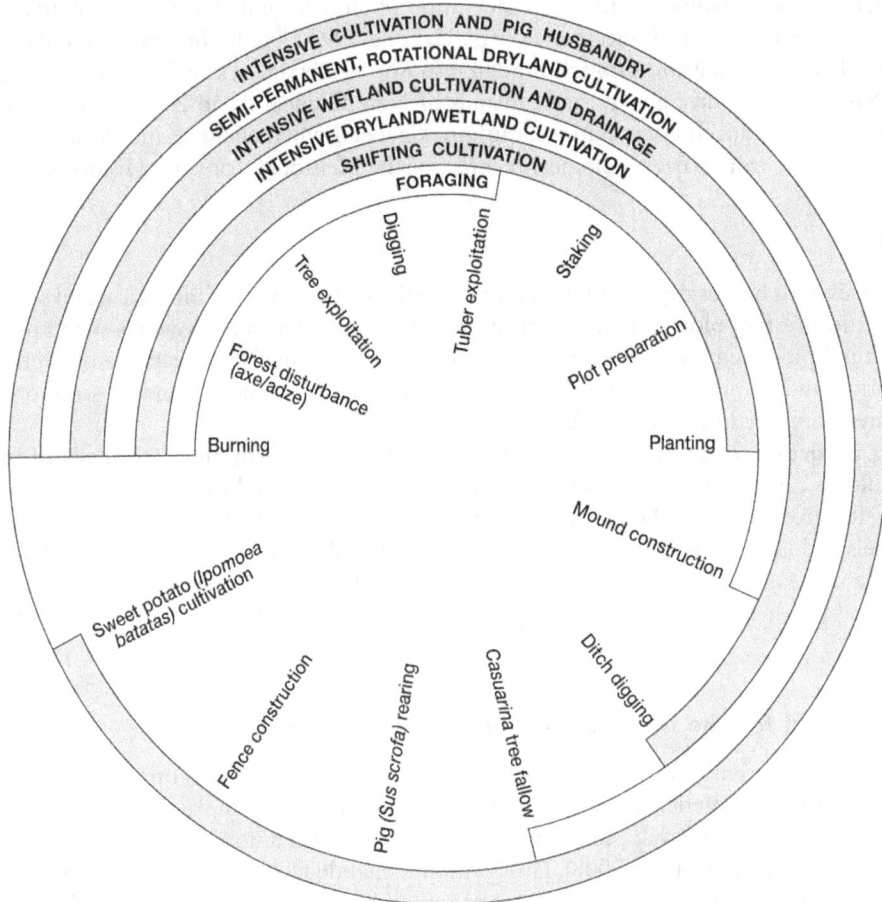

Figure 4.18 A schematic diagram illustrating the practices (inner spokes) constitutive for different types of plant exploitation and cultivation (outer rims) in the highlands.

Note: Notice the alternate shading of the outer rims is solely for ease of reference and to differentiate types of plant exploitation and cultivation

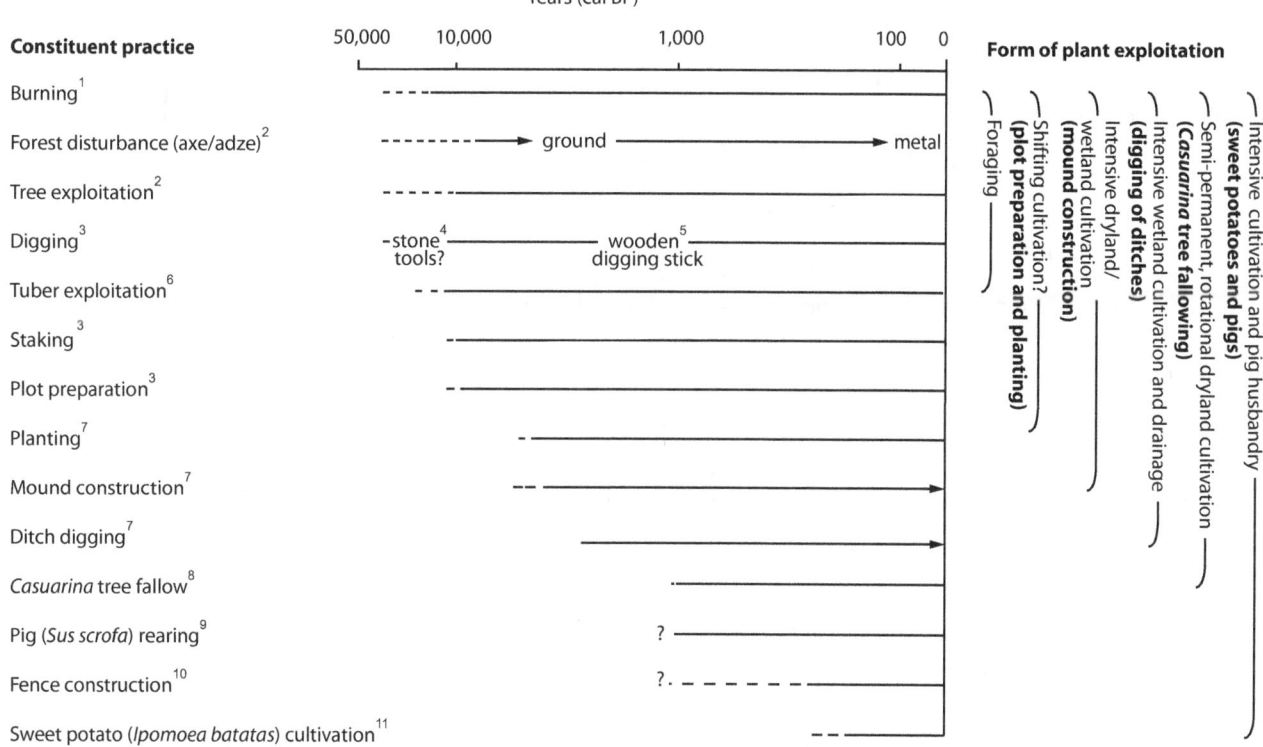

Figure 4.19 Chronology of practices and forms of plant exploitation in the Upper Wahgi Valley. Notes: [1]Denham, Haberle et al. 2004; [2]Christensen 1975a; [3]Denham 2005a; [4]S. Bulmer 2005; [5]Golson et al. 1967; [6]Fullagar et al. 2006; [7]Denham et al. 2003; [8]Haberle 2007; [9]Sutton et al. 2009; [10]Golson 1977a; [11]Golson 1982a.

Source: Denham 2013d: Figure 2

- geomorphological investigations indicate slope erosion/deposition rates for the last 10,000 years at Kuk and Manim 2, respectively; and,
- archaeological excavations of rockshelters along an altitudinal gradient in a tributary valley (Wurup Valley) – Manim 2 (1770 m), Kamapuk (2050 m), Etpiti (2200 m) and Tugeri (2450 m) – supplemented by house sites on the wetland margin at Kuk, collectively indicate changes in material culture over the last 10,000 years.

The antiquity and temporal extension of constituent practices have been reconstructed using a combination of archaeobotany (plant use), archaeology (cultivation) and palaeoecology (environmental transformation). For example, anthropic burning is inferred from elevated charcoal frequencies and prolonged decreases in primary forest taxa with concomitant increases in secondary forest taxa and grasslands (Haberle 1994; Hope 2009).

These types of palaeoecological and geomorphological data are primarily chronological; they depict change through time. In terms of plant exploitation, though, there is a need to go beyond the cultivated plot to consider how people engaged in a variety of practices across the landscape. It is thus necessary to 'spatialise' this chronological information, namely, to move beyond site-based data and consider the spatial extension of these records across a landscape (Figures 4.20 and 4.21). In concert with more spatially discrete archaeological data, it is thus possible to 'populate pollen diagrams' and move beyond the one-dimensional portrayals of human agency in them, as represented by changing pollen and microcharcoal frequencies.

The historical and geographical specificity of each line of evidence in these reconstructions is secure because each has been dated and they are derived from a single landscape. A focus on one landscape, rather than on a larger geographical region, avoids problems of conflating practices from different places that never actually co-occurred in the past (Denham 2011: S383). For instance, there are diverse plant exploitation practices across New Guinea today. If various practices from across the lowlands and highlands were brought together, they could construct an artificial agricultural system that

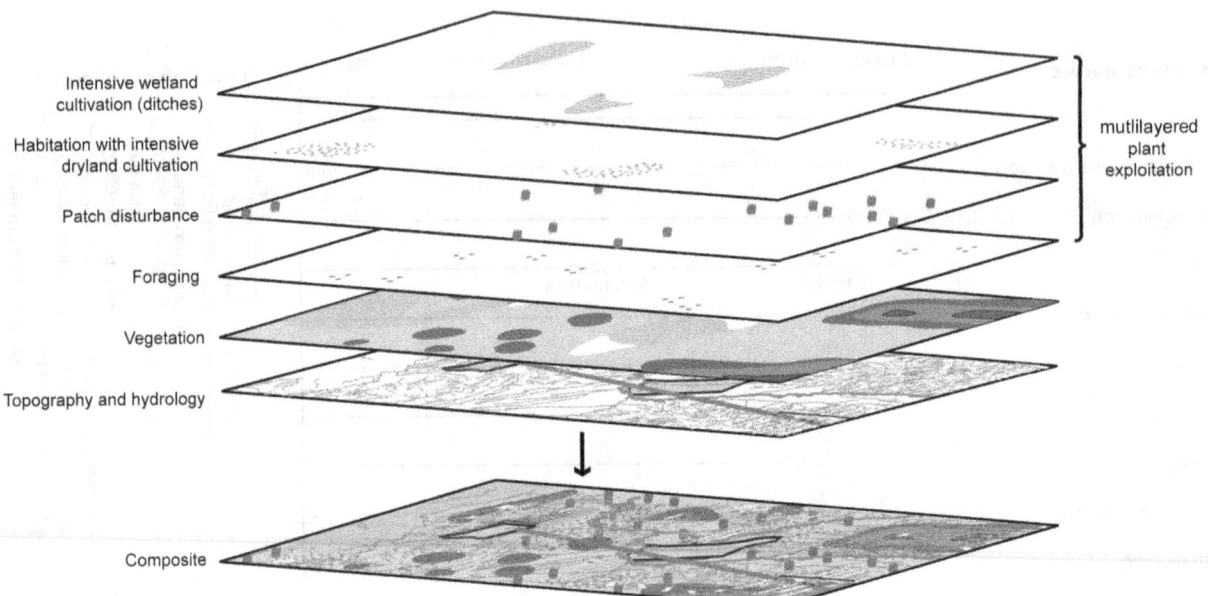

Intensive wetland
cultivation (ditches)

Habitation with intensive
dryland cultivation

Patch disturbance

Foraging

Vegetation

Topography and hydrology

Composite

mutlilayered
plant
exploitation

Figure 4.20 Schematic, multilayered spatialising of plant exploitation practices for the Upper Wahgi Valley at c. 2500 cal BP.

Source: Denham and Haberle 2008: Figure 8

is not practised anywhere on the island. Similar problems pertain when drawing evidence from diverse socio-spatial and temporal contexts together to potentially create a mistaken impression of agriculture in the past.

Bundling practices in time and place

Once a chronology has been established, the constituent practices that co-occurred in a given locale can be brought together, or 'bundled', to infer different forms of plant exploitation in the past (Figure 4.22; Denham 2005c, 2009, 2011, 2013b). The charting of individual practices and 'bundling' draws explicitly on the time geography of Hägerstrand (1970; see Gregory 2000). The interpretation of forms of plant exploitation from the temporal bundling of constituent practices is not just reliant on co-occurrence. It is necessary to have some interpretative insights and for multidisciplinary information to be grounded in archaeological evidence of past practices.

For example, the differentiation of the earliest cultivation in the Upper Wahgi Valley, as opposed to wild plant resource exploitation or intensification (Gott 2005), focuses on bananas (*Musa* spp., based on anomalously high frequencies of banana phytoliths; Denham et al. 2003), the bases of former mounds (excavated archaeologically; Denham, Haberle et al. 2004) and major environmental transformations, including the degradation of montane forest on the valley floor to grassland (from microcharcoal and pollen records; Haberle et al. 2012). It is neither just the types of banana phytolith that are significant, nor the landscape clearing; rather, the association of these archaeobotanical and palaeoecological findings with archaeological remains of former mounds are collectively suggestive of cultivation. The highly specific archaeological evidence of past practices grounds the interpretation of archaeobotanical and palaeoecological findings.

Transposing plants and practices

Even though the evidence is currently lacking, we can integrate information regarding past practices and archaeobotanical data to generate informed hypotheses regarding the stages of domestication for some food plants at specific locales in the past (Denham 2009: 665). Namely, scenarios for the exploitation, cultivation and domestication of individual food plants can be mapped onto the practice-based reconstruction (Figure 4.23). The approach can be illustrated for certain food plants that were plausibly transformed through time following incorporation into different forms of plant exploitation, exemplified here for bananas in the Upper Wahgi Valley (*Musa* spp.; following Denham 2009: 665; Figure 4.23).

Although *Musa* spp. grew in the Kuk vicinity during the Terminal Pleistocene and early Holocene, *Musa acuminata* ssp. *banksii* is the most significant taxon for the history of banana domestication (De Langhe and De Maret 1999; Carreel et al.

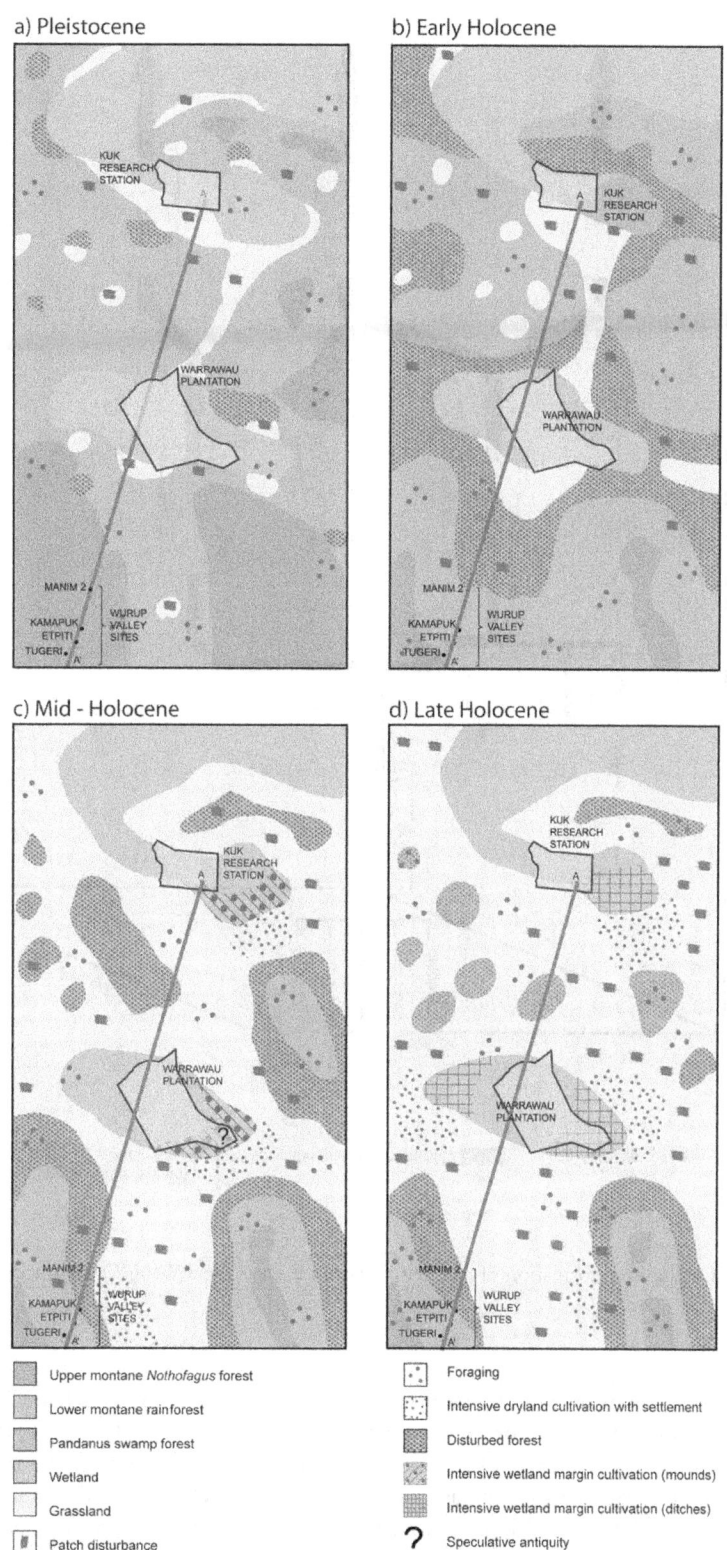

Figure 4.21 Spatial scenarios of landscape change through time

Note: (a) Terminal Pleistocene, c. 12,000 cal BP; (b) Early Holocene, c. 9000 cal BP; (c) Mid-Holocene, c. 6500 cal BP; (d) Late Holocene, c. 2500 cal BP

Source: Denham and Haberle 2008: Figure 6

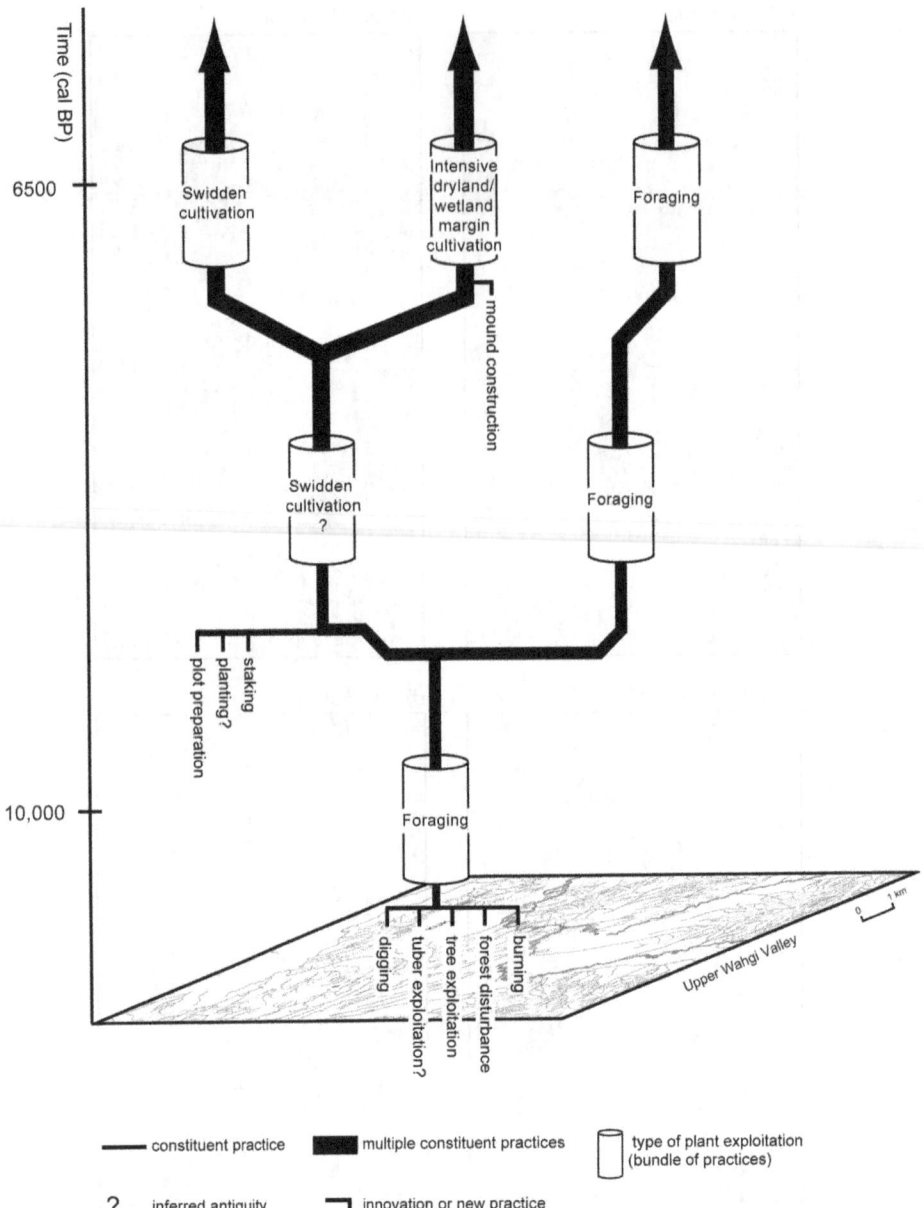

Figure 4.22 Bundling of practices and transformation of plant exploitation in the Upper Wahgi Valley during the early-to-mid Holocene
Source: Denham 2009: Figure 2

2002; Perrier et al. 2011). Bananas were initially exploited by foragers as part of broad spectrum diets, although people increasingly began to focus on starch, fat and protein-rich plants (Denham and Barton 2006). During the early Holocene, or earlier, people began to move bananas around the landscape, either through the planting of seed (sexual reproduction) or through the vegetative propagation of suckers (asexual reproduction). *Musa acuminata* ssp. *banksii* could have spread to the Upper Wahgi Valley during the early Holocene as a result of human translocation, or wetter and warmer climates, or a combination of both. It is not known whether the plants documented archaeobotanically at this time retained the potential to sexually reproduce (Lentfer and Denham 2017).

By c. 7000–6400 cal BP multiple lines of evidence indicate cultivation, including of bananas, on the wetland margin at Kuk (Denham et al. 2003; Denham, Haberle et al. 2004; Haberle et al. 2012). Genetic research suggests that the persistent cultivation of *Musa acuminata* ssp. *banksii* diploids, especially if vegetative, may have favoured the anthropic selection of parthenocarpy (production of mature fruit without fertilisation), seed suppression (reduction in size of seeds and increased

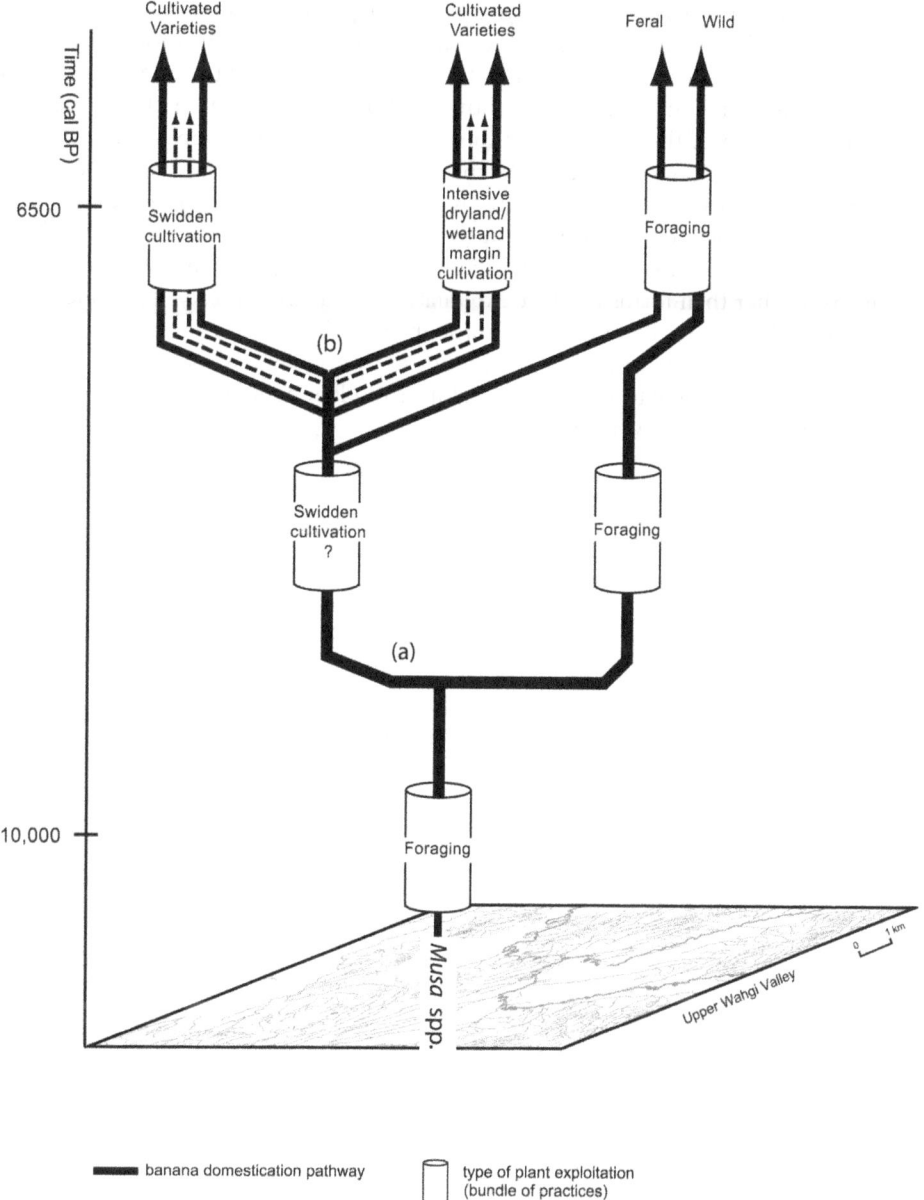

Figure 4.23 A hypothetical scenario of banana domestication superimposed upon forms of plant exploitation in the Upper Wahgi Valley during the early-to-mid Holocene

Note: (a) Represents initial planting of wild stock (cultiwild), most probably via vegetative propagation; (b) Represents more systematic vegetative propagation of cultivated stock leading to the creation of an array of cultivars

Source: Denham 2009: Figure 3

proportion of pulp within fruit), and sterility (inability of plants to reproduce sexually) (De Langhe et al. 2009; Perrier et al. 2011). Continued anthropic selection, via vegetative propagation, of cultivated plants created an array of cultivars.

Transformation through time

The sequential history of agricultural emergence and transformation in the highlands has previously relied on Golson's agricultural phases for Kuk and resultant inter-site comparison (Golson 1977c, 1982a; Denham 2003b, 2005a, 2007a; Bayliss-Smith 2007; Golson et al. 2017). Here, the use of phases as an organising principle for discussion is abandoned in favour of viewing agricultural history as a series of innovations and introductions that are then adopted and incorporated into, thereby transforming pre-existing forms of plant exploitation. Significantly, the timing of innovation and introduction may be much earlier than the timing of widespread adoption. Further, constituent practices and forms of plant exploitation do not just occur within a set period only to then be abandoned; rather, they may persist, disperse and transform through time. Within the method advocated here, there is an interpretative shift from a sequential and staged passing of thresholds to a transformative, expanding repertoire of cultivation practices.

The articulation of constituent practices into different forms of plant exploitation should not be inferred to represent a unilinear developmental trajectory. If the Upper Wahgi Valley is considered in isolation, or if New Guinea is considered as a whole, multiple forms of plant exploitation occurred at the same time in different places, as they still do. Different forms of plant exploitation should be viewed as an expanding repertoire deployed by people in different ways in different social and environmental contexts. Rather than forms of plant exploitation being sequentially abandoned, most were retained and had different levels of significance in different locales through time. Yet some forms of plant exploitation may have been abandoned in the past, whereas other formerly more significant forms may now be marginal (Terrell 2002; cf. Balée 1994). Theoretically, it may even be possible to identify novel forms of plant exploitation, namely those that occurred in the past for which there are no contemporary analogues.

5   The plants of highland cultivation

Until about twenty years ago, New Guinea was considered a relatively minor area of plant domestication (Sauer 1952; Zhukovsky 1962; Hawkes 1983; Vavilov 1992). Although a range of indigenous plant domesticates were known (Barrau 1955; Yen 1973; Simmonds 1976a, 1976b), most major staples were inferred to have dispersed in ancient times to the island from Southeast Asia and in post-Magellan times from the Americas. Recent archaeological, agronomic, genetic and linguistic research is beginning to reveal more complex histories of plant domestication, primarily through vegetative propagation, in the New Guinea region (following Lebot 1999; Grivet et al. 2004; Perrier et al. 2011). A new picture is emerging, in which the island is likely to be a major centre of plant domestication for a range of globally significant subsistence and cash crops.

Here, the domesticatory relationships between people and food plants in the highlands are considered in the present and hypothesised for the past. The ecology of major and minor crop plants are reviewed, with particular attention to the ways people exploit them. The last section considers the archaeobotanical visibility of cultivation practices and domestication processes for vegetatively propagated plants in the highlands.

Domesticatory relationships, degrees of domestication and cultivation mosaics

Domesticatory relationships refer to the ways in which people engage with plants in their world. The idea of characterising people-plant relationships in these terms is to avoid simplistic distinctions of 'wild' versus 'cultivated', or 'domesticated', plants (Harris 1990). These botanical distinctions are often prioritised at the expense of understanding the social-biological complexities of human-plant domesticatory relationships in the present and the past. Within the New Guinea context, the binary differentiation of wild: domesticated has limited relevance for many plants under traditional forms of plant management and cultivation (Yen 1990, 1991; Kennedy and Clarke 2004). Rather, it may be more appropriate to view different cultivated varieties and subspecies of a crop plant, as well as different species of crop plant, as existing along degrees or gradients of domestication (Denham 2005c: 300).

In the highlands, some crops are planted from 'wild' (cultiwild) and cultivated varieties (cultivars) (De Langhe et al. 2009). Plants considered 'wild' may include formerly cultivated plants that are now feral, plant lineages that were never cultivated and, potentially, spontaneous sexual offspring. The resultant degrees of domestication reflect the cumulative effects of human interference in the life cycle and dispersal of plants over the long term.

The character of phenotypic and genotypic domestication under vegetative propagation is not fully understood and may be qualitatively different to that under sexual forms of reproduction. Cultivators select plants based on desired phenotypic traits, including morphology (e.g., size, shape, colour, taste) and behaviour (e.g., respective timings of key stages in life cycle, environmental tolerance, disease resistance). Through vegetative propagation, people are able to genetically isolate cultivated stock and thereby prevent in-breeding with wild populations and subsequent dilution of selected traits. People may be able to establish greater control over phenotypic traits within the natural range of the plant under vegetative propagation than is possible under sexual reproduction; the latter entails continual interbreeding with wild populations and other cultivated varieties that dilute selected traits. Additionally, vegetative propagation enables reproduction of a plant not only outside of its natural range, but also outside the range in which it produces viable seed or has suitable pollinators. The clonally reproduced stock is subject to somatic mutation.

The traditional twin pillars of domestication – namely, directed selection (or human selective pressure) and genetic isolation – provide an organising framework for understanding the domestication of plants under prolonged vegetative propagation. Domestication is often characterised in terms of a balance between these two sets of mutually reinforcing processes, with the previously under-appreciated influence of genetic isolation now being better understood (Jones and Brown 2007; Larson et al. 2014; Marshall et al. 2014). These pillars are not mutually exclusive processes; many practices through which genetic isolation is achieved represent direct human selection of favoured phenotypes, whereas others may be inadvertent.

Different degrees of human selective pressure are exerted on plants through various practices in different forms of plant exploitation (Figure 5.1). Stepped changes in the degree of human intervention in the life cycle of plants can be hypothesised to include selective exploitation of favoured ecotypes; management of favoured groves and stands through tending, weeding and ownership; resource intensification through burning and transplantation; resource production through cultivation; and so on through to the technology of modern agribusiness. Additional selective pressures are exerted over the long term by the types of location within which plants are grown, cropping regime, degree of tillage, soil enrichment, pest protection and so on. Significantly, these selective pressures need not be unilinear, or cumulative; they can vary in time and place. For instance, plants could be propagated from wild-types or long-cultivated stock, then abandoned and left to go feral, from which they can then be propagated or just gathered. As a result, plants of individual species within a single landscape, and even within a single cultivated plot, can vary greatly in terms of the genotypic and phenotypic inheritance attributable to human selective pressures.

Vegetative propagation enables a greater degree of control over the cultivated gene pool within the natural range of the plant than is commonly achieved under sexual reproduction. Different degrees of genetic isolation are enabled in New Guinea for vegetatively propagated plants primarily through movement beyond natural range, elimination of wild populations through extensive environmental degradation, the development of sterile forms and cultivation practices that inhibit sexual reproduction (Figure 5.2). These practices can all result in the reproductive (i.e., genetic) isolation of cultivated plants. Movement beyond natural range prevents fertilisation of cultivated plants by wild plants of the same species, although sexual reproduction with other cultivated plants of the same species may still occur in some circumstances. Large-scale environmental transformations, such as forest clearance or climate change, may remove wild plants from the vicinity of cultivated stock and pollinators (Denham 2007d). Continuous vegetative propagation beyond the natural range isolates the gene pool of favoured varieties, while the development of sterile forms also isolates cultivated stock genetically. Harvesting prior to flowering or seed-set hinders the ability of a plant to reproduce sexually.

If New Guineans have been altering their environment from initial colonisation, it would be anticipated that plants even within forest environments will have adapted, albeit in a limited way, to the deliberate and inadvertent selective pressures exerted upon them (Yen 1989; Hather 1996). For considerable periods of the past, green leafy vegetables and herbs were likely propagated from 'wild' sources. These wild sources included plants growing in the vicinity, self-sown offspring, wild 'bush' stock and garden escapees. Other plants, particularly starch-rich staples, were plausibly planted from more highly conserved, 'cultivated' stock over prolonged periods, such as various species of banana (*Musa* cvs), sugarcane

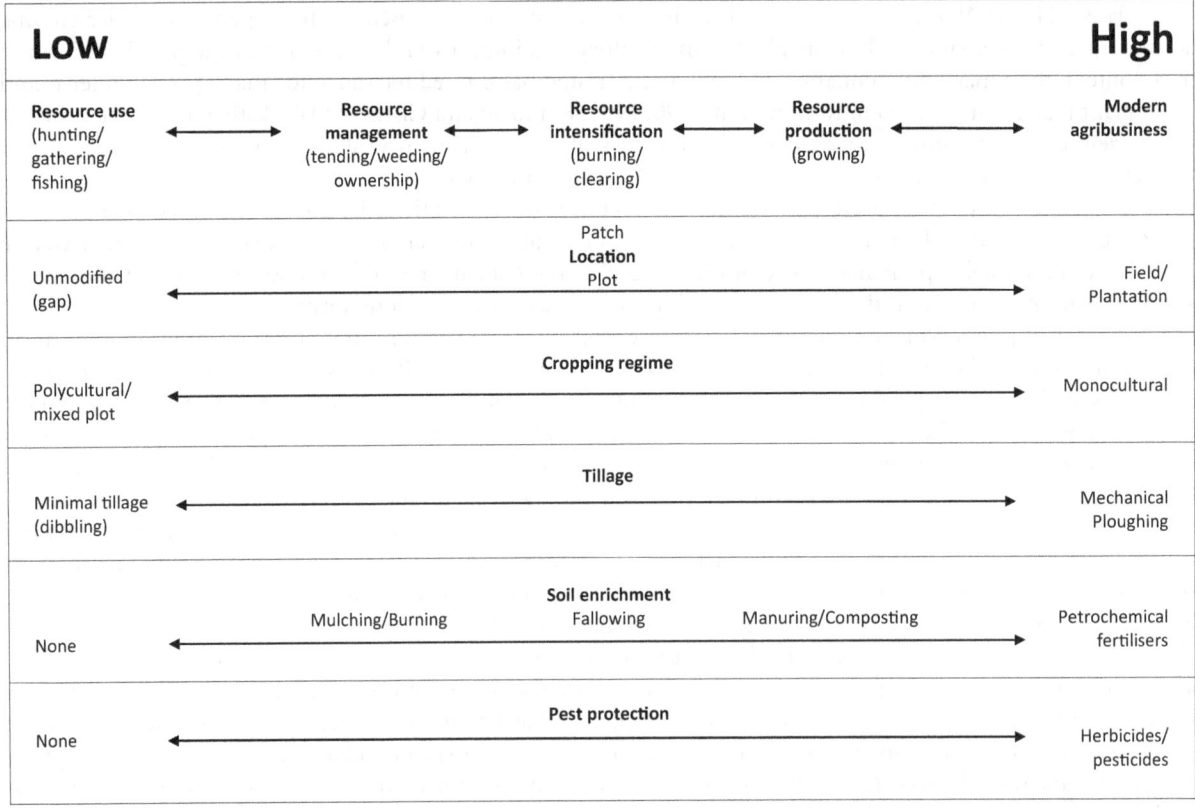

Figure 5.1 Relative influence of different cultivation practices on human selective pressures, as relevant to understanding plant domestication

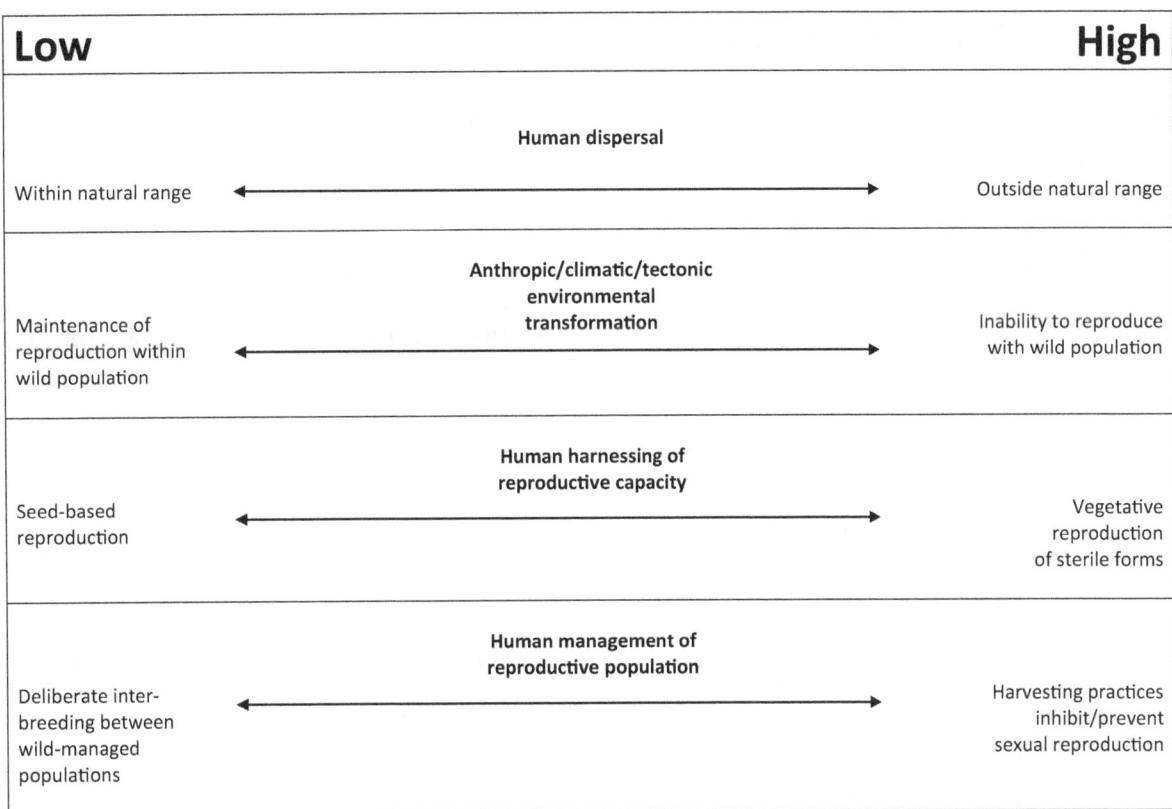

Figure 5.2 Relative influence of different cultivation practices on genetic isolation, as relevant to understanding plant domestication

(Saccharum officinarum), yams *(Dioscorea* spp.) and taro *(Colocasia esculenta)*. These more intensively managed staples are likely to have undergone greater directed selection and genetic isolation.

Through time, the cumulative effects of direct and indirect selection on plants and plant parts within the human niche (Smith 2012) will have been highly variable; some plants will have undergone only slight transformation from wild-types, whereas other plants became so heavily selected and hybridised that multiple domesticated cultivars were created. These species-specific effects vary from place to place, as well as through time, in response to the types of practices within which they are, and have been, enmeshed. Consequently, there are gradients or, more accurately, mosaics of varying degrees of domestication for different species, subspecies and varieties of plant grown within individual gardens, as well as for plants under different types of exploitation regime across the landscape (Table 5.1).

Despite prolonged histories of exploitation, wild and cultivated plants of some species do not seem to have developed clear or permanent morphogenetic diagnostics of domestication. As a result, some cultivated plants have been referred to using concepts of semi-domestication (Yen 1991) and degrees of domestication (Caballero 2004; Denham 2005c). Degrees of domestication exist between the extremes of rarely utilized 'wild' plants in largely unmanaged environments and intensively managed 'cultivated' plants in intensively managed landscapes. The reasons for the uncertain domestication status of many plants can be summarised as follows:

- interbreeding or gene flow between wild and cultivated populations within the natural range of plants, with subsequent incorporation of sexually reproduced offspring into the vegetatively propagated stock (cf. Clement et al. 2010);
- recurrent propagation of many vegetables and greens from wild or feral (abandoned or garden escapees) plants growing within disturbed forest and old garden sites, as opposed to solely from vegetatively cultivated stock (Powell 1970b, 1982b);
- prolonged clonal reproduction of plants that are selected on the basis of phenotype and which accumulate genetic traits through somaclonal mutation rather than through sexual combination; and,
- the phenotypic plasticity of many plants under cultivation, most notably documented among tuberous crops such as taro and yams, that leads to a lack of clear correspondence between phenotypic and genetic diagnostics for wild and cultivated plants, thereby hindering the ready identification of domestication traits (Dumont and Vernier 2000).

Table 5.1 Hypothetical degrees of domestication for cultivated plants in the highlands

Domestication Status	Scientific Name	Common Name	Modes of Cultivation
'Wild' Grows by itself	*Pandanus antaresensis* *Pandanus brosimos* *Castanopsis acuminatissima*	Wild *karuka* Wild *karuka* Castanopsis	Gathered, rarely transplanted by people
	Colocasia esculenta *Commelina diffusa* *Ficus copiosa* *Ficus dammaropsis* *Musa* cvs *Oenanthe javanica* *Pandanus julianettii* *Setaria palmifolia*	Diploid taro Wandering Jew *Kumu musong* Highland *kapiak* Diploid banana Oenanthe *Karuka* Highland *pitpit*	Cultivated intensively: mostly propagated from cultivated plants – with wild, feral or self-sown plants variably exploited and cultivated
	Abelmoschus manihot *Psophocarpus tetragonolobus* *Rorippa schlechteri* *Rungia klossii* *Saccharum edule* *Saccharum officinarum*	Aibika Winged bean – Rungia Lowland *pitpit* Sugarcane	Cultivated intensively: primarily only cultivated plants grown in highlands
'Domesticated' Grown by people	*Colocasia esculenta* *Dioscorea alata* *Musa* cvs *Pandanus conoideus*	Triploid taro Greater yam Triploid banana *Marita*	Cultivated intensively: asexual reproduction of cultivated plants only

Note: 'Wild' or feral types of some bananas, yams and taro, as well as various green vegetables are known in highland valleys (e.g., in the Wahgi Valley; Powell et al. 1975; also see Sillitoe 1983; Cook and Webster 2016). Common names in italics are pidgin terms

Loss of sex in vegetatively propagated plants

Prolonged asexual reproduction in a plant, whether humanly or environmentally induced, can lead to the suppression of characteristics important for sexual reproduction, thereby necessitating continued reliance on vegetative propagation. This has been noted for clonally reproducing plants generally, even though the precise mechanisms are poorly understood (Eckert 2002). Following Eckert's observation, little is known about the effects of prolonged vegetative cultivation and human-directed 'altitudinal forcing' on the loss of sexual reproductive capacity in the highlands.

One scenario could be that people initially began to cultivate plants at the 'edge-of-the-range', where plants potentially had a diminished, or had already lost, sexual reproductive capacity. People could have locked on to this phenological characteristic during the earliest forms of plant management and cultivation in the highlands. Such a scenario suggests people initiated agriculture in the highlands where plants were at or near sexual reproductive limits (following Denham, Haberle et al. 2004), rather than in the lowlands (see Hope and Golson 1995; Yen 1995).

Based on a consideration of a range of New Guinea cultivars, some appear to have lost sexual reproductive capacity – including marita *(Pandanus conoideus)*, greater yam *(Dioscorea alata)*, some banana cultivars – whereas others have developed phenotypic traits suggesting a partial loss of reproductive capacity. Where beneficial, these often adventitious traits were selected by people; for example, reduced seed size and increased pulp size enhance the mastication and caloric yield of some fruits, such as bananas. Some of the important changes to the phenology of vegetatively propagated plants in New Guinea are briefly discussed here (extracted, expanded and amended from Denham and Barton 2014 and Denham 2017: 42–43).

Parthenocarpy is the development of fruit to maturity without fertilisation. It is a genetic mutation arising spontaneously in some plants. Parthenocarpic fruits are preferentially selected due to smaller and softer, vestigial or embryonic seeds, such as in bananas (*Musa* cvs.; De Langhe and de Maret 1999) and figs (*Ficus carica*; Condit 1947). Parthenocarpic plants can often reproduce sexually if fertilised, such as occurs with some diploid banana cultivars (Perrier et al. 2011) and all cultivated figs (Denham 2007d). The development of parthenocarpy frees a plant from its natural range; climates within which sexual reproduction is viable in terms of reproductive phenology, pollen sources and pollinators; and degraded environments in which wild populations have been eradicated. A parthenocarpic plant confers a major agronomic advantage: it is still productive beyond the range in which it can sexually reproduce.

Sterility denotes the inability of a plant to reproduce sexually and is often achieved through polyploidy, such as the development of triploids in bananas (Perrier et al. 2011) and taro (Matthews 2004), and various polyploids in yams (Arnau et al. 2010). Ordinarily, plants and animals are diploids, namely made up of two sets of chromosomes derived from each parent. Plants have an ability for polyploidy, in which multiple sets of chromosomes become incorporated into the genome. Even-numbered polyploids can sexually reproduce, whereas plants with odd-numbered sets of chromosomes are unable to sexually reproduce. Triploid cultivars, although sterile, are often favoured because of greater starch production, as well as increased robustness and tolerance to environmental stress, pests and disease. Sterile cultivars are reliant on humans to be moved geographically, such as the movement of AAB plantains across the Indian Ocean to Africa (Perrier et al. 2011), although some can self-spread vegetatively given the right environmental conditions. Most banana and taro cultivars in the highlands were traditionally diploids (Kennedy and Clarke 2004).

Effective sterility is achieved through asynchronic flowering in male and female plants. Pollen production and fertilisation is feasible and can occur in laboratory conditions; however, it does not ordinarily occur because male and female plants are sexually active at different times of the year. This behaviour has been noted in the greater yam (*Dioscorea alata*; Abraham and Gopinathan Nair 1991), and it is not known whether asynchronous flowering is a result of human selection or not.

Seed suppression, namely the reduction in size of seeds so that they are no longer viable, is selected for by people because it increases the relative proportion of edible pulp or starchy part of the fruit. Seed-suppressed cultivars are presumably already parthenocarpic, no longer able to reproduce sexually and are reliant on vegetative reproduction. For example, vestigial seeds are present in most of the major cultivar groups of banana (*Musa* spp.). In this respect, breadfruit (*Artocarpus altilis*) and breadnut (*Artocarpus camansi*) are unusual, although the domestication history of these two crops in the New Guinea region is not fully understood (Zerega et al. 2004). Most breadfruit cultivars in the Indo-Pacific have been selected for edible pulp, with a concomitant reduction in size of the seeds that are not eaten, whereas on the mainland of New Guinea, people selected breadnut for the consumption of seeds, namely the seeds have increased in size and are eaten while the pulp is discarded.

Hypothetical domestication scenarios for the highlands

Based on the foregoing discussion, hypothetical scenarios can be proposed for plant domestication under different forms of plant exploitation in the highlands of New Guinea. The scenarios schematically depict the ways in which plant domestication relates to the emergence of agriculture from pre-existing plant exploitation practices. The focus is upon the hypothesised rates of change in genotypic and phenotypic attributes, hereby referred to by the degree of domestication of plants under different forms of plant exploitation. The first scenario sketches the degrees of directed selection under different types of plant exploitation and associated degrees of domestication (Figure 5.3). The second two scenarios consider the differences in the nature of the domestication process within and outside of the natural range of plants, namely, looking at the effects of genetic isolation (Figures 5.4 and 5.5).

The domestication episode refers to the period over which there is directional change in genotypic and phenotypic traits resulting from human exploitation – including cultivation – until they become 'fixed', namely, until they become prevalent in cultivated populations. Genotypic traits accumulate in clonally propagated plants largely through somaclonal mutation, whereas phenotypic traits associated with increased yield, decreased toxicity and acridity and so on may be more plastic; namely, plants that exhibited domesticated phenotypes may revert to 'wild-type' once the environment of growth reverts to a more 'natural' state. The clearest examples of phenotypic plasticity are tuberous plants, in which large, starchy tubers under cultivation can revert to small, fibrous tubers when left to grow feral (Denham 2008b). The fixation of phenotypic traits in vegetatively propagated plants may work differently to that in sexually reproduced plants, such as cereals and legumes (Lebot 2009; Fuller et al. 2014); potentially, the fixation of domestication traits is a relative concept. As noted for cereals, the fixation of different domestication traits – such as seed size and non-shattering – need not occur at the same time (Fuller et al. 2014). Primary domestication traits are fixed prior to the generation of numerous varieties that share these traits.

For plants within their natural range, pre-cultivation directed selection occurs under extensive forms of plant management, resource intensification and transplantation within the landscape (Figure 5.4). The rates of change and accumulation of phenotypic traits under these practices are variable, intermittent and non-cumulative. The increased influence of directed selection within the natural range is largely offset by interbreeding with wild populations. Even with the initiation of planting in especially prepared plots, clear domestication signatures may not rapidly emerge in cultivated stock because of persistent interbreeding between wild and vegetatively cultivated plants, the periodic propagation of wild and feral plants, and the incorporation of spontaneous offspring. The rate of change to 'fixation' is dependent, in part, upon the degree to which people persist in the selection and the vegetative propagation of specific phenotypes. As

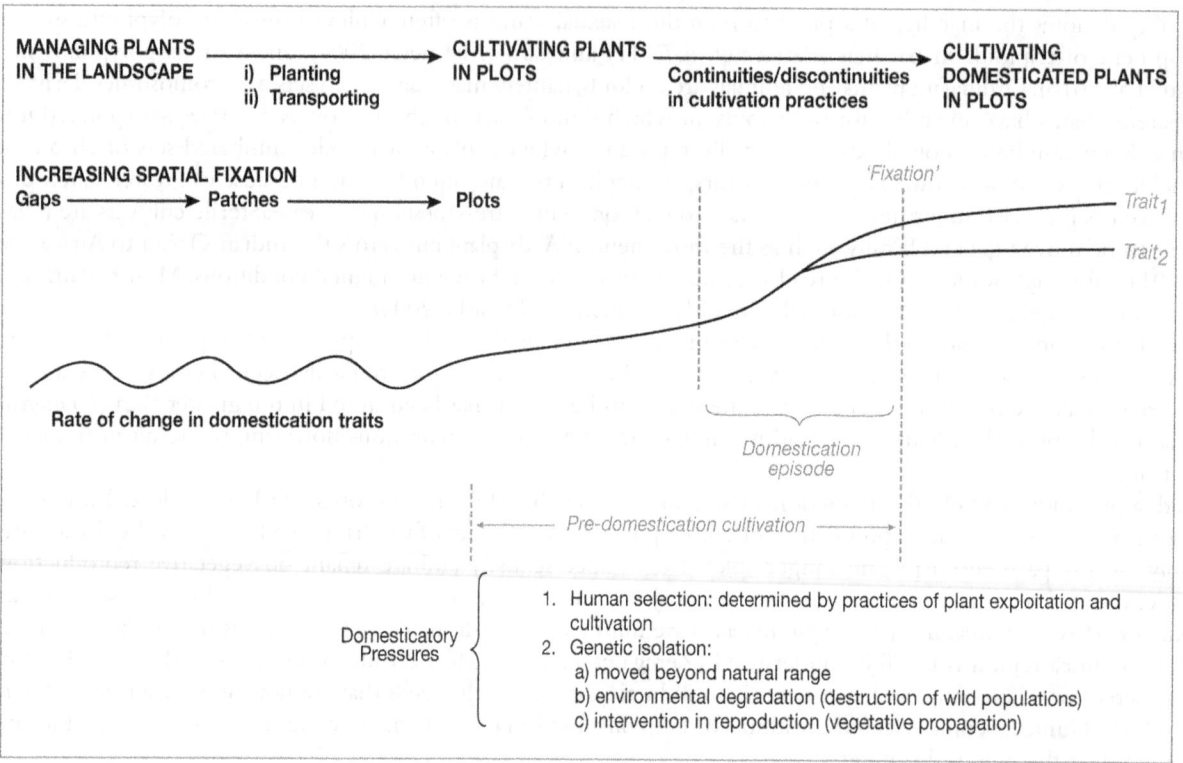

Figure 5.3 Hypothetical scenario of plant domestication in New Guinea highlands, which links practices of cultivation (in black) to the twin domesticatory pressures of human selection and genetic isolation (in grey)

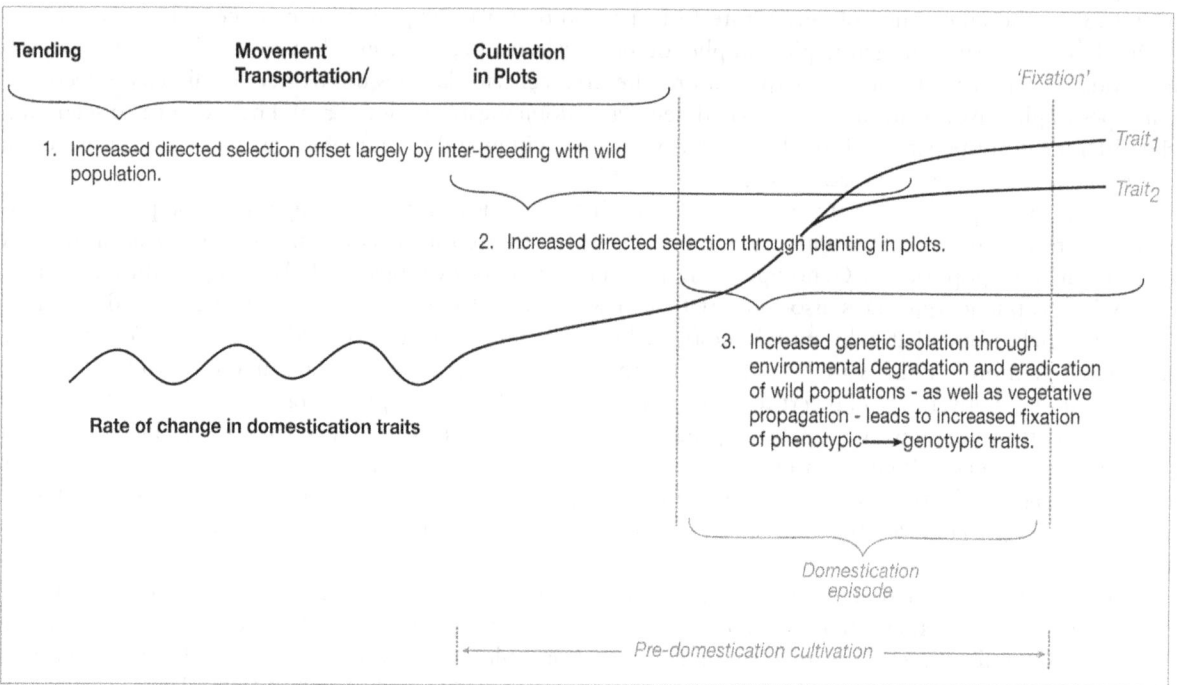

Figure 5.4 Hypothetical scenario of plant domestication within the natural range

Note: Period of pre-domestication cultivation is longer than the domestication episode. The accumulation of phenotypic→genotypic traits is directed by cultivation practices and the degree to which they restrict interbreeding with wild/feral populations. The rate of accumulation of domestication traits increases with landscape degradation and eradication of proximal wild/feral populations

such, plants may be subject to short or long periods of pre-domestication cultivation. The emergence of domestication traits – through the partial fixation of distinctive phenotypes and the accumulation of somaclonal mutations – may follow prolonged cultivation and is dependent on the genetic isolation of cultivated stock from wild and feral plants.

Within the natural range of the plant, fixation is achieved through continued vegetative propagation, the loss of sexual reproductive capacity and widespread environmental degradation. Under these conditions, cultivated stock is no longer able to interbreed with wild or feral plants, leading to the accumulation of somaclonal mutations and more pronounced phenotypic differentiation in cultivated stock. In the New Guinea highlands, many cultivated plants were probably not effectively isolated genetically until after widespread forest clearance and the establishment of grasslands on the floors of some of the main inter-montane valleys, thereby eliminating local populations of wild and feral stock. The domestication episode within the natural range of the plant could be relatively short and does not correlate with the initiation of cultivation; rather, domestication may greatly post-date the inception of cultivation and is more dependent upon when genetic isolation is achieved.

By contrast, distinctive genotypic and phenotypic traits associated with human transplantation and manipulation may accumulate and emerge relatively rapidly for plants taken outside of their natural range (Figure 5.5). In this scenario, there is a bottleneck effect in which a sub-population is taken to a new location and the effects of human exploitation, transplantation and management are not offset by interbreeding with genetically diverse cultivated, feral or wild plants. Divergence between the translocated 'curated' stock and the distant parent population could be initiated relatively quickly. Such divergence may be offset by spasmodic interbreeding between managed and feral plants near the 'edge-of-the-range' where sexual reproduction is only periodically viable. The domestication process starts relatively early and its duration depends upon the character of human-directed selection, namely management or cultivation. More distinctive phenotypic variations become more pronounced and 'fixed' following more intensive directed selection of specific phenotypes under prolonged cultivation.

In many ways, the degree of directed selection is similar in both scenarios because the plants are enmeshed in similar types of practice. However, the rate and duration of the domestication episode vary greatly between the scenarios largely due to differences in the degrees of genetic isolation for plants within and outside of their natural range. For

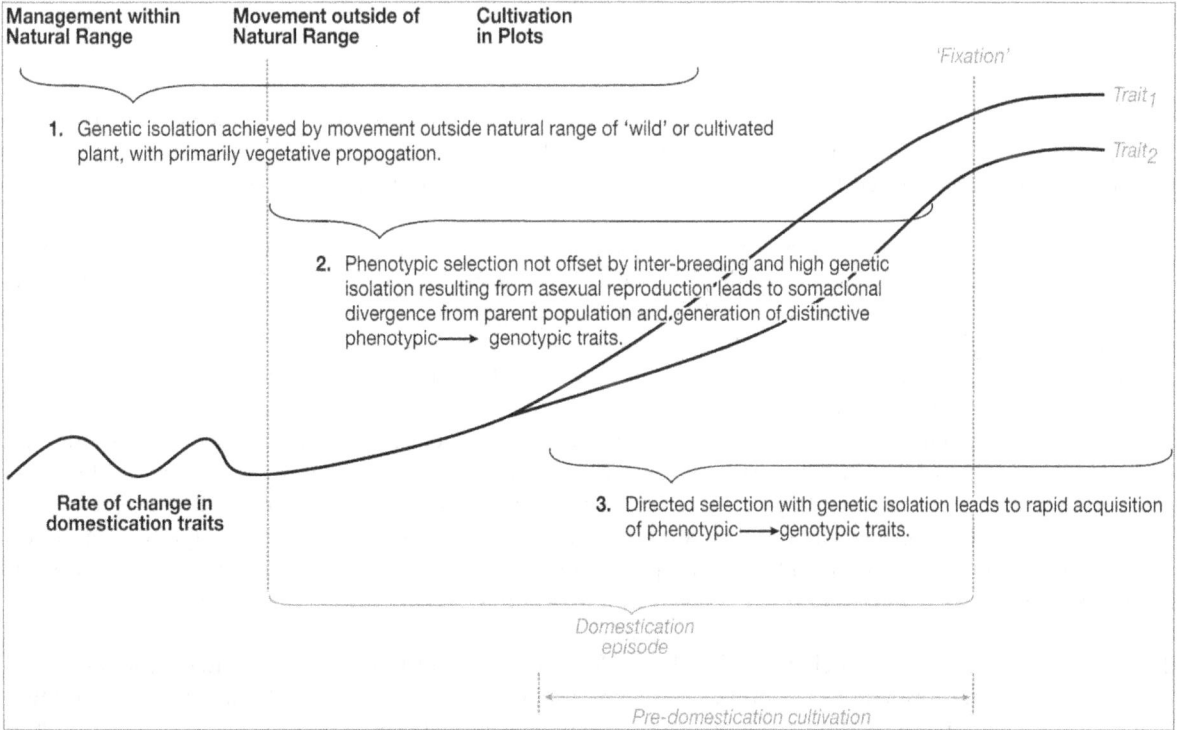

Figure 5.5 Hypothetical scenario of plant domestication outside the natural range

Note: Period of pre-domestication cultivation is shorter than the apparent domestication episode. Plants begin to accumulate phenotypic→genotypic traits once taken outside of the natural range, whether taken for cultivation or inadvertently spread. A faster rate of domestication is anticipated for plants outside the natural range than for those within, under comparable selection pressures

plants within their natural range, a prolonged period of pre-domestication cultivation precedes a shorter 'domestication episode', when genotypic and phenotypic attributes become more fixed following genetic isolation. For plants outside of their natural range, a prolonged period of pre-cultivation domestication may occur due to their genetic isolation and plausibly forms part of a longer domestication episode. These processes contribute to the widely varyng degrees of domestication exhibited by cultivated plants (in plots) and exploited plants (in the landscape) in most highland locales.

These are idealised scenarios, which serve as guides for understanding plant domestication in the past. Currently, there is no clear morphological (phenotypic) transformation of plant microfossils or macrofossils that might be expected to accompany the domestication process for any food plant in the highlands (cf. Haberle 1995; Yen 1996). The absence of a robust, archaeobotanically derived chronology for domestication does not necessarily reflect a lack of domesticatory relationships; rather, it reflects the nature of those relationships and the low archaeobotanical visibility of likely domestication traits in many vegetatively propagated crop plants.

Crop plants in the highlands

Communities across New Guinea traditionally exploited hundreds of different plants for a variety of different purposes (following Powell 1976; Sillitoe 1983; Bayliss-Smith 1985, 1988; French 1986; Kocher Schmid 1991; Kennedy and Clarke 2004; Cook and Webster 2016; Bourke 2017). Plants were used for food, fodder, bedding and construction, decoration and ornamentation, medicinal and psychoactive properties, ritual and ceremonies, boundary markers and so on (Powell 1976). The emphasis here is on major food plants of highland agriculture (Table 5.2).

Staple crops

Staple crop assemblages vary considerably across New Guinea today, and are likely to have done so in the past (Figure 5.6). Prior to the introduction and widespread adoption of the sweet potato, traditional staples – such as banana, taro and yam – would have been more significant for highland cultivation than they are today (Figure 5.7). Here, the primary uses and methods of cultivation for each plant, major domestication traits resulting from prolonged cultivation and likely history of domesticated relationships are indicated (extracted, amended and expanded version of Denham 2017: 43–48).

*Bananas (*Musa *cvs.)*

Bananas are usually cultivated for edible fruit, although a variety of others uses are known (Kennedy 2009). Carbohydrate-rich varieties can be eaten raw as fruit and starchy varieties can be cooked as vegetables. In general terms, domestication has transformed wild, large-seeded forms with little edible pulp into cultivated, vestigial-seeded forms with abundant edible pulp.

Bananas are usually propagated vegetatively through the transplantation of suckers or shoots growing at the base of the stem of the plant. The sucker or shoot is composed of leaves and roots that are trimmed prior to planting. Bananas are important for diet and also as a ceremonial food, for example, accompanying brideprice payments. Bananas are ordinarily considered to favour well-drained soils, though some varieties grow in saturated conditions.

Although hundreds of banana cultivars are known (Stover and Simmonds 1987; Arnaud and Horry 1997), the most significant in terms of global food production are descended – at least in part – from *Musa acuminata* ssp. *banksii* (Perrier et al. 2011). Traditional banana cultivars in the highlands were hybrids descended, to some degree, from *Musa acuminata* ssp. *banksii* (section Musa). Edible diploids of this subspecies are thought to have undergone initial domestication on New Guinea, including the development of parthenocarpy (De Langhe and de Maret 1999). Major diploid and triploid cultivar groups are thought to have emerged within the maritime landscapes of Island Southeast Asia-Melanesia through interspecific and inter-subspecific hybridisation to produce parthenocarpic, seed-suppressed and sterile forms (De Langhe and de Maret 1999; Kennedy 2008; Denham and Donohue 2009; Perrier et al. 2011; De Langhe et al. 2015).

Other bananas on New Guinea include the highland *Musa ingens* (Ingentimusa), and cultivated varieties of Callimusa section bananas (formerly Australimusa section). The starchy pith of the pseudo-stem of *Musa ingens* is sometimes eaten, although it is only a very minor food plant (French 1986, 2006). Of Callimusa bananas, some Fe'i bananas are cultivated in the highlands for fruit, but they are more important in lowland New Guinea, Island Melanesia and the Pacific (Kennedy 2009). Another member of the Musaceae family, *Ensete glaucum* grows on New Guinea and the pseudo-corm at the base of the stem is eaten as a famine food (Powell 1982b).

Table 5.2 Altitudinal range of important crop plants in the highlands (above 1200 m AMSL) of Papua New Guinea

Scientific Name	Common Name	Mean Usual Altitudinal Range (m)	Extreme Altitudinal Range (m)
Starch-rich plants			
Colocasia esculenta	Taro	0–2400	0–2760
Dioscorea alata	Greater yam	0–1900	0–2100
Dioscorea bulbifera	Potato yam	0–1900	0–2110
Dioscorea nummularia	Nummularia yam	0–1900	0–2050
Dioscorea pentaphylla	Five leaflet yam	0–1500	0–1620
Ipomoea batatas[1]	Sweet potato	0–2700	0–2850
Musa cvs	Fe'i banana	0–1750	0–2060
Musa cvs	Diploid banana	0–1800	0–2030
Musa cvs	Triploid banana	0–2150	0–2580
Pueraria montana var. lobata	Pueraria/kudzu	0–2300	0–2740
Saccharum officinarum	Sugarcane	0–2600	0–2760
Vegetables			
Abelmoschus manihot	Aibika	0–1900	0–2110
Amaranthus tricolor	Amaranth	0–1950	0–2050
Commelina diffusa	Wandering Jew	0–?	0–2390
Ficus copiosa	Kumu musong	0–2200	0–2450
Ficus dammaropsis	Highland kapiak	800–2750	0–2820
Lagenaria siceraria	Bottle gourd	0–?	0–2670
Oenanthe javanica	Oenanthe	1050–2700	0–3400
Psophocarpus tetragonolobus	Winged bean	0–1900	0–2070
Rorippa schlechteri	–	750–2700	180–2850
Rungia klossii	Rungia	950–2700	0–2760
Saccharum edule	Lowland pitpit	0–1800	0–2270
Setaria palmifolia	Highland pitpit	0–2700	0–2760
Fruit- and nut-bearing plants			
Castanopsis acuminatissima	Castanopsis	700–2350	570–2440
Pandanus antaresensis	Wild karuka	1000–2350	850–2460
Pandanus brosimos	Wild karuka	2400–3100	1800–3300
Pandanus conoideus	Marita	0–1700	0–1980
Pandanus julianettii	Karuka	1800–2600	1450–2800
Rubus rosifolius	Red raspberry	950–2800	700–2900
Plants marginal in highland environments			
Artocarpus altilis	Breadfruit	0–1250	0–1450
Cocos nucifera	Coconut	0–950	0–1310
Dioscorea esculenta	Lesser yam	0–1550[2]	0–1670
Metroxylon sagu	Sago	0–1150	0–1250

Source: Data from Bourke 2017

Notes

1 Sweet potato was introduced to New Guinea within the last 400 years
2 *Dioscorea esculenta* yam is not commonly grown above 900 m

Taro (Colocasia esculenta)

Taro is grown primarily for its edible corm, or enlarged underground storage organ. In many parts of the world its leaves are cooked as a green vegetable, although this does not seem to have been a common practice in the highlands. The corm is abundant in starch and contains other compounds, including oxalates, that make processing and cooking necessary prior to consumption (Paull et al. 1999).

Taro is propagated by corm division and replanting of smaller side suckers or side corms that grow around the larger central corm, or by planting of the upper part of the larger central corm with trimmed petiole, that is, the 'taro top'. Taro is planted and grown in dry and wet soils in the highlands. Rituals are often associated with the planting of taro (Powell et al. 1975: 21). Taro is rarely grown in the same garden twice in succession, because of the rapidly decreasing yields caused by infestations of taro beetle (*Papuana* spp.) and other pests (Bayliss-Smith 1985).

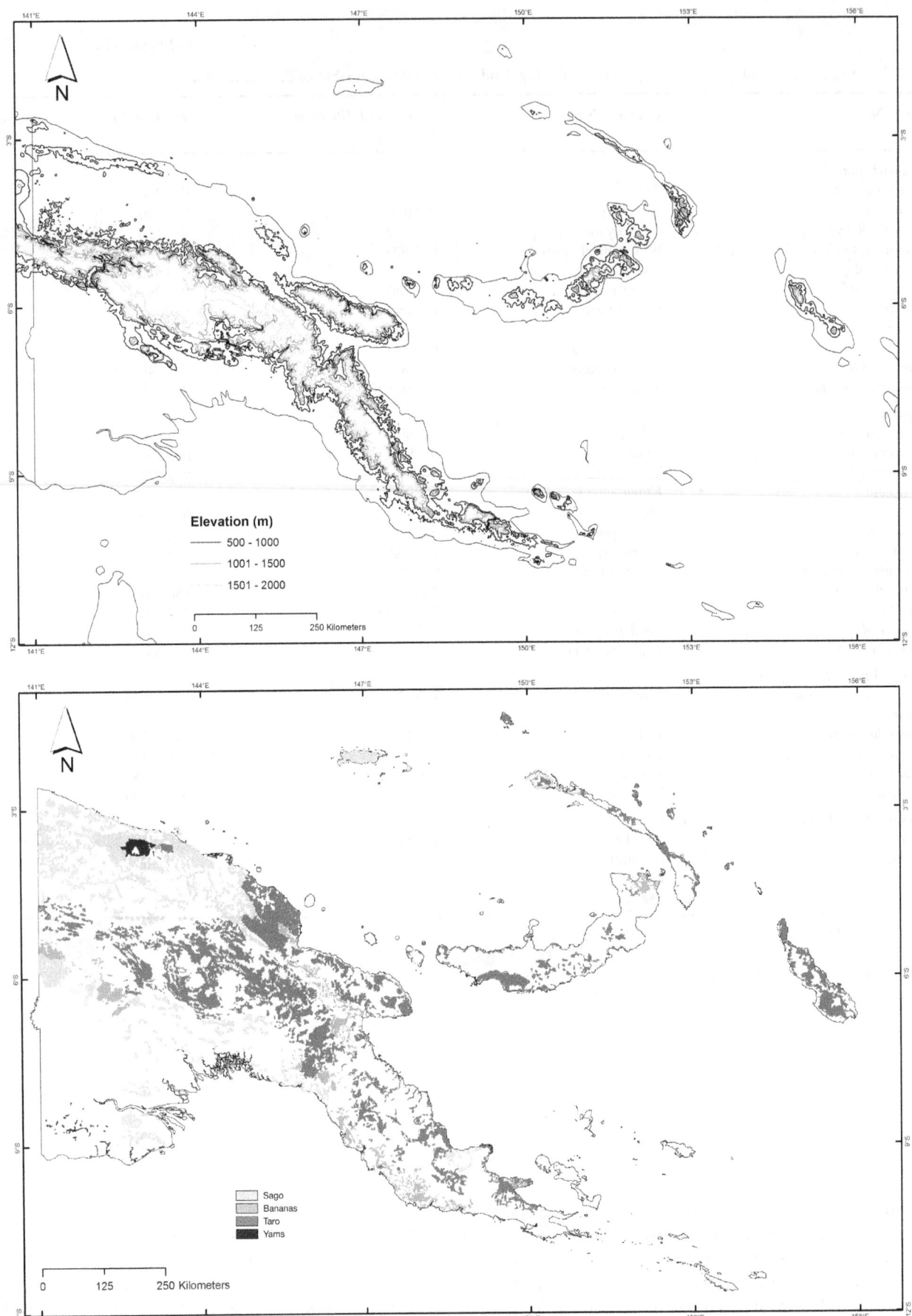

Figure 5.6 Maps depicting the variability in plant exploitation across Papua New Guinea

Note: (Upper) contour map of Papua New Guinea; (Middle) an interpretation of the geographical distribution of major crop plants with recently introduced crops excluded. Major crop plants likely to have been introduced within the last 500 years comprise sweet potato (*Ipomoea batatas*), cassava (*Manihot esculenta*) and 'Chinese' taro or taro kongkong (*Xanthosoma sagittifolium*); (Lower) geographical distribution of major crop plants today.

Source: Original data from RMAP, ANU; Denham 2011: Figure 3

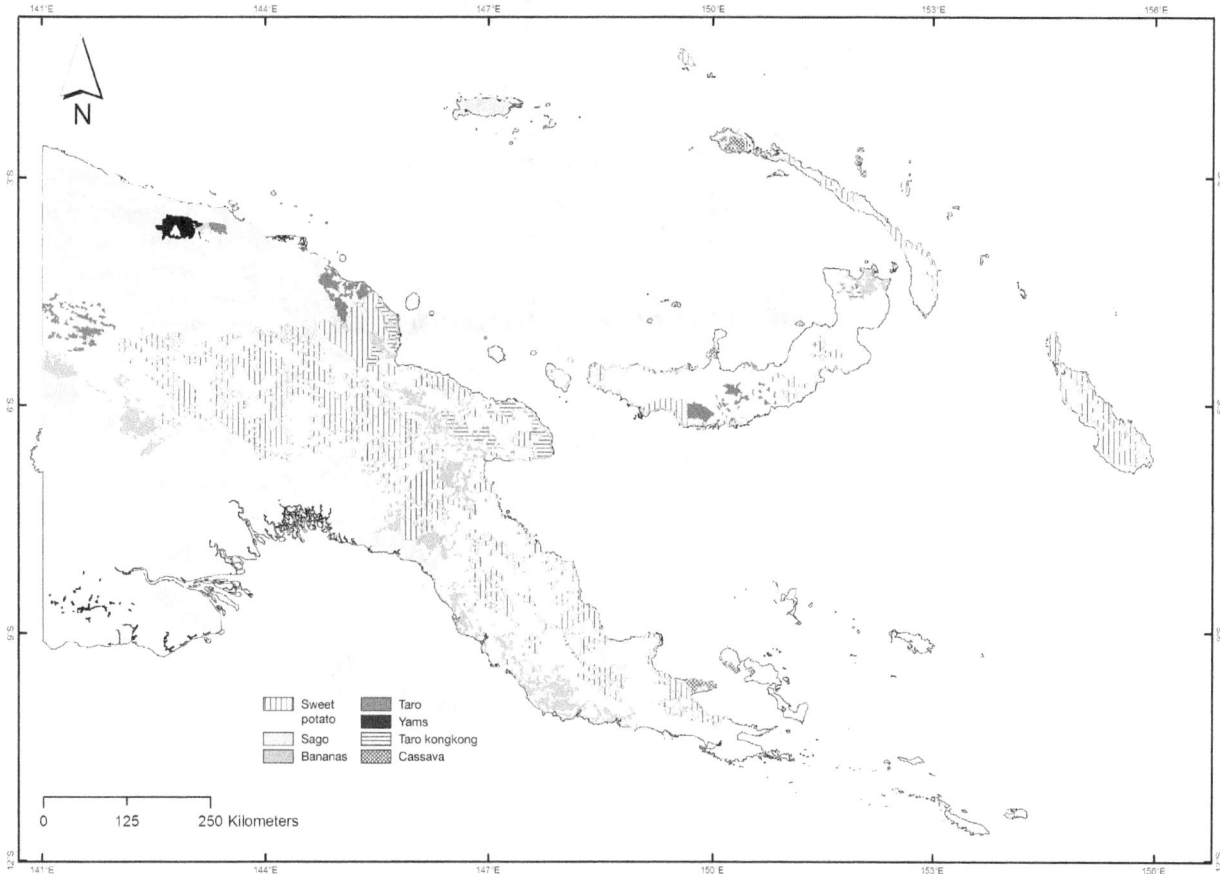

Figure 5.6 (Continued)

The transition between wild and domesticated forms is usually marked by increased corm size and decreased acridity. In part, these phenotypic changes are a function of growth environment; namely, they are plastic. Diploid cultivars can reproduce sexually with each other and with wild plants, whereas triploid cultivars are sterile. Traditional taro cultivars in New Guinea were predominantly diploids, whereas those in Southeast Asia included diploids and triploids (Kennedy and Clarke 2004; Matthews 2004).

Many taro cultivars are differentiated by highlanders on the basis of colour of corm flesh, colouration of stalk and so on (Sillitoe 1983: 37–42; Cook and Webster 2016). Over the last few decades, numerous traditional cultivars have been lost or grow in feral stands, whereas others are deliberately grown by those seeking to preserve crop diversity and the traditional knowledge that goes with it. Although sweet potato has usurped the role of taro as a primary staple over the last few hundred years across swathes of the highlands, taro retained its role as an important ceremonial crop in many societies at least into the late twentieth century (e.g., R. Bulmer 1982).

Wild-type taro *(Colocasia esculenta* var. *aquatilis)* occurs over a vast area from northern India to the New Guinea region and northern Australia (Matthews 1991, 1995, 2014; Ramanatha et al. 2010). Taro is likely to have undergone multiple, possibly independent domestications within its natural range, including Southeast Asia and New Guinea (Lebot 1999, 2009). It is not known where taro was first cultivated or domesticated on New Guinea (Yen 1995; cf. Denham, Haberle et al. 2004). Initial molecular analyses suggested limited genetic admixture, or gene flow, between geographical regions, such as between New Guinea and regions to the west (Lebot et al. 2004), although more complex scenarios are now being inferred (e.g., Chair et al. 2016). The continual replacement of old varieties with new ones is a major problem for inferring origins based on present-day cultivar distributions (Figure 5.8). As with many tropical crops, ongoing genetic analyses may radically alter current understandings of the geodomestication history for taro.

Figure 5.7 Traditional starch-rich crops in the highlands

Note: (upper, this page) banana (Source: Tim Denham, Upper Sepik, 2010); (lower, this page) sugarcane (Source: Tim Denham, Karimui, 2007); (upper, next page) taro (Source: Peter Matthews, Kokoda, 2008); greater yam (Source: Mike Bourke, Milne Bay, 1995)

Figure 5.7 (Continued)

*Yams (*Dioscorea *spp.)*

Yams are exploited for starch-rich tubers, although some species produce above-ground, or aerial, bulbils that can be eaten. Multiple yam species can be cultivated on the floors of the main highland valleys, including *Dioscorea alata*, *D. bulbifera* and *D. nummularia*, whereas *D. esculenta* and *D. pentaphylla* approach the upper altitudinal range of cultivation around 1500–1600 m. Prior to the introduction of sweet potato, yams in combination with bananas and taro were the

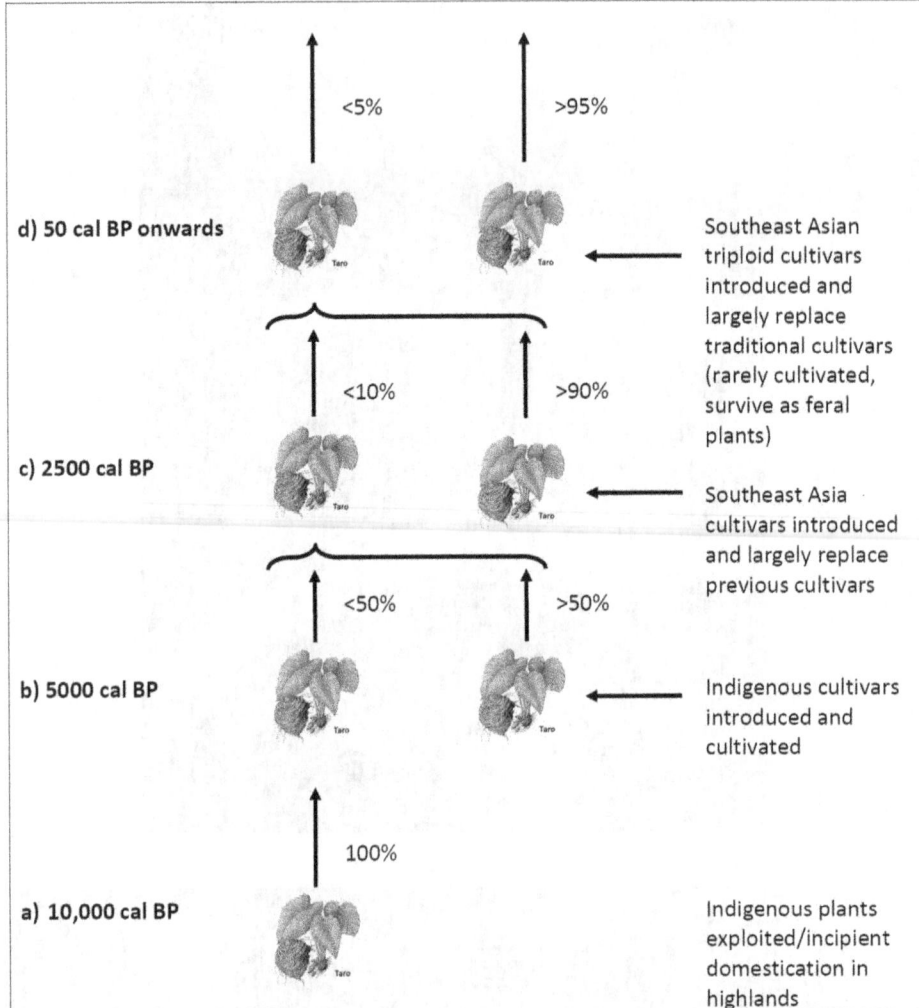

d) 50 cal BP onwards
<5%   >95%

Southeast Asian triploid cultivars introduced and largely replace traditional cultivars (rarely cultivated, survive as feral plants)

c) 2500 cal BP
<10%   >90%

Southeast Asia cultivars introduced and largely replace previous cultivars

b) 5000 cal BP
<50%   >50%

Indigenous cultivars introduced and cultivated

a) 10,000 cal BP
100%

Indigenous plants exploited/incipient domestication in highlands

Figure 5.8 A cautionary tale of genetic reshuffling in cultivated taro lineages

Note: The hypothetical scenario begins at 10,000 cal BP with the exploitation and incipient domestication of indigenous highland taro populations, with sequential introductions of taro cultivars from elsewhere on New Guinea and from Southeast Asia leading to the resultant dilution and eventual loss of indigenous cultivar diversity

most significant crop complex across much of the highlands. Today, yams are cultivated in the main highland valleys as minor crops (although see Sillitoe 1983), and were probably more important in the past. Wild yams of unknown species and derivation have been recorded growing wild in the Upper Wahgi Valley (Powell et al. 1975: 35), among other places.

The antiquity of different yam species on New Guinea has been subject to much speculation, largely on the basis of phytogeography and ethnohistory (Coursey 1976). De Candolle claimed New Guinea to be a possible source of yam (1884: 79), whereas on the basis of the plurality of forms, Coursey proposed New Guinea as a secondary centre of dispersal for *D. alata* and *D. esculenta* (Coursey 1972: 226, 1976: 71; also see Alexander and Coursey 1969: 417; Hather 1992: 71). Minor food species are considered to be indigenous including *D. bulbifera*, *D. nummularia*, *D. pentaphylla* and *D. hispida* (Gagné 1982: 247; Powell 1976: 119; Lebot et al. 2017). Inferences of antiquity for any yam species, however, should be treated cautiously in the absence of robust archaeobotanical and ancient DNA data.

Globally, the greater yam, or water yam *(D. alata)* is the most significant yam species. Although the precise locus of greater yam origin and domestication is not known, and a wild-type or precursor has not been identified (Lebot et al. 1998), multiple lines of genetic and morphological evidence have been drawn on to suggest the New Guinea region as a source (following Denham 2010: 4). AFLP-fingerprinting profiles show that *D. alata*, *D. nummularia* and *D. transversa* are closely related and belong 'to a Southeast Asian-Oceanian genepool which is rather confined to the former Sahulian and Wallacean regions' (Malapa et al. 2005: 928). Previously, several authors proposed New Guinea as the source region because it has the greatest genetic diversity (Lebot 1999: 625); a large diversity of varieties differentiated by flesh colour and the shape of leaves and tubers (Sillitoe 1983: 46); and, primitive cultivar types, 'most bizarre and least improved types', and most types found elsewhere (Martin and Rhodes 1977: 5). The greater yam flowers naturally in Melanesia (Lebot 1999: 625) and has been reported to yield fertile seed (albeit rarely) in the highlands (Sillitoe 1983: 44), although such claims require independent verification.

Even taken together, these multiple lines of evidence only circumstantially suggest the New Guinea region as the place of *D. alata* origin and domestication, from which cultivar clones have dispersed widely across the globe (see Burkill 1935 for an alternative view). As De Candolle (1884: 13) cautioned, diversity is no guarantor of origin. The genetic analysis of new accessions from Southeast Asia, especially eastern Indonesia, may radically alter current scenarios of yam domestication.

The greater yam, like other yams, is propagated vegetatively. Ordinarily, a tuber top with a length of stem is replanted. The tuber top may still be viable after storage for years (Powell et al. 1975: 24). Although today the plant is effectively sterile due to asynchronic flowering, genetic analyses have inferred sexual reproduction in the past. However, the majority of genetic variability among greater yam populations represents somaclonal variation resulting from prolonged asexual, or vegetative, reproduction (Lebot et al. 1998; Malapa et al. 2005). Thus, geographically dispersed populations of greater yam became established through human translocation.

Most yams exhibit considerable phenotypic plasticity, namely, the morphology of yam tubers may represent growth environment, as much as specific genetic traits. In New Guinea, various cultivation methods are used to alter the size and shape of yam tubers (Lea 1966; Coupaye 2013). A working hypothesis is that the domestication of yams in New Guinea, like other parts of the world, has sought to increase tuber size, decrease vine and tuber spininess, as well as to promote culturally specific and idiosyncratic traits associated with colour, edibility, shape and taste.

Sugarcane (Saccharum officinarum)

Of the various cane grasses domesticated in New Guinea region, only sugarcane is globally significant. In the highlands, sugarcane is generally considered a snack food, or refreshment, although it was potentially a staple in drier areas of the eastern highlands and elsewhere in the past (Rappaport 1968; Daniels and Daniels 1993). Sugarcane is cultivated for the sugar-rich sap that is sucked or masticated from the stalk.

For vegetative propagation, the tops of mature canes are cut, together with two or three mature and reproductively viable inter-nodes, and then planted erect in the ground (Sillitoe 1983: 86). Reproduction can also occur through division of the root stock. In the highlands, flowering and seed-set are often inhibited. Numerous varieties are differentiated on the basis of cane colour and thickness, as well as taste. Sugarcane is intolerant of saturated soils.

The geodomestication history of sugarcane is not well known (Simmonds 1976a; Daniels and Daniels 1993; Lebot 1999; Grivet et al. 2004). Simmonds (1976a) considered anthropogenic selection and domestication to have occurred in New Guinea from a wild ancestor of *S. robustum*, principally for its sugar-rich sap. Lebot concluded that *S. officinarum* and *S. edule* domestication occurred in New Guinea from introgressions of wild forms of *S. robustum* with *S. spontaneum* and from wild gene pools of *S. robustum*, respectively (Lebot 1999: 623). Alternative scenarios exist that shift the locus of domestication further towards Asia (Daniels and Daniels 1993). Archaeobotany sheds little light on the history of sugarcane domestication on New Guinea: a putatively early claim for sugarcane at Yuku (S. Bulmer 1975; Horrocks et al. 2008) is unverified (Yen 1998; Denham 2016a), and the earliest robust record dates to only a few hundred years ago at Kuk (Lewis et al. 2016).

Even though the initial stages of the domestication of bananas and sugarcane plausibly occurred in the New Guinea region, globally dominant cultivar groups are thought to derive from interspecific or inter-subspecific hybridisations in Island Southeast Asia before reintroduction to the island (Grivet et al. 2004). Many indigenously derived cultivars of sugarcane, like banana, are still cultivated on the island (Kennedy and Clarke 2004) and were probably more significant in the past. The understanding of sugarcane domestication will probably change in the light of more robust genomic studies and more rigorous archaeobotany.

Other traditional staples

Kudzu *(Pueraria montana var. lobata)* is generally considered to be an Asian domesticate. Kudzu is rarely cultivated in the highlands for its tuber, which is slow maturing and a ceremonial, reserve or famine food (Bowers 1964; Watson 1964, 1968; Strathern 1969; Sillitoe 1983: 46–48). French (1986: 18) notes that wild forms can self-seed and usually grow between 30–1860 m, whereas cultivated forms are vegetatively propagated from stem cuttings and are common at higher altitudes 'up to 2700 m'. A long time depth for kudzu cultivation in the highlands is implied by its ceremonial use, which suggests considerable cultural embeddedness. The antiquity of kudzu introduction to New Guinea is not known, but it was plausibly an important starch-rich staple plant in the past (Barrau 1965).

Several important staple crops on New Guinea do not grow in the main highland valleys. Of greatest significance are sago *(Metroxylon sagu)* and breadfruit *(Artocarpus altilis)*, both of which were domesticated in the New Guinea region (Kjaer et al. 2004; Zerega et al. 2006, respectively). Although central highlanders do not cultivate these plants, some groups maintain rights of access to exploit stands and groves at lower altitudes. For example, some Huli retain rights over stands of sago at lower altitudes, which they periodically visit to tend, harvest and process.

Sweet potato (Ipomoea batatas)

Sweet potato is a South American domesticate that is primarily grown in tropical and subtropical locations for its edible, starch-rich subterranean storage roots and its leaves are eaten in some places (Yen 1974). In Papua New Guinea, sweet potato is probably the most widely eaten food and fodder (for pigs) plant. Under cultivation, sweet potato is propagated vegetatively by taking a cutting of the vine, stem or slip, rather than using part of the underground storage organ. The plant can also reproduce and disperse from fertilised seed, as suggested for the interior of New Guinea (R. Bulmer 1966) and cultivators incorporate spontaneously arising self-sown varieties into their cultivated stock (Powell et al. 1975). There is an enormous diversity of sweet potato cultivars in the highlands, with varieties being differentiated on the basis of leaf shape and colour, as well as the shape, colour, taste and texture of the tuber (Yen 1974).

Sweet potato is the dominant staple crop for people and pigs across most of the highlands today. Despite being frost and flood intolerant, sweet potato has replaced taro across much of the highlands because it provides higher yields on poorer soils. Even though the plant is probably a post-Magellan introduction to New Guinea, namely within the last 550 years (Roullier et al. 2013), the plant has enabled major social and environmental transformations. Sweet potato is accredited with increasing populations of pigs and people, with attendant intensification of exchange practices and the development of the 'big man' institution (e.g., Watson 1965; Modjeska 1982; Denham 2013a).

Manioc, or cassava (Manihot esculenta)

Manioc or cassava *(Manihot esculenta)* is increasingly being cultivated for food and fodder in the highlands (Figure 5.9). Manioc adapts well to marginal environments, is drought resistant, has a flexible growth cycle and has relatively high energy yields per unit area of land (Lebot 2009). Manioc is usually reproduced vegetatively under cultivation; a portion of

Figure 5.9 Manioc *(Manihot esculenta)* growing in the Upper Sepik

Source: Tim Denham, 2010

the stem is cut and replanted. Cultivated manioc can produce viable seed, which is considered important for the domestication process and the generation of cultivar diversity in South America (Clement et al. 2010).

Although the antiquity of manioc on New Guinea is uncertain, it was introduced ahead of direct European contact in some parts of the interior (Hays 2005). People do not want to say that they plant or eat manioc, which is often considered an inferior food, even though it is inter-cropped in their gardens (Mike Bourke pers. comm. 2010). The agronomic importance of manioc will increase in the future as a source of food for people and fodder for pigs because of its relatively high yields in marginal environments.

Vegetables

A variety of minor food plants were domesticated on New Guinea, some of which are indigenous to the highlands. Of these, several are still widely cultivated including edible cane grasses – *Setaria palmifolia* and *Saccharum edule* – bamboo, and several leafy greens, including aibika *(Abelmoschus manihot)*, *Ficus copiosa*, rungia *(Rungia klossii)* and *Oenanthe javanica*, as well as winged bean *(Psophocarpus tetragonolobus)* and potentially other legumes (Figure 5.10).

Other minor crops have been introduced to New Guinea at various times in the past. Some of these introductions are ancient (see Wilde and Duyfjes 2010), such as wax gourd dating to at least 3000–2000 years ago and bottle gourd *(Lagenaria siceraria)* potentially earlier (Powell 1970b; Golson 2002). Others – like the sweet potato and manioc – were probably introduced following European exploration of the Indo-Pacific. The precise antiquities of most remain a mystery.

Assorted green leafy vegetables

Numerous green leafy vegetables are cultivated in New Guinea gardens (Powell 1976; Sillitoe 1983; French 2006). Numerous varieties are differentiated using leaf size, shape and colour, stem colour and so on. Most underwent some form

Figure 5.10 Assorted vegetables and greens

Note: (upper) highland pitpit *(Setaria palmifolia)* with rungia *(Rungia klossii)* and taro *(Colocasia esculenta)* (Source: Tim Denham, taken by Alice Bedingfield, Karimui, 2007); (middle) variegated highland pitpit *(Setaria palmifolia)* (Source: Tim Denham, taken by Kale Sniderman, Karimui, 2007); (lower) sweet potato *(Ipomoea batatas)*, yam *(Dioscorea alata)* and winged bean *(Psophocarpus tetragonolobus)* growing (from left to right) with a clump of manioc *(Manihot esculenta)* planted above (Source: Mike Bourke, Kainantu, 1982)

Figure 5.10 (Continued)

of domestication on New Guinea – specifically rungia *(Rungia klossii)*, *Oenanthe javanica* and crucifer spinach (*Rorippa* sp.). Some plants have wide tropical distributions and probably underwent independent domestications in different regions, such as *Commelina diffusa*.

The majority of vegetables are propagated from cuttings or rooted stems, although some are planted from seed – such as *Amaranthus tricolor* and *Rorippa* sp. Vegetative cultivation entails planting stem cuttings into a dibble hole. Often multiple cuttings may be planted together in hope that at least one survives. Some plants are known in wild and cultivated forms, and may be propagated from both, for example, *Commelina diffusa*, whereas others are known in cultivated form, for example, *Abelmoschus manihot*, *Amaranthus tricolor* and *Rorippa* sp. Yet others are 'semi-domesticated', being self-sown in gardens and burned areas, for example, *Solanum* spp. (Symon 1985; Edmonds and Chweya 1997), and a few are successful weeds that readily establish feral populations, for example, *Oenanthe javanica* (Powell et al. 1975: 29–30). Leafy vegetables are adventitiously exploited in feral or wild form by hunting and foraging parties (Majnep and Bulmer 2007; also see Croft 1982 for exploitation of ferns).

Edible cane grasses (pitpit)

Of the various cane grasses, two have been domesticated as vegetables, *Saccharum edule* and *Setaria palmifolia* (Berding and Koike 1980; Plarre 1995). These two crops are colloquially referred to as lowland *pitpit* and highland *pitpit*, respectively. *Setaria palmifolia* is a widely distributed plant throughout Asia, which has been domesticated as a vegetable in New Guinea. Altitudinal distributions ordinarily overlap from sea level to 1800 m, and in extreme circumstances up to c. 2300 m, although *Setaria palmifolia* is more commonly planted above 500 m (Bourke 2017). Both plants are considered to have been domesticated in New Guinea – and together with sugarcane – represent a rather unique assemblage of grass domestication through vegetative propagation. Highland *pitpit* is propagated using leafy lateral shoots and lowland *pitpit* using stalk cuttings.

Today, both vegetables are relatively minor contributors to diet. Lowland *pitpit* is cultivated for its unopened flower, or edible inflorescences, and highland *pitpit* for thickened young shoots and stem hearts (Powell et al. 1975: 27–28; Sillitoe 1983: 81; French 1986: 96–97). Feral and abandoned stands of highland *pitpit* are known and exploited by people, as are the wild plants, which have thinner stalks. The Wola say that cultivated and wild forms of highland pitpit are distinct; the cultivated form does not revert to the wild self-propagating form if abandoned (Sillitoe 1983: 81). No wild-types of lowland *pitpit* are known in the highlands.

Winged bean (Psophocarpus tetragonolobus)

The antiquity and origin of winged bean on New Guinea is uncertain. Like recently introduced beans and peas, the winged bean is cultivated from seed. As well as bearing a highly nutritious seed (or bean), the flowers, leaves, young pods and a subterranean tuber can be eaten (Powell et al. 1975: 25). The plant usually only produces a tuber between 1200–1850 m altitude, and French (1986: 34) states: 'Very important for tubers in the Western and Eastern Highlands. Moderately common for beans in other places'.

Gourds

Only two gourd species are known to have been cultivated in the interior of New Guinea, wax gourd *(Benincasa hispida)* and bottle gourd *(Lagenaria siceraria)*, even though bitter gourd *(Momordica charantia)* is cultivated today. Both traditional gourd species are cultivated from seed and eaten as a vegetable, although neither is a significant food plant today (French 1986: 107–108). The young fruits of bottle gourd are boiled and the young tips and leaves eaten, whereas the old fruits are primarily used as a container. The flesh, seeds, young leaves and flower buds of cultivated wax gourd are eaten. The cultivated form of wax gourd is larger, with a smaller, wild form potentially indigenous to Australia-New Guinea (Whistler 1990).

Bottle gourd is a widely dispersed plant, which appears in human-associated contexts dating from the early Holocene in Asia and the Americas, although the mechanisms of dispersal are open to debate, including human agency and oceanic drift (Erikson et al. 2005; Clarke et al. 2006; Kistler et al. 2014). The introduction of bottle gourd to New Guinea could be associated with the original arrival of people to the island, although there is no archaeological or genetic evidence to support such an assertion. In contrast, wax gourd is generally taken to be an Asian domesticate, even though the antiquity, cultural associations and location of its domestication may include New Guinea and northern Australia (Wilde and Duyfjes 2010). Matthews (2003) has proposed a more westerly origin for wax gourd in India, or potentially even Africa, because *Benincasa* is a monotypic genus, whose closest taxonomic relatives are *Citrullus* (water melons) and *Lagenaria* (bottle gourds), both presumed of African origin.

Nut and fruit trees

The majority of nut and fruit-bearing trees and pandanus growing in New Guinea, as well as sago, are restricted to lowland altitudes (Table 5.2), where they are major contributors to diet (Bourke 1996). In the highlands, *karuka* pandanus, as well as formerly castanopsis nuts, is a contributor to diet on a seasonal or periodic basis (Figure 5.11).

Castanopsis (Castanopsis acuminatissima)

Castanopsis chestnut is a producer of abundant highly nutritious nuts that grows throughout New Guinea and was formerly targeted by highland populations (R. Bulmer 1964). Castanopsis can form dense stands in lower montane forest environments (French 2006). The tree regenerates sexually, with subsequent transplantation of germinated seedlings.

*Pandanus – karuka (*Pandanus julianettii/iwen/brosimos *complex) and marita* (Pandanus conoideus)

There are two main types of Pandanus, or screwpine, cultivated on the island of New Guinea today (Stone 1982a, 1984; Hyndman 1984; French 1986): highland *karuka* (members of the *Pandanus julianettii/iwen/brosimos* complex) is grown

Figure 5.11 Photographs depicting cultivated pandans

Note: (upper and lower, this page) marita (*Pandanus conoideus*; Source: Tim Denham, Karimui, 2007); (upper and lower, next page) *karuka* (*Pandanus julianettii*; Source: Tim Denham, Simbai, 2007)

Figure 5.11 (Continued)

for its caloric-, oil- and protein-rich nuts; and *marita (Pandanus conoideus)* is grown for its energy and oil content. The classification of *karuka* is problematic, with multiple authors suggesting that a cline exists between 'cultivated' *P. julianettii* and 'wild' *P. brosimos*, with *P. iwen* as a possible intermediate form (Cook 1999; Cook and Webster 2016). Both *karuka* and *marita* are periodic – and in some places seasonal – sources of food for groups in the highlands and lowlands of New Guinea (Bourke 1996).

Karuka and *marita* are both propagated by people vegetatively: the crowns and crown-cuttings of old trees are replanted, as well as suckers of *marita*. *Karuka* can grow from self-sown seeds, whereas *marita* is one of the few cultivated plants in the New Guinea region that has been considered to only reproduce asexually. Until recently, *marita* had no known wild progenitor, although one has now been identified in western New Guinea (Keim 2012). It is not known whether prolonged vegetative propagation caused the loss of sexual reproduction in *marita*, or whether it was already only asexually reproducing before people began to cultivate it vegetatively. Thus, the role of people in the development of parthenocarpic and asexual *marita* is unknown, although prolonged vegetative propagation could be a contributory factor.

These pandans have relatively discrete altitudinal ranges: wild *karuka* ordinarily occurs between 2400–3100 m, whereas cultivated *karuka* ordinarily occurs between 1800–2600 m; by contrast, *marita* is ordinarily cultivated between 0–1700 m, but is not important below 500 m and is common in mid-altitudes, with an upper altitudinal extreme of 1980 m (see Table 5.2). Potentially, prolonged cultivation may have 'forced' the cultivation of *karuka* to lower altitudes, whereas *marita* was 'forced' into higher altitudes. As a result, an altitudinally intermediate pandanus, *P. antaresensis* that ordinarily grows between 1000–2350 m, may have been replaced as a food source. Ralph Bulmer remarked that the kernels of *antaresensis* nuts 'require considerable effort to extract, for a rather small return' (Majnep and Bulmer 2007: 314; cf. Hyndman 1984: 296). Today, *P. antaresensis* is a very minor food, whereas archaeological excavations in the Wurup Valley show that this species was intensively utilised during the early Holocene (Christensen 1975a; Donoghue 1989). Altitudinal forcing during the Holocene may have led to the replacement of *P. antaresensis* as a food source in altitudes that were intermediate between the former altitudinal ranges of *karuka* (above 2400 m) and *marita* (below 1000 m).

Archaeobotanical visibility of vegetative domestication and cultivation

A domestication syndrome has been proposed to characterise phenotypic changes between wild and domesticated plants (Table 5.3). Domestication traits are largely derived for cereals, legumes and other sexually reproduced plants, even though only a few of these traits are directly visible in the archaeobotanical record, such as grain size and non-shattering. The domestication episode for visible traits can be tracked and reconstructed using archaeobotanical assemblages from multiple sites (Fuller et al. 2014).

As yet, the comparison of plant domestication traits and syndromes for sexually reproduced and vegetatively propagated plants has not been undertaken. For vegetatively propagated plants, unlinear concepts of domestication may not be applicable in the same way: rates of change may be variable, domestication episodes may be drawn out and there may be periods of reversion to 'wild-type' when growing feral. Further, the sequences and synchronicity for the emergence of domestication traits may be different.

A priori, the uncertain domestication status of many crop plants in the highlands of New Guinea makes it difficult to determine a set of domestication syndrome traits, as well as to track the emergence and fixation of these traits in archaeobotanical assemblages. Certainly, several similar types of trait have been selected for and have become fixed, with varying degrees of permanence, in crop plants under vegetative propagation (Table 5.3). Clear phenotypic differences are known between wild and cultivated types for many plants, but for others there are grey, in-between areas that were probably more prevalent in the past.

There are several reasons why domestication traits of vegetatively propagated plants may be difficult to track in the archaeobotanical record. First, some traits may not be genetically determined but reflect gene expression and growth environment; namely, they are epigenetic and plastic. For example, yam tuber and taro corm phenotypes may reflect growing conditions and revert to 'wild-type' once feral. The implications of phenotypic plasticity for reconstructing the fixation of traits during a domestication episode in the past are unclear. Potentially, parenchyma and starch granule morphologies could be similarly plastic thereby complicating the charting of domestication in the archaeobotanical record using these proxies. Thus, the investigation of domestication in vegetatively propagated tuberous plants may be difficult to characterise based on phenotypic changes, especially size, in plant macrofossils and microfossils. Similarly, people selected for decreased acidity and lower oxalate content in some tubers, just as they have selected for taste and texture in other plants. With more prolonged cultivation and increased domestication, toxins, acrid chemicals and oxalates would be anticipated to become less prevalent – if preserved – in the archaeological record. Second, people selected for seed-suppressed forms in some plants, namely smaller seeds and larger edible pulp in bananas and breadfruit. As a result, seeds will drop out of the archaeobotanical record through time – or are less likely to be preserved, collected and identified in vestigial form – even though the plant may be more intensively cultivated.

Table 5.3 Comparison of domestication traits in sexually reproduced plants (such as cereals, legumes and some fruits) and vegetatively propagated plants (such as root crops, grasses and vegetables in the New Guinea region)

Trait	Domestication in Sexually Reproducing Plants	Domestication in Asexually Reproducing Plants	Example from New Guinea
Size of edible portion	Increased seed size in cereals, legumes, nuts	Increased size of edible portion, most marked in starch-rich plants, decreased seed size/suppression in some fruits	Increased edible pulp and decreased seed size in bananas and breadfruit
Ease of harvesting	Development of non-shattering seeds	Potential development of bunching equivalent	Adherence of individual banana fruits within a bunch to enable harvesting, also possible for *karuka* and *marita* pandanus
Timing of fruit production	Synchronous food production within plant and between plants	Asynchronous and more continuous food production, with in-ground storage for some root crops	Root crops – including taro and yams – can produce edible tubers/corms over a prolonged period and be 'stored' growing in the ground for a short period
Plant architecture	Apical dominance	Apical dominance in some plants	Artificially maintained in numerous vegetatively propagated plants, through management and removal of suckers to encourage growth and food production
Plant life cycle	Shift towards annual life cycle based on sexual reproduction from seed	Shift towards perennial life cycle based on vegetative reproduction using suckers, shoots, subterranean storage organs and other viable plant parts	Domesticated grasses – including sugarcane and edible pitpits – grow as vegetatively propagating perennials rather than as annuals
Photoperiod sensitivity	Loss of photoperiod sensitivity	Maintenance of photoperiod sensitivity?	Plant maturation and production heavily reliant on photoperiod exposure – in the highlands partially a product of cloud cover
Defensive adaptations	Loss of defensive adaptations (spines, hard seed casings, toxicity, acridity) to enhance harvesting, processing and consumption	Loss of defensive adaptations (spines, hard seed casings, toxicity, acridity) to enhance harvesting, processing and consumption	Reduction in oxalate crystals (raphides) in taro through human selection, loss of spines in cultivated yams and sago
Environmental tolerance	Easily adapted to new environments	Triploidy and vegetative propagation enable cultivation in wider environmental (altitudinal and latitudinal) ranges	Vegetative propagation enables movement of cultivars beyond natural range, enabling altitudinal extension of production in highlands
Disease resistance	Reduced resistance to disease and pests due to human selection following continued sexual reproduction of sub-population	Dramatic reduction in resistance to disease and pests due to low genetic variability in clonally reproduced cultivars	Susceptibility of AAA banana cultivars (Gros Michel and Cavendish) to Panama disease
Palatability	Increased selection for desired traits	Increased selection for desired traits	Reduction in acridity and oxalate content in cultivated taro Reduction in fibrous stems and leaves in edible grasses and green vegetables

Note: Drawing on Gepts 2002; Zohary et al. 2012; Allaby 2014; Barton and Denham 2017; author's fieldwork observations

Third, prolonged vegetative propagation may assist in the loss of sexual reproductive capacity (following Eckert 2002), potentially including flowering and seed-set. Further, plants under cultivation are usually harvested before flowering and seed-set (Powell 1970b). Given the phenology, reproductive habits, infrequent flowering at altitude and harvesting prior to flowering of these plants under cultivation, a muted palynological signal is not entirely surprising (after Powell 1970a: 199).

In general, seeds may be less relevant to understand domestication processes in the wet tropics because domesticatory processes of selection and cultivation have not largely focussed on the seeds themselves. Consequently human-directed evolutionary pressures on seed morphology probably would not be comparable to cereals and legumes, in which the seed is the primary goal of exploitation and cultivation.

The archaeobotany of domestication under vegetative cultivation is different to that of seed-based cultivation, especially in the wet tropics (Harris 1996b: 568; Piperno and Pearsall 1998: 8; Smith 2001: 16–17; Yen 1985). As noted by others (Golson and Ucko 1994), the hardy and better-preserved macrobotanical remains – such as charred seeds, husks and processing debris – are often missing from archaeobotanical assemblages in the wet tropics. Nonetheless, three advances in tropical archaeobotany have aided the investigation of early agriculture and plant domestication: phytoliths (Piperno 2006); starch granules (Torrence and Barton 2006) and archaeological parenchyma (Hather 2000). The application of this suite of techniques has raised the visibility of early cultivation practices in the lowland neotropics (Piperno and Pearsall 1998), West African rainforest (Mbida Mindzie et al. 2001) and the Australia-New Guinea region (Denham et al. 2003; Denham, Atchison et al. 2009). Despite these advances, agricultural practices in many tropical regions remain invisible because such techniques are not routinely applied during archaeological investigations.

Given problems in identifying the domestication of vegetatively propagated plants in the past, how is it possible to identify vegetative cultivation in the past? As discussed earlier, the domestication episode may not be clear, because the beginning and end of phenotypic changes in plants may be plastic, and may not be clearly discriminated in the archaeobotanical record; for instance, does a larger tuber size or starch granule size represent cultivation of a domesticated plant or a plastic response to a more favourable growth environment? Movement beyond the natural range may not be clear because little is known about the phytogeography and ecology of most crop plants in New Guinea, and so it is often not possible to determine when a plant is outside or inside its natural range. Furthermore, people may advertently or inadvertently translocate plants outside of their natural range without domestication or cultivation, either through deliberate transplantation or through plant regrowth from discarded plant parts, respectively.

In the highlands, the cultivation of vegetatively propagated plants has been inferred using elevated frequencies of pollen and phytoliths of crop plants and weeds beyond that anticipated for landscapes without cultivation. This type of record needs to be interpreted cautiously. High frequencies of crop plants may indicate cultivation, but could equally indicate resource intensification in the landscape. To infer cultivation, any archaeobotanical remains need to be linked directly to archaeological evidence of plant exploitation and cultivation practices preserved as features, artefacts and palaeosols. Archaeological excavations provide the essential multidisciplinary evidence for interpreting plant exploitation practices in the past. As such, the archaeology of practices is not only ontologically prior, it is also methodologically prior to archaeobotanical and palaeoecological lines of evidence for the identification of tropical cultivation in the past. Archaeological evidence of past practices provides the evidential foundation to ground the consilience of: archaeobotanical data on the presence, cultivation and use of plants; palaeoecological reconstructions of forest clearance and disturbance; and, stratigraphic evidence of soil formation and plot preparation.

Part III

Practices in the past

6 Exploiting diversity in the Pleistocene

According to a long-held view, the mountainous interior of New Guinea is thought to have been occupied on a temporary basis for hunting and the exploitation of seasonally-producing high-altitude *Pandanus* spp. *(karuka)* during the Pleistocene (see S. Bulmer 1977a: 69; Hope et al. 1983: 40–41, 43–44; Golson 1991a: 87; Hope and Golson 1995: 822–823). More recent archaeological findings at Kosipe Mission and other sites in the Ivane Valley, Owen Stanley Range (Fairbairn et al. 2006; Summerhayes et al. 2010), in concert with the palaeoecological record from adjacent wetlands (Hope 2009), have breathed new life into this scenario, albeit with a variation on the traditional theme: these authors acknowledge the potential for broader-based subsistence and settlement continuity in the interior during the Pleistocene, including the Last Glacial Maximum (LGM). In this chapter, multidisciplinary evidence and inferences indicate that plant exploitation across the whole of the highlands during the Pleistocene would not solely have been reliant on high-altitude *Pandanus* (Denham 2007c). The scenario developed here provides alternative historical horizons for contextualising the emergence of agriculture in the highlands during the Holocene.

The starting point for this investigation is the colonisation of Sahul and commonalities of practice that existed across Island Southeast Asia, New Guinea and northern Australia during much of the Pleistocene (Denham 2008b; Denham, Fullagar et al. 2009; Denham, Donohue et al. 2009; Barton and Denham 2011). Common practices include landscape and plant management using fire; broad spectrum plant exploitation encompassing tree (nuts, fruits and forms of sago) and tubers; broad spectrum animal exploitation; and high mobility. These practices are characteristic of long-term human occupation in many tropical rainforests globally (Barton et al. 2012) and have formed the backdrop for understanding how agriculture emerged in tropical rainforest environments on different continents (Piperno 1989; Golson 1991a; Piperno and Pearsall 1998).

As a methodological note, there is very little archaeological work that has yielded direct archaeobotanical evidence of plant exploitation in the highlands of New Guinea during the Pleistocene. The notable exception is the excavation of multiple sites in the Ivane Valley, Owen Stanley Ranges, Papua New Guinea (Fairbairn et al. 2006; Summerhayes et al. 2010). Most interpretations of plant exploitation in the Pleistocene are inferential and derived from palaeoecology, plant ecology and functional evaluations of stone tool assemblages (e.g., S. Bulmer 1977b; Groube 1989; Golson 1991a).

To Sahul

The first settlement of New Guinea occurred at least 55,000–50,000 years ago (Groube et al. 1986; O'Connell and Allen 2007; Summerhayes et al. 2010; Clarkson et al. 2017). At that time, New Guinea was joined to Australia to form the continent of Sahul (Figure 6.1). The two land masses began to separate with post-glacial sea level rise and the formation of the Arafura Sea and Gulf of Carpentaria approximately 12,000 years ago. They were fully separated only following sea level rise and the formation of the Torres Strait within the last 8000 years (Lambeck and Chappell 2001).

Archaeological investigations suggest that Sahul was rapidly colonised by people following an ocean crossing from Island Southeast Asia. On the basis of evidence from Island Southeast Asia and Australia, people had a range of technologies that were employed in diverse environments including composite projectiles (Barton et al. 2009), edge-ground stone tools (Geneste et al. 2012; Hiscock et al. 2016), pelagic fishing (O'Connor et al. 2011), seed grinding (Fullagar et al. 2008) and toxic plant processing (Barton and Paz 2007). Even though pre-40,000-year-old sites exhibit a coastal and hinterland bias, they occur in varied climatic and biological zones, including wet tropical forests of lowland and highland New Guinea (Groube et al. 1986; Summerhayes et al. 2010), monsoonal tropical forests of northwestern Australia (Clarkson et al. 2017), savanna and grasslands of subtropical northwestern Australia (Balme 2000; Fifield et al. 2001), temperate southwestern Australia (Turney et al. 2001) and the arid and semi-arid interior of southeastern Australia (Bowler et al. 2003; Hamm et al. 2016). By 30,000 years ago, people inhabited every type of environment on the continent including ephemeral occupation of the tropical rainforests of northwestern Queensland (David et al. 1997), intermittent occupation

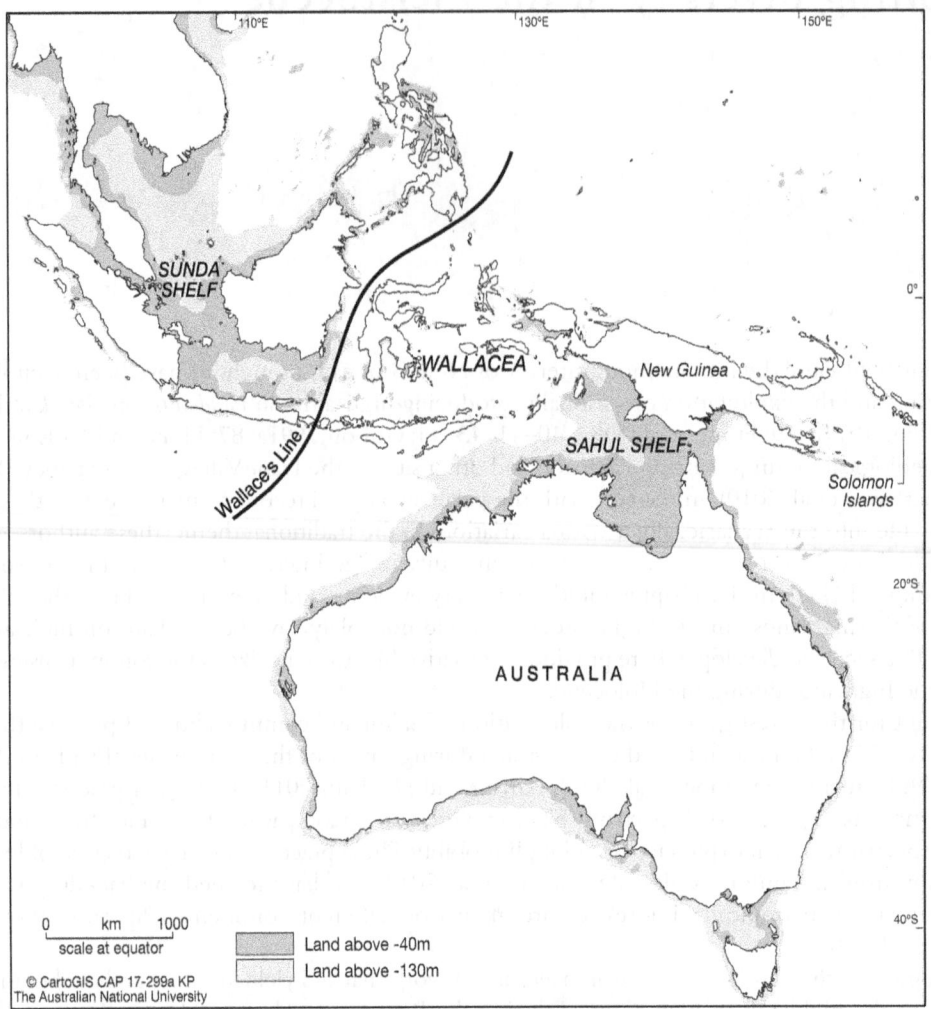

Figure 6.1 Map of Sunda and Sahul depicting sea levels at −40 m and −130 m

of the arid interior of Australia (Smith et al. 2001) and permanent occupation of cool temperate southern Tasmania (Cosgrove 1995).

The relatively early occupations of Yombon on New Britain (c. 35,000 years; Lentfer et al. 2010) and Buang Merabak on New Ireland (pre-40,000; Leavesley et al. 2002) indicate that the first crossings to Sahul were not accidental. These islands were never connected to each other or to the mainland of New Guinea. Maritime technology was maintained in some form throughout the Pleistocene to enable: the colonisation of the Solomon Islands (Wickler 2001) and Manus (Fredericksen et al. 1993), the potentially deliberate introduction of marsupials from New Guinea to the Bismarck Archipelago (Heinsohn 2010), the potentially deliberate introduction of plants from New Guinea to the Bismarck Archipelago (Yen 1996) and inter-island trade within the Bismarck Archipelago (Summerhayes 2003).

Various models have been proposed to characterise the dispersal of modern humans eastwards from Eurasia across Southeast Asia to Sahul; each focuses on different types of environment, whether coastal (Stringer 2000), estuarine (Bulbeck 2007), savanna (Bird et al. 2005) or tropical rainforest (Barker 2014). In contrast to these interpretations, the archaeological evidence from Sahul suggests that early colonists were not specialists – whether coastal, estuarine, savanna or rainforest – rather, they were generalists. People rapidly dispersed across and around land masses due to their ability to apply a set of technologies to the exploitation of animal and plant resources in the diverse environments they encountered: 'people exhibited generalist behaviour and adaptive flexibility in the ways they exploited resources in diverse environments' (Denham, Fullagar et al. 2009: 31).

Adaptive flexibility was enabled by several practices that appear to have been common across Sahul during the Pleistocene. These common practices are discussed below with a focus on the highlands. These practices are inferred from the archaeological record of settlement, lithic technology and animal exploitation; palaeoecological indicators of burning and forest disturbance; and multidisciplinary evidence and interpretations of broad spectrum plant exploitation.

Archaeological traces

The earliest occupation sites for New Guinea are at Bobongara on the Huon Peninsula in the lowlands (Groube et al. 1986) and multiple locations in the Ivane Valley, Owen Stanley Ranges in the highlands (White et al. 1970; Summerhayes et al. 2010). At these lowland and highland locales, archaeological evidence has been interpreted to represent human presence over 40,000 years ago. At Bobongara, waisted blades were documented below and embedded within a tephra on an upraised coral terrace (Groube et al. 1986). In the Ivane Valley, multiple and multi-period occupation deposits contain stone artefacts, including waisted blades; starch residues representing plant exploitation, including potentially of yams (*Dioscorea* spp.); and charred macrobotanical remains of high-altitude *Pandanus* spp. characteristic of cooking in hearths (White et al. 1970; Fairbairn et al. 2006; Summerhayes et al. 2010).

Characteristically for early sites in Sahul (O'Connell and Allen 2007), there are problems of association between dated and archaeological material, exacerbated by the reporting of provenances in very coarse terms. Although the Bobongara and Ivane Valley sites are undoubtedly 'early', the antiquity of human presence on New Guinea is not known precisely. Other early (pre-30,000-year-old) evidence interpreted to represent human activity occurs at two other open locations in the highlands: burning horizons on Supulah Hill in the Upper Baliem Valley (Hope 1998), sometimes referred to as a hearth (Hope and Haberle 2005: 544), and a possible hearth at Kuk Swamp in the Upper Wahgi Valley (Jack Golson pers. comm. in Hope 1998: 459).

Two open sites with putative structural remains have been radiocarbon dated to the Pleistocene in the highlands: Wañelek (c. 19,400–17,300 years ago; S. Bulmer 1977a, 1991) and NFX (c. 23,600–20,000 years ago; Watson and Cole 1977). However, there are problems with the dating and interpretation of features at both (Golson 1991a: 86, 1996: 168; Denham 2016a). Given that the next oldest structural remains in the highlands indicative of settlement date to around 4500–4000 years ago (Denham 2014), claims for Pleistocene house sites are highly anachronistic and should not be accepted at face value (Denham and Ballard 2003: 131).

The earliest occupation of Nombe rockshelter dates to at least 25,500–19,600 cal BP and comprises sparse cultural materials (Mountain 1991a, 1991b; Evans and Mountain 2005; Denham and Mountain 2016). Small quantities of cultural material pre-date 15,000 years ago, including an edge-ground axe, a waisted artefact, ochre, burnt bone and faunal remains that indicate human exploitation of forest and alpine grassland environments (Figure 6.2; Mountain 1991a: 4.17). Sporadic occupation occurred at Nombe for the remainder of the Pleistocene. Four other occupation sites in the highlands date to the Pleistocene/Holocene transition (Table 3.1): Manim 2, Kiowa, Kafiavana and Yuku; although re-dating at Yuku raises questions regarding the archaeostratigraphic integrity of that site (Denham 2016a).

The oldest stone tools on New Guinea comprise variously flaked implements representing adventitious use of available local sources, most probably stream beds (Swadling 1983). In the Ivane Valley excavations, stone artefacts are made from a wide range of igneous and metamorphic rocks, most of which are available locally (namely, within a day's walk; Summerhayes et al. 2010). Other than relatively crude implements made from cobbles, the most distinctive artefact form is the waisted blade, including waisted axe (S. Bulmer 1964, 1977b, 2005; Muke 1984), which occurs in the earliest human-associated deposits at Bobongara (Groube et al. 1986). The artefacts are named after the distinctive notches that form a 'waist' for the hafting of the implement (Figure 6.2). S. Bulmer (2005: 399) notes that some types are more appropriately termed 'stemmed' or 'tanged', and it may be simpler to follow Golson's advocacy of 'butt-modified' as a generic term (Figure 6.3; Golson 2001, 2005). Waisted artefacts persist in the archaeological record of the highlands throughout the Pleistocene and into the mid-Holocene.

Waisted tools are multifunctional and general-use implements. Groube (1989) initially proposed that waisted axes were used for habitat modification, including ring-barking, cutting branches and smaller trees and root clearance. His arguments have been extended to suggest deliberate habitat modification by humans since initial colonisation (Denham and Barton 2006; Fairbairn et al. 2006). It is now generally accepted that these tools could not have been used to cut down and fell trees, or clear forest. S. Bulmer (2005: 411) has proposed that the primary role of waisted axes was for food procurement: 'harvesting sago, splitting cycad trunks, thinning pandanus stands to ripen their fruits' as well as 'splitting logs to harvest grubs, chopping into hollow logs or trees to capture sleeping and hiding animals and digging up edible roots'. Whatever the precise function, waisted blades were used by the earliest colonisers of Sahul, including in the highlands, and these people were engaged in forms of habitat modification, most plausibly to encourage the growth of useful plants in edges and gaps of the montane rainforest.

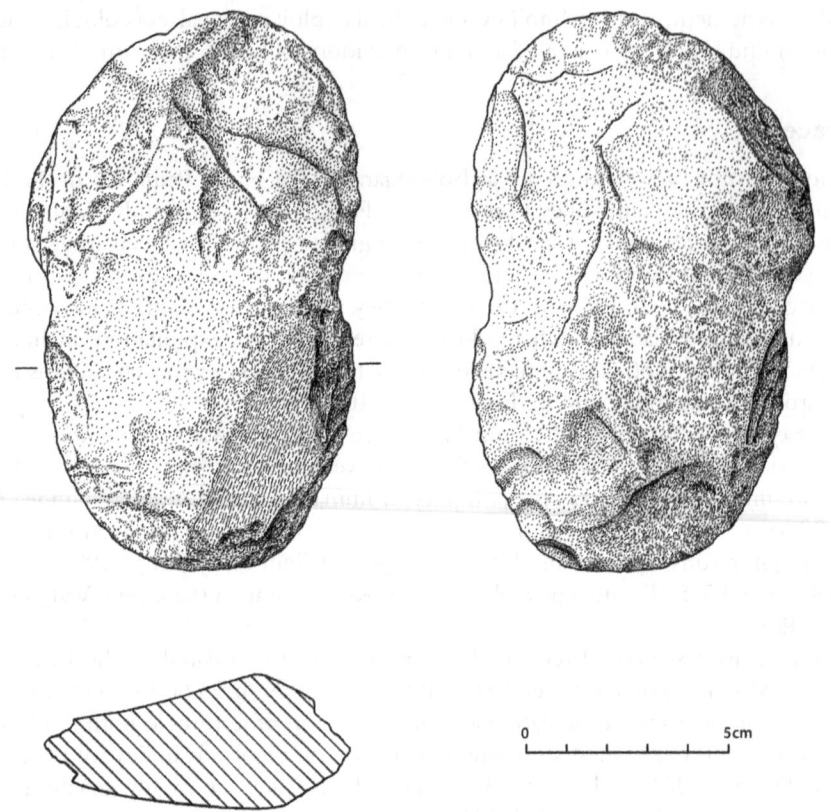

Figure 6.2 Waisted blade excavated from c. 30,000-year-old context at Kosipe, Ivane Valley, Papua New Guinea

Source: Anderson and Summerhayes 2008: Figure 5

Various theories have been proposed to characterise the faunal exploitation of early colonists, although most debate has centred on the timing and human role in the extinction of megafauna (e.g., Roberts et al. 2001; Field et al. 2008; Wroe et al. 2013). Long records of faunal exploitation from New Ireland in Island Melanesia and the Bird's Head in lowland New Guinea show heavy reliance on individual taxa: cuscus *(Phalanger orientalis)* from Buang Merabak (Leavesley 2004) and wallaby *(Dorcopsis muelleri)* from Toé Cave, Birds Head (Pasveer 2004), respectively. By contrast, there are no robust records of faunal exploitation in the highlands during the Pleistocene; Nombe provides only sporadic, sparse and admixed information (Mountain 1991b; Sutton et al. 2009: 47–48, 50; Denham and Mountain 2016).

The record at Nombe is complicated by site formation processes (Denham and Mountain 2016). Pleistocene contexts at the cave have yielded remains of four extinct faunal species (Flannery et al. 1983; Mountain 1991a): three faunal species are unique to Nombe – *Protemnodon nombe, Protemnodonn tumbuna* and *Dendrolagus noibano*, with an unidentified diprotodontid also present (Flannery et al. 1983). The age and cultural associations of these megafaunal remains are uncertain. Five additional species of extinct megafauna have been recovered from poorly dated contexts in the highlands: from river clays along the West Baliem River (c. 40,000–15,000 years ago; Hope et al. 1993) and at Pureni swamp (c. 38,000 years ago; Flannery and Plane 1986), as well as from an old stream channel within Kelangurr cave (c. 20,000 years ago; Flannery 1992; Hope et al. 1993). Taken as a group, these remains suggest diverse faunal exploitation, including of megafauna, persisted in the highlands long after human colonisation (e.g., Hope 1998). Moreover, they suggest that people played a role in the extinction and local extirpation of faunal species in highland rainforests during the Pleistocene, just as they have during the Holocene and up to the present day (Flannery 1995; Sutton et al. 2009).

Archaeological sites tend to preserve hard calcareous materials, such as bone and shell, which distorts interpretations of faunal exploitation in the past. Ethnographic inference suggests a wide range of mesofauna and invertebrates would have

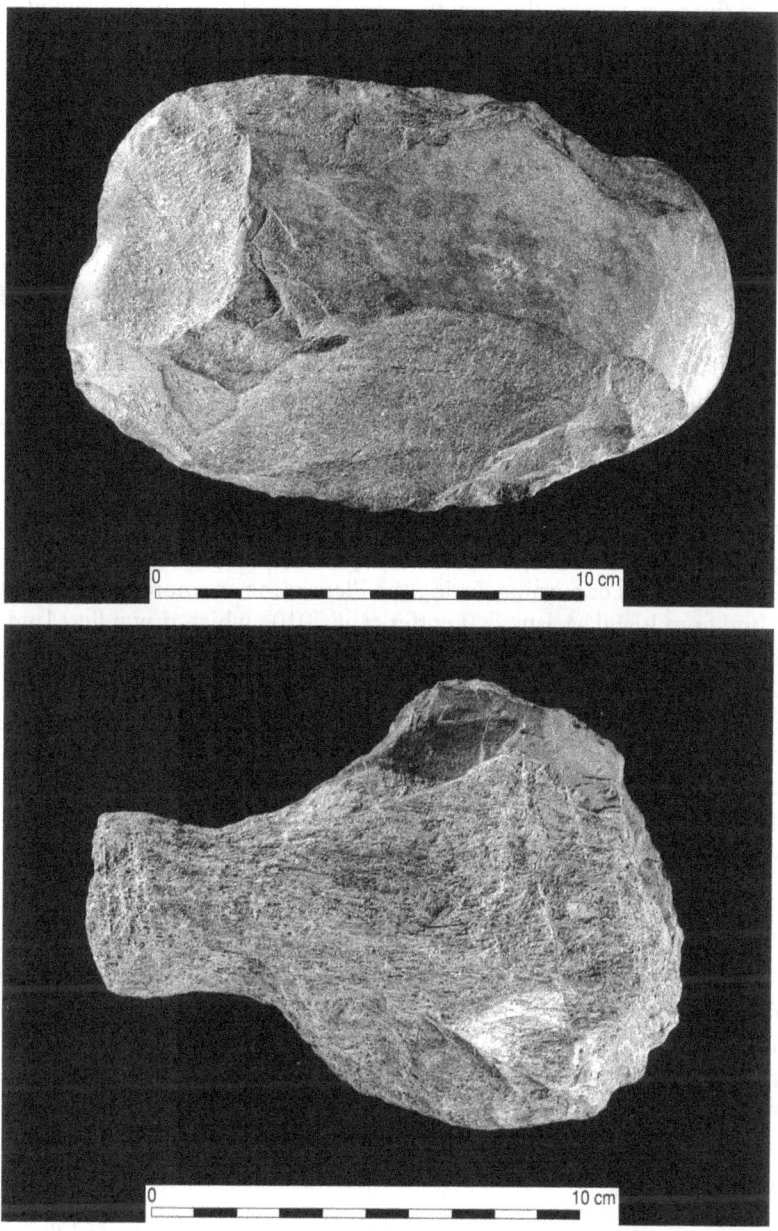

Figure 6.3 Hand axe (upper) and tanged, or butt-modified, blade (lower) excavated from Pleistocene levels at Nombe
Source: Mary-Jane Mountain, photograph taken by Dragi Markovic

been gathered and formed a large percentage of protein in everyday diets (R. Bulmer 1968; R. Bulmer and Menzies 1972, 1973; R. Bulmer et al. 1975). For instance, in the highland interior, the hunting of large to medium animals – including birds, mammals and reptiles – would provide periodic meals, whereas the gathering of invertebrates and small animals would provide regular, albeit individually small, sources of protein to supplement diets. Zooarchaeological remains of most mesofauna and invertebrates are unlikely to preserve.

Except for the Ivane Valley sites, archaeology sheds relatively little light on the chronology and subsistence practices of early human occupants of the highlands during the Pleistocene. As a result, interpretations of human occupation and land use have relied heavily upon palaeoecological reconstructions of fire and vegetation histories (e.g., Haberle and Chepstow-Lusty 2000).

Palaeoecological inference

The colonisers of Sahul used fire to encourage the growth, production and density of favoured plants (Jones 1969; Gott 2005), as well as to hunt and manage game (Bowman et al. 2001; Vigilante and Bowman 2004). The purpose of burning was probably activity and environmentally specific, for example, to encourage the growth of yams in tropical rainforests and to hunt in semi-arid/or monsoonal grasslands and savanna. The cumulative effects of burning on landscapes are indicated in vegetation histories for the continent; they were regionally variable and appear to be greatest in climatically sensitised environments (Kershaw et al. 2006). Only a few sites demonstrate evidence for burning coeval with colonisation (Turney et al. 2001); most marine and terrestrial records exhibit considerable variability in the levels of burning and effects on vegetation communities (Kershaw et al. 2006).

In the highlands of New Guinea evidence of burning occurs sporadically throughout the period of Pleistocene occupation (Hope and Haberle 2005). The apparent intensity, timing and duration of burning, as well as of associated vegetation changes, varied considerably and the human contribution to the burning signal is not always clearly differentiable from climatic causes. The human origin of these disturbances can be inferred from the long-term decline in primary forest and concordant rises in secondary forest, grassland and charcoal frequencies (Haberle 1994). Hope (2009: 2274) states: 'The relationship of fire, inferred from charcoal, and human activity is complex, but in a very wet montane setting consistent or extensive fire is largely the result of human activities'.

Relatively early burning occurs from c. 32,000 cal BP in the Baliem Valley to the west (Haberle et al. 1991) and from c. 40,000 cal BP in the Owen Stanley Ranges to the east (Hope 2009). Various indicators of Pleistocene burning occur in the major intervening inter-montane valleys: Telefomin (Hope 1983), Haeapugua (Haberle 1998) and Wahgi Valley (Denham, Haberle et al. 2004). Records of burning during the Pleistocene are more limited from lowland New Guinea (Hope and Haberle 2005) and from Island Melanesia (Lentfer et al. 2010), which may reflect land management practices; the greater difficulty of burning in wetter, lowland tropical rainforest; and potentially lower climatic stressing in the lowlands relative to highland environments.

In the highlands, fire records and vegetation histories are highly variable spatially and temporally. Whereas some valleys show relatively frequent human disturbance from the Pleistocene to the present, for example, Baliem Valley (Haberle et al. 2001), others show little evidence of burning and forest disturbance until the last few hundred years, for example, Nurenk Swamp (Hope et al. 1988). Anthropic disturbance to montane rainforests during the Pleistocene is usually episodic, only in a few locations is there continued and slightly increased disturbance towards the Holocene (Hope and Haberle 2005; Haberle 2007).

The disentangling of anthropic and climatic contributions to vegetation disturbance and change is complex. The effects of similar types of practices, such as landscape management to create gaps and patches in the rainforest for useful plants, will be greater in climatically sensitised environments. As a result, the intensity of human rainforest management cannot be directly read from the intensity of burning (evident in charcoal records) or from the degree of disturbance to vegetation communities (evident in pollen and phytolith records). Rather, the reconstruction of human impacts on the environment is interpretative and requires careful consideration of multiple lines of evidence and multiple potential contributions (i.e., climatic, human and tectonic).

Not just trees and tubers . . .

Taken together, the archaeological (settlement, stone tool and faunal) and palaeoecological (burning and vegetation history) evidence indicates that people have been agents of change in the interior rainforests of New Guinea since the Pleistocene. What was the purpose of these practices? In general terms, people are considered to have maintained or created edges, gaps and patches in the rainforest to encourage the growth of favoured species (Denham and Barton 2006; Denham and Haberle 2008). For example, bananas (*Musa* spp.) are gap colonisers and yams (*Dioscorea* spp.) are vines that can thrive in disturbed rainforest environments. Additionally, people may have cleared competing vegetation around favoured nut, fruit and sago-yielding plants – such as palms, pandanus and trees – thereby encouraging the growth of sexually and vegetatively reproduced stands and groves (discussed in Barton and Denham 2011).

The only sites to provide archaeobotanical evidence of plant use in the highlands during the Pleistocene are those in the Ivane Valley (Fairbairn et al. 2006; Summerhayes et al. 2010). Multiple sites in this locale have yielded macrobotanical evidence for the exploitation of *Pandanus* spp. and starch residues extracted from stone tools indicate the exploitation of yams (*Dioscorea* spp.) and other unidentified plants. Directly dated highland *Pandanus* kernels are 37,000–35,000 years old at Kosipe (Fairbairn et al. 2006) and are noted for older deposits at other sites in the Ivane Valley (Summerhayes et al. 2010). Similarly, plant residues including those of yams are inferred to pre-date 40,000 years and to reflect lowland plant use by mobile groups who also utilised high-altitude pandans. Previous claims that drupes from *marita (Pandanus conoideus)* at Yuku were of Terminal Pleistocene antiquity (Bulmer 2005: 392–393) can be discounted; not only is there evidence for site disturbance, but also putatively ancient drupes have been radiocarbon dated to within the last few hundred years (Denham 2016a). Currently, the Ivane Valley has yielded evidence of plant use more than 25,000 years older than anywhere else in the highlands.

Methodological biases in archaeobotanical recovery provide a bimodal impression of plant exploitation in Island Southeast Asia and New Guinea, with seeming foci on palms, pandanus and tree products on the one hand and tuberous plants on the other (Barton and Denham 2011; Denham, Fullagar et al. 2009). Archaeobotanical remains of nuts and fruits from trees occur in samples analysed at Niah Cave (including Pleistocene contexts; Barton and Paz 2007), at multiple sites in Island Melanesia (Fredericksen et al. 1993; Fairbairn 2005) and highland New Guinea (at least 37,000–35,000 years ago; Fairbairn et al. 2006). Comparable microfossil evidence regarding early tuber/root use emerges for a variety of plants from Borneo (pre-45,000 years ago; Barton and Paz 2007), the Solomon Islands (c. 30,000 years ago; Loy et al. 1992) and highland New Guinea (pre-40,000 years ago; Summerhayes et al. 2010). Very little information is available on the exploitation of other types of plant and plant parts during the Pleistocene, particularly when compared with archaeological assemblages of seeds in Australia, for example, Kimberly Ranges (c. 40,000 years ago; McConnell and O'Connor 1997; McConnell 1998).

Following from the previous, there have been claims for the nascent domestication of trees and tubers in rainforest environments of New Guinea extending back into the Pleistocene (from Groube 1989 and Yen 1989 onwards). These claims are based on the inferred long-term management of plant resources during the Pleistocene leading to the selection of preferred types, including of sterile forms, and movement out of natural range. The development of distinctive phenotypic and genotypic traits, namely domesticated forms, could have occurred due to a combination of anthropic selection and genetic isolation (Chapter 5). Genetic isolation of managed populations could have occurred through the preferential selection of sterile forms and through geographical movement to areas in which wild populations of the plant did not occur, namely to new islands or new altitudes. Certainly, there is evidence for the inter-island movement of marsupials between New Guinea and islands to the north (Bismarck Archipelago) and to the west (Wallacea) during the Pleistocene (Heinsohn 2010). So the Pleistocene inhabitants of New Guinea were oriented to their world in such a way that the movement of things, plausibly plants as well as animals, made sense.

Yen (1996) claimed that canarium nut *(Canarium indicum)* was domesticated and transported between New Guinea and the Bismarck Archipelago during the Terminal Pleistocene. His claims were based on the occurrence of canarium nut dating up to 20,000 years ago in archaeobotanical assemblages at Pamwak on Manus, which he considered to have been deliberately transplanted from the New Guinea mainland (Yen 1996). Two issues cloud any definite conclusions: uncertainty over the ability to discriminate domesticated and wild canarium in archaeobotanical assemblages (Fairbairn 2005) and the uncertain role of humans in the dispersal of canarium to Manus and other islands. On the basis of pollen records, Haberle (1995) inferred the potential domestication of taro *(Colocasia esculenta)* and pandanus *(Pandanus* spp.) on the New Guinea mainland during the Terminal Pleistocene. However, there is likely to have been human-aided dispersal and use of these plants for millennia on New Guinea and neighbouring regions, with resultant genetic isolation and differential selection leading to domestication. Currently, there is a lack of robust archaeobotanical evidence to track these processes across time and place.

Haberle's (1995) claims for the early domestication of taro *(Colocasia esculenta)* and pandans *(Pandanus* spp.), like Yen's (1996) claims for the Pleistocene domestication of canarium, should be treated cautiously as hypotheses. Once introduced, whether to new islands or new altitudes, genetic isolation could have led to the development of distinctive phenotypic and genotypic traits among isolated and managed populations. In this sense, early translocation led to domestication.

Pandanus species in the highlands

The ancient *Pandanus* in the Ivane Valley has been taken to substantiate White's earlier hypothesis that the occupation of the region was based on the exploitation of wild edible *Pandanus* in the vicinity (White et al. 1970: 168–169). Like several other researchers before them, Fairbairn et al. (2006: 379) and Summerhayes et al. (2010: 81) imply a generalised model of highland subsistence based on hunting and the exploitation of high-altitude *Pandanus*. Although contributing greatly to a fragmentary archaeological record, their interpretations echo long-held assumptions regarding the nature of plant exploitation during the Pleistocene in the highland interior (this section reproduced in amended form from Denham 2007c: 42–43).

A re-evaluation of this recurrent model of highland subsistence requires an in-depth consideration of *Pandanus* phenology, vegetation history and resource availability in the interior rainforests of New Guinea, including the highlands, during the Pleistocene. This alternative interpretation highlights the spatial and temporal variability in the seasonality of *Pandanus* production; the persistence of mixed *Castanopsis-Lithocarpus* lower montane forest on the lower slopes and floors of some highland valleys during the Pleistocene, including during the Last Glacial Maximum (LGM); the resultant variability in abundance and diversity of plant food resources across space and through time; and, the highly variable food procurement strategies adopted by people inhabiting the interior rainforests (Denham 2007c).

High-altitude *Pandanus* species are members of the *P. brosimos/julianettii/iwen* complex, which are all nutritious nut-bearing trees referred to as *karuka* in New Guinean pidgin. Given the numerous cultivated varieties of *P. julianettii* and types of *P. brosimos*, as well as the uncertain taxonomic status of *P. iwen*, the classification and anthropic associations of members of this high-altitude *Pandanus* spp. complex remain to be fully elucidated (Denham 2007c). In historical records, these *Pandanus* spp. are noted as major sources of fat and protein for some highland groups; they can be collected, dried/smoked and stored for consumption later in the year, for example, Baruya (Lemonnier 2002). *P. julianettii* is often considered to be the cultivated form of *P. brosimos* (Stone 1982a; Donoghue 1988, 1989; Haberle 1995; cf. Lemonnier 2002 in which both are referred to as 'semi-domesticated'). Cook raises the possibility that *P. iwen* is 'in-between' the cultivated *P. julianettii* and the wild *P. brosimos*, since the Amungme of Irian Jaya consider it to be the 'uncle' of both (1999: 96; cf. Stone 1984).

Under present climatic conditions, Bourke (2017; see Table 5.2) described the normal altitudinal ranges of *P. brosimos* and *P. julianettii* as 2400–3100 m and 1800–2600 m, respectively, and the extremes as 1800–3300 m and 1450–2800 m, respectively. Cook notes that cultivation of *P. julianettii* occurs as low as 1200 m among the Amungme in West Papua (1999: 95) and Stone (1982b: 116–117) reports that L.J. Brass found fruiting *P. julianettii* at 1300 m. The fruit-producing *Pandanus conoideus (marita)* has a normal altitudinal range of 0–1700 m, an extreme range of 0–1980 m (Bourke 1989), and is an important dietary item between 500 m and 1400 m (French 1986: 210; Bourke 1996).

Today, there is an altitudinal 'gap' in the availability of nutritious and highly ranked, wild *Pandanus* species between highland *P. brosimos* (usually from 2400 m upwards) and lowland *P. conoideus* (usually from 1700 m downwards). *Pandanus antaresensis*, also referred to as *karuka*, fills this altitudinal and ecological gap; it usually grows from 1000–2350 m and has an extreme altitudinal range of 850–2460 m (Bourke 2017; see Hyndman 1984). Today, *P. antaresensis* is only eaten by some communities (Donoghue 1988: 49; Mike Bourke pers. comm. 2006). Of note, *Pandanus antaresensis* was heavily exploited during the early Holocene at Manim 2 (Donoghue 1988: 71–78).

The 1700–2400 m *Pandanus* gap has not remained constant in the highlands through time. First, considerable variability in the altitudinal distribution of all *Pandanus* species can be anticipated in response to various climatic changes during the Pleistocene and Holocene. Due to lower altitudinal vegetation boundaries during the Pleistocene, most marked during the LGM, the *Pandanus* gap would have been narrower than today. Second, the effects of long-term management on the altitudinal ranges of *karuka* and *marita* are uncertain. The antiquity of *Pandanus* spp. domestication is unknown and subject to speculation (e.g., Haberle 1995: 207–208). On the basis of her studies of the Wurup valley assemblages, Donoghue (1988: 94) concluded that morphological variations in *Pandanus* drupes between the Wurup valley sites 'represents the exploitation of several wild species, namely, *P. antaresensis, iwen* and *brosimos*, with the species procured relating to the altitude of the site'. There was no indication of morphological transformations characteristic of domestication (cf. Golson in Christensen 1975a: 24).

Even though *Pandanus* spp. are the only macroremains of an economic plant at Pleistocene-aged occupation sites in the highlands, this is largely a product of preservation and archaeological visibility. *Pandanus* kernels are relatively large and preserve well when charred. Furthermore, excavations at most occupation sites have focussed on chronology, fauna and stone tools rather than on archaeobotanical recovery. The systematic application of archaeobotanical techniques to recover plant macrofossils and microfossils (including parenchyma, phytoliths and starch grains), such as employed in the Ivane Valley (Summerhayes et al. 2010), is likely to reveal a greater range of plants were available, eaten and otherwise exploited at occupation sites in the highlands during the Pleistocene.

More than hunting and the seasonal exploitation of *Pandanus*

Recurrent accounts portray pre-agricultural occupation of the highlands in the Pleistocene as temporary and based on hunting and the exploitation of seasonally-producing *Pandanus* spp. (S. Bulmer 1977a: 69; Hope et al. 1983: 40–41, 43–44; Golson 1991a; Hope and Golson 1995: 822–823; Fairbairn et al. 2006; Summerhayes et al. 2010; this section reproduced in amended form from Denham 2007c: 43–45). Putting aside issues related to hunting, which are less contentious, these scenarios are based on three major assumptions. First, *Pandanus brosimos* is considered to be seasonally producing and the main plant food at high-altitudes, or in the highlands generally, during the Pleistocene (e.g., Golson 1991a; Fairbairn et al. 2006: 378–379, 382). Second, highland valleys were carpeted in *Nothofagus*-dominated forests during the altitudinal depression of vegetation communities during the LGM and other periods of the Pleistocene (e.g., Golson 1991a: 86; Fairbairn et al. 2006: 375, 381). Third, beech *(Nothofagus)* forests are comparatively resource-depauperate, so highland valleys during the LGM and other periods of the Pleistocene were relatively devoid of edible plant resources other than *Pandanus* (see Golson 1991a: 87; Hope and Golson 1995: 827). The relevance of each of these assumptions is spatially and chronologically contingent, and they certainly do not apply to the highlands as a whole during the Pleistocene or LGM.

First, Mike Bourke has shown that the production of nuts in *Pandanus julianettii* is climate dependent, and largely a product of water stress (Bourke 1996: 49–50, 54; Bourke et al. 2004: 39–40, 154, 188–189). There are considerable variations in *P. julianettii* nut production across the highlands (see Donoghue 1988: 50):

> In the Eastern Highlands where rainfall distribution is seasonal, production is more regular and approaches a regular seasonal pattern. In the western part of the highlands where production is much less regular, rainfall seasonality is very weak or non-existent.
>
> (Bourke et al. 2004: 40)

The information for *P. brosimos* is more limited and indicates that in regions with weakly seasonal climates, such as the central highlands generally (McAlpine et al. 1983: 69–70), 'production is discontinuous and non-seasonal' and the producing season in *P. brosimos* may or may not coincide with that of *P. julianettii* at lower altitudes within a given region (Bourke et al. 2004: 41). Indeed, *P. brosimos* is noted as fruiting in most months of the year and '[c]laims in the literature for annual or biennial bearing are not supported by the available data' (Bourke et al. 2004: 41).

In contrast, for *marita (P. conoideus)* there are clear relationships between length and timing of fruiting season, seasonality of fruit production and altitude; there are no relationships between the timing, periodicity or duration of fruiting and rainfall seasonality or latitude (Bourke et al. 2004: 29–30). At higher altitudes, the main fruiting season is later, shorter, discontinuous and very seasonal, whereas at sea level, production is non-seasonal and continuous (Bourke et al. 2004: 29; cf. Bulmer 2005: 392–393). Seasonal variations in *marita* availability at different altitudes are exploited by people living along the highland fringes and in the lowlands, thereby ensuring greater continuity in access to this highly nutritious food (Bourke et al. 2004: 29–30); a practice likely to have occurred in the past.

Given variations in the seasonality of nut (and fruit, in the case of *marita*) production for cultivated and wild *Pandanus* varieties depend upon variations in climate, high-altitude pandans should not be assumed to have provided a predictable seasonal crop during the Pleistocene. The variations in *P. brosimos* nut production across the highlands today are likely to characterise similarly variable, albeit different climates in the past (after Brookfield 1989). Inhabitants of the central highlands with weakly seasonal climates could not have based their occupation on the exploitation of an unpredictable resource during the Pleistocene, whereas those in the more seasonal eastern highlands and Owen Stanley Ranges are more likely to have relied on seasonal *karuka* production.

Second, the altitudinal depression of lower montane forests during the LGM was neither even across the highlands nor as extreme as sometimes suggested. Fairbairn et al. (2006: 381 modifying Hope and Haberle 2005: 545–546), claim 'a forest of beech *(Nothofagus)* remained through the LGM' at '1200–1800m in the Wahgi, Tari and Telefomin valleys'. Although this may be true for Haeapugua (1650 m) in the Tari Basin (Haberle 1998), it is doubtful for other intermontane valleys at similar altitudes, such as the Upper Wahgi Valley.

A pollen diagram through the LGM at Kuk (c. 1560 m) in the Upper Wahgi Valley (Figure 6.4a; Powell 1984) indicates that *Nothofagus*-dominated mid-montane forests were restricted to the upper valley walls and higher reaches of the valley; mixed *Castanopsis-Lithocarpus* lower montane forest persisted on lower slopes and valley floors. The Kuk 5A diagram clearly shows the persistence of mixed lower montane forests before, during and after the LGM. In contrast, at the slightly higher altitude site of Draepi-Minjigina within the same valley (c.1890 m; Figure 6.4b; Powell 1982b: 219), mixed oak forests drop markedly and *Nothofagus* forests predominate during the presumed LGM. The effects of climate change on vegetation and, by implication, resource availability vary from one highland valley to another (compare Kuk in the Upper Wahgi Valley to Haeapugua in the Tari Basin; Haberle 2007), as well as between reaches of the same valley (compare Kuk to Draepi-Minjigina in the Upper Wahgi Valley; Denham and Haberle 2008).

Furthermore, the period of Pleistocene occupation of New Guinea witnessed variable climates; high magnitude variability overlay general cooling and warming trends (Turney et al. 2004). In highly general terms, LGM conditions persisted for several thousand years, whereas human occupation during other periods of the Pleistocene witnessed milder conditions. In the Upper Wahgi Valley, a core (DR 29) at Draepi-Minjigina exhibits more mixed forests for the majority of the Pleistocene period represented and before the peak in *Nothofagus* forest inferred to correspond to the LGM (Powell 1982b: 219). Similarly, the Pleistocene portion of a core (AM6) at Lake Ambra dated to between c. 27,000 to 32,000 uncal. BP indicates admixed forests in the vicinity (Powell 1982b: 221). Given the temporal, as well as spatial, variability in climatic conditions and resultant distributions of vegetation and food resources in different locales, care should be taken to be chronologically and geographically specific when referring to subsistence practices during the Pleistocene in the highlands.

Third, *Nothofagus*-dominated forests are considered to inhabit cool misty conditions and to be relatively resource poor (Golson 1991a: 87; Hope and Golson 1995: 827; Read et al. 1990). In contrast, *Castanopsis-Lithocarpus* forests are considered to be relatively resource rich (after Bulmer and Bulmer 1964: 49; Clarke 1971: 51–66). Regarding the availability of resources, Golson has stated:

> the mixed oak forest is a favourable environment for plant-food procurement. Besides *Castanopsis* itself, which produces prolific quantities of small nuts, Bulmer and Bulmer (1964: 69) list a number of other trees with edible nuts and seeds, including *Elaeocarpus, Sloanea, Finschia, Sterculia* and especially *Pandanus* . . . vines with edible fruits; many trees, shrubs and ferns with edible foliage; many kinds of edible fungi; and wild edible yam-like tubers, apparently of the genus *Dioscorea*. Of this corpus of edible plants only *Pandanus* is at home in the beech [*Nothofagus*-dominated] forest.
> (Golson 1991a: 87)

There are three reasons to infer that a wide variety of edible food plants were available to inhabitants of the Wahgi Valley at altitudes of c. 1560 m and below throughout the Pleistocene, including during the LGM:

1   The floor and lower slopes of the valley were carpeted with relatively resource-rich, mixed *Castanopsis-Lithocarpus* montane forests throughout the Pleistocene;

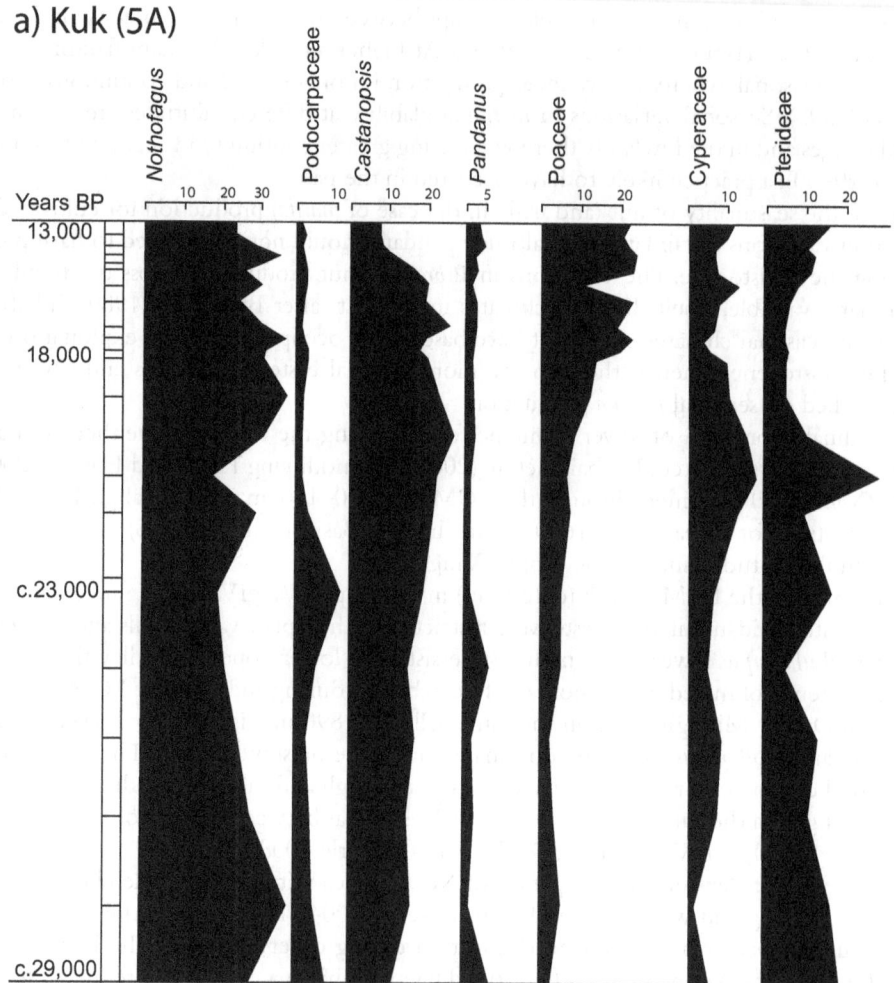

Figure 6.4 Summary of key pollen taxa during the Pleistocene for Kuk (core 5A) and Draepi-Minjigina (core DR 29)

Note: The taxa frequencies in the two diagrams have not been calculated in the same way, and so are not directly comparable: for Kuk 5A, taxa frequencies are given as a percentage of total pollen minus aquatics and unknowns (Powell 1984), whereas for Draepi-Minjigina DR 29, taxa frequencies are given as a percentage of total woody taxa (Powell 1982a). Despite these differences, the diagrams enable comparisons of the relative frequencies of key taxa at each site during the Pleistocene. Uncalibrated radiocarbon ages are given on both diagrams

Source: Denham 2007c: Figure 1

2   Based on altitudinal ranges today, many food plants would have been available on the valley floor and along the lower reaches of the Wahgi Valley during the LGM; and

3   Archaeobotanical and palaeoecological records for the Pleistocene at Kuk Swamp document a broad range of edible plants, including potential major (Musaceae) and minor (cf. *Setaria palmifolia*) sources of starch, as well as nut and fruit-bearing trees (such as *Castanopsis* sp., *Ficus* spp. and *Pandanus* spp.) and leafy vegetables *(Oenanthe javanica, Solanum nigrum)*. Unfortunately, the exact age of these food plant remains is uncertain; they were collected from organic deposits that pre-dated 10,000 years cal. BP.

Powell (1982b: 210–211) conjectured that the richness and diversity of edible plants in New Guinean rainforests may have fostered broad spectrum subsistence. Many communities utilised a broad range of plants for food and other uses until the recent past (Powell 1976). The availability of floral resources on the floor of the Upper Wahgi Valley opens up the possibility for mobile groups to be sustained by broad spectrum diets and to live permanently in the interior of the island during the Pleistocene, including throughout the LGM.

The earlier considerations of *Pandanus* seasonality, the altitudinal migrations of vegetation communities in response to climate change and resource availability through time, are further complicated by recently identified changes in climate during the Pleistocene. These include millennial-scale period ('flickers') of rapid warming and cooling (Turney et al. 2004), as well as variations in El Niño Southern Oscillation (EÑSO) periodicity and intensity (Tudhope et al. 2001),

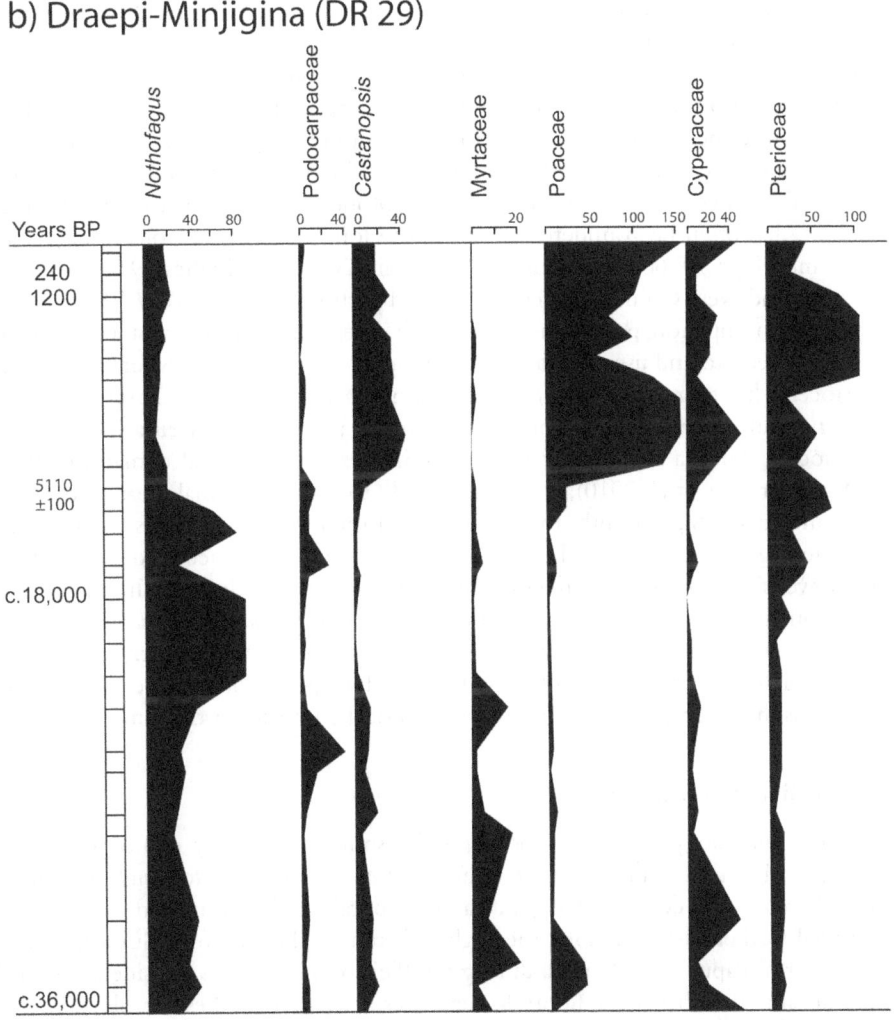

Figure 6.4 (Continued)

in rainfall (Dewar 2003) and in seasonality (Haberle and David 2004: 166–169). The implications of these developing interpretations of climatic variability during the late Pleistocene on faunal and floral resource availability and, hence, on human occupation have yet to be seriously considered for the highlands of New Guinea.

Rethinking occupation of the interior during the Pleistocene

Archaeological evidence of human occupation is not sufficient to determine the character of occupation in the highlands during the Pleistocene and early Holocene, namely, whether it was permanent or seasonal. Open sites dating to the Pleistocene have been identified in the three locations in the highlands: multiple sites in the Ivane Valley (including Kosipe; Summerhayes et al. 2010), NFX (Watson and Cole 1977) and Wañelek (S. Bulmer 1977a, 1991). NFX and Wañelek are claimed to have structural remains of former houses, but there are uncertainties over the dating and archaeological significance of the published evidence at both sites (Golson 1991a: 86, 1996: 168; Denham and Ballard 2003: 131). An additional two 'hearth-type' sites have been dated in the Upper Baliem and Upper Wahgi valleys, respectively (Hope and Haberle 2005) and Nombe rockshelter contains sparse cultural material of Pleistocene age.

Despite a paucity of evidence, it is hypothesised here that highly mobile communities inhabited the interior of the island of New Guinea permanently during the Pleistocene (this section reproduced in amended form from Denham 2007c: 45–46). Since first colonisation, people engaged in environmental manipulation to enhance resource availability, as well as in extensive hunting and foraging. People targeted relatively resource-diverse gaps in the forest and along ecotones, such as those along wetland margins and riparian corridors (Denham and Barton 2006; cf. Groube 1989), and the high-altitude grassland/forest boundary (Hope and Hope 1976a; cf. Golson in Hope et al. 1983: 44).

To take this proposition seriously, we need to disinherit ourselves of a preconception about New Guinean history. We need to recognise that permanent occupation of a region does not necessarily imply sedentism or the continuous use of a single area or site; it can refer to the permanent occupation of a region by highly mobile groups. People moved around the landscape making temporary use of rockshelters and caves, as well as temporary camps, and are likely to have begun to target particular types of resource in different habitats and ecotones across a broad range of altitudes (see Denham and Barton 2006). Mobility has often been underestimated in the long-term history of the highlands of New Guinea (exceptions being Hope and Hope 1976b and Evans and Mountain 2005; see Hughes 1977), whereas mobility is a key and familiar concern in lowland New Guinea (Roscoe 2002) and parts of Australia (Veth 1993).

Throughout the history of occupation, people moved through diverse habitats and vast altitudinal ranges across New Guinea thereby broadening access to and use of resources. Although a very limited basis for understanding human occupation during the Pleistocene, hunting and *Pandanus* exploitation are applicable to understanding intermittent resource utilisation of some (but not all) mid-montane environments during the LGM, as well as higher areas throughout the Pleistocene and the Holocene. Such a model may well apply in the more seasonal climates of the eastern highlands (Fairbairn et al. 2006; Summerhayes et al. 2010); it does not apply to more aseasonal areas of the central highlands (cf. Feil 1987 for east-west differences). Significantly, the seasonality of these regions today is likely to have differed during the Pleistocene. Furthermore, groups utilising high-altitude resources would have accessed a much broader array of food plants on the floors and lower slopes of some inter-montane valleys, such as the Upper Wahgi Valley, as well as neighbouring lowland valleys that contained additional resources, including *marita (Pandanus conoideus)*, and other fruit-, nut- and starch-bearing palms, pandanus and trees (after Bourke 1996). Thus, it does not make much sense to say that people were living permanently in the highlands or lowlands of New Guinea during the Pleistocene, rather they were engaged in activities across these regions that enabled permanent occupation of the interior at this time.

Common practices in diverse contexts

Considerations of *Pandanus* phenology, vegetation history and resource availability demonstrate that the exploitation of major plant foods in the highlands could not have been seasonal or restricted to *Pandanus* during the Pleistocene. Rather there was, just as there is today, considerable spatial and temporal variability in food plant availability; climates and resources are more seasonal further eastward along the highland spine of Papua New Guinea. Thus, any consideration of occupation in the interior of Papua New Guinea during the Pleistocene requires consideration of the significance of place, especially of the mid-altitudes, that is, the floors, lower slopes and lower reaches of valleys. These were more abundant in food plant resources during the Pleistocene than is often acknowledged. In the eastern highlands, people adjusted their rhythms of plant exploitation to more seasonal resources, whereas in the central highlands more permanent occupation was possible by mobile groups exploiting a broad array of aseasonally producing food plants.

Significantly, there is little in the Pleistocene history of the interior of New Guinea that prefigures the emergence of agricultural practices in the highlands during the early Holocene. The long-term records of the highlands have recurrently been read to lead to agriculture and the long-term records of the lowlands have similarly been read to lead to arboriculture. There is no long-term trajectory deep in the Pleistocene that tracks, or inexorably leads to, either scenario. Rather, evidence for plant exploitation in the highlands during the Pleistocene is comparable with that for Australia and Island Melanesia (Denham, Fullagar et al. 2009) and Island Southeast Asia (Barton and Denham 2011). Across these regions, people selectively deployed a common suite of practices to exploit diverse plant resources, including trees and tubers (Denham, Fullagar et al. 2009). Only during the Holocene does regional diversity in plant exploitation practices emerge as differences of emphasis in plant exploitation practice gradually accumulate to become differences of kind.

7 Ambiguities of practice during the early Holocene

The timing of the transition to agriculture in the highlands of New Guinea is ambiguous. There are no definite archaeological traces of cultivation in the highlands during the early Holocene (from approximately 11,500 to 7000 years ago). Although archaeological evidence dated to c. 10,000 years ago on the wetland margin at Kuk Swamp in the Upper Wahgi Valley has been interpreted to indicate manipulation for cultivation, alternative interpretations are plausible. Furthermore, and despite assertions regarding a lowland derivation and human-mediated dispersal to the highlands for some key staples, such as bananas (*Musa* sp.), taro (*Colocasia esculenta*) and yams (*Dioscorea* spp.) (Yen 1995), the 'natural' ecological ranges of these plants possibly extended into the highlands at this time. Here, the ambiguous nature of the multidisciplinary evidence dating to c. 10,000 years ago at Kuk is argued to be consistent with transitions to nascent forms of cultivation, plausibly representing a novel form without modern analogue.

Multidisciplinary evidence at Kuk, c. 10,000 years ago

> Obviously there are many difficulties with the case for [10,000]-year-old agriculture at Kuk.
>
> (Golson 1991b: 489)

Archaeological evidence unique to Kuk was claimed to represent agriculture dating to approximately 10,000 years ago (Golson 1977c, 1989, 1991a; Golson and Hughes 1980; Hope and Golson 1995: 824). These finds were referred to as 'Phase 1' within the phaseology for Kuk and represented the earliest archaeological remains of cultivation (Golson 1977c). The interpretation of early Holocene agriculture at Kuk was always controversial and viewed with scepticism and uncertainty even by the original excavator (Golson 1977c: 613–614). Since then, the agricultural interpretation has been referred to as 'indirect and unusual' (Golson 1981: 56), 'possible' (Golson 1991b: 484), based on analogies with more recent evidence (Golson 1991b: 485; Hope and Golson 1995: 824), and made with 'somewhat less confidence' than for later archaeological evidence (Golson 2000: 232). Archaeological evidence of past cultivation during the early Holocene at Kuk originally comprised a palaeochannel and palaeosurface (Golson 1977c: 613–614, 1981: 55–56, 1985: 308–309, 1991b: 485, 1991c: 48, 2000: 232; Golson and Hughes 1980; Hope and Golson 1995: 824).

Earlier interpretations of the archaeological evidence have been revised following re-excavation at Kuk (Denham 2003a, 2004a, 2005a, 2007b; Denham et al. 2003; Denham, Golson et al. 2004, 2017a; Denham, Haberle et al. 2009; Denham, Sniderman et al. 2009; Denham and Grono 2017). Originally, a palaeosurface and palaeochannel were considered to be artificial. Based on re-excavation and a re-evaluation of the archaeological and geomorphological evidence, the palaeochannel is considered to be natural here (Denham 2003a, 2004a; although see debates in Golson 2007; Denham, Golson et al. 2004, 2017a).

Despite the revision, there are chronological and functional associations between the palaeosurface and palaeochannel. Even though radiocarbon dating and biostratigraphic signatures provide only a general guide of contemporaneity, they both date to around 10,000 years ago and are sealed by a distinctive stratigraphic unit – grey clay – at Kuk (Denham 2003a, 2004a; Denham, Sniderman et al. 2009). The palaeosurface is located on an area of higher ground, potentially a levee, adjacent to and plausibly drained by the palaeochannel; there is no evidence that the latter was dug by people. Even though a natural watercourse, the sediments and artefacts within the palaeochannel shed light on the activities of people in the vicinity at this time. Hence, interpretations of plant exploitation practices dating to c. 10,000 cal BP at Kuk are reliant on the archaeology of features on the palaeosurface, archaeobotanical remains of plants, palaeoecological reconstructions and stratigraphic evidence for soil formation and admixed feature fills (Denham et al. 2017a).

A gap, patch or plot on the wetland margin?

The 10,000-year-old palaeosurface was exposed to any large degree only in one area successively excavated over several years (Figures 7.1 and 7.2). Five feature complexes and numerous dispersed discrete depressions were exposed (Denham 2004a: 53). Three composite, curvilinear or sinuous runnels were uncovered, each composed of deeper basins connected by shallower depressions to form an irregular curvilinear suggestive of drainage (labelled A-C in Figure 7.1). Two complexes of upraised areas defined by surrounding inter-cut depressions were recorded (labelled D-E in Figure 7.1). There

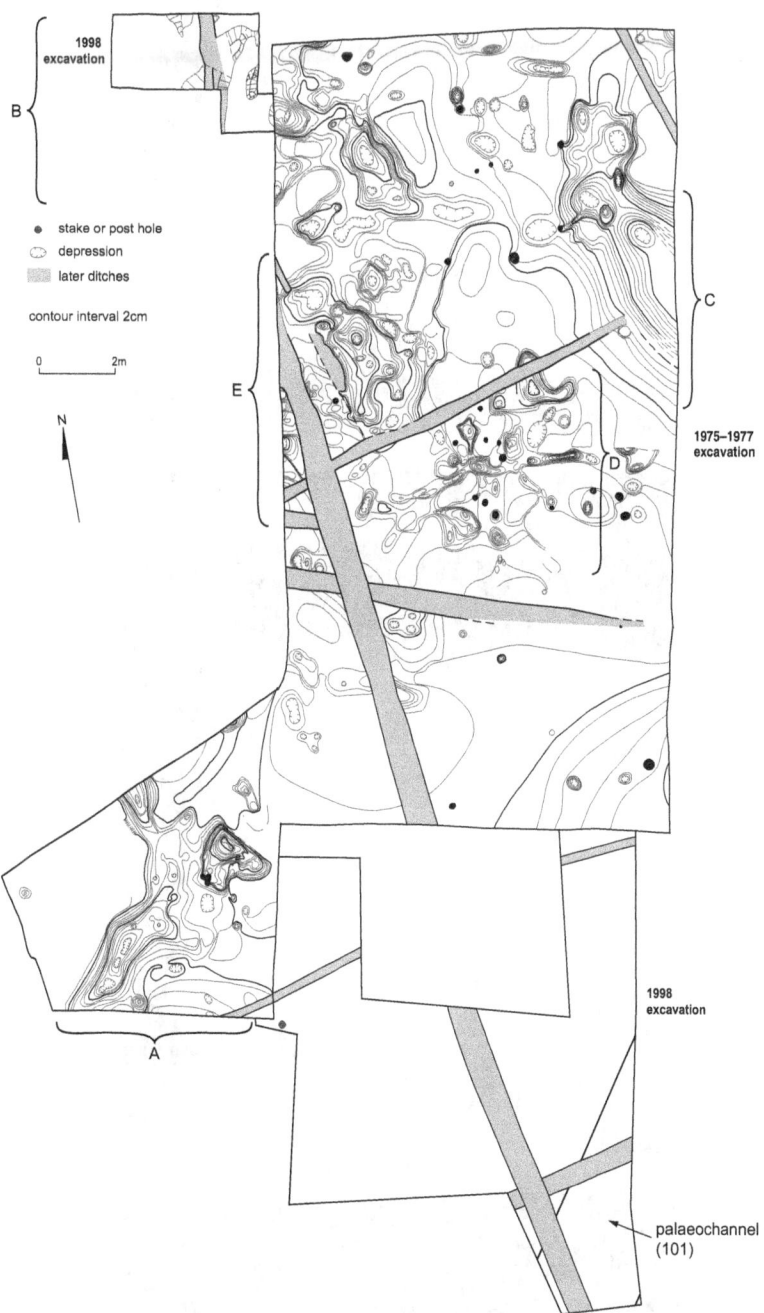

Figure 7.1 Plan of areas sequentially excavated in 1975–1977 and 1998 to expose the c. 10,000-year-old palaeosurface at Kuk Swamp

Source: Denham 2006: Figure 8.2

Figure 7.2 Photographs of the c. 10,000-year-old palaeosurface at Kuk swamp

Note: (upper) close-ups of discrete features; (lower) runnel and stakeholes; (next page) overview of 1975–1977 excavations

Source: Kuk archive

Figure 7.2 (Continued)

were recurrent feature types, as some of the more defined and deeper depressions were associated with postholes and stakeholes. A few stone artefacts were collected from the fills of these features and associated deposits.

The scale and nature of activities represented by the features are uncertain. Golson and Hughes (1980) reviewed a number of alternative, natural processes to account for formation, such as, deflation, vegetation and surface drainage. They rejected each in favour of a human origin, although some features were plausibly formed by localised micro-drainage, tree throw, root activity and tree boles (Denham 2004a: 53–54).

Upon re-examination, two elements of the archaeological record are unequivocally anthropogenic. First, at least twenty stakeholes and postholes were distributed across the palaeosurface, a number of these were associated with depressions. The repetitive association of feature types is suggestive of digging (depressions) and possibly staking and supporting (postholes and stakeholes) of plants. Second, the artefacts and manuports associated with the palaeosurface indicate a human presence in this locale approximately 10,000 years ago. These include flakes (76/S28, 76/S29A+B) associated with the palaeosurface, as well as a pestle or grinding/pounding stone (75/S178) collected from the basal fill of the adjacent palaeochannel (Figure 7.3). The paucity of lithic artefacts is expected within a utilised patch, or plot, and represents the use of wooden tools to harvest and process plants prior to consumption.

The edges of most features on the palaeosurface suggest that they were dug, because they are smooth and clearly defined against underlying strata (Denham et al. 2017a: 193). Edge definition is more suggestive of features dug by people than would be anticipated for more irregularly edged features derived from tree bowls, roots and tree throw depressions. The presence of a chert flake fragment (76/S28) at the base of a palaeosurface feature reinforces the impression that they are, in large part, artificial.

The mode of formation of the three runnels is less certain; they were plausibly dug adventitiously to enhance drainage within a cleared plot, or equally plausibly represent natural rills formed on exposed surfaces. These microtopographical drainage features are less likely to form under closed canopies of established vegetation communities than on recently cleared land. They either represent deliberate microtopographic drainage or indirectly represent localised forest clearance.

The features on the palaeosurface contain fills that are more heterogeneous than underlying and overlying strata (Denham and Grono 2017). This heterogeneity is interpreted to represent pedogenesis associated with ripening and limited pedoturbation of underlying clays, as well as mechanical admixture. Radiograph, thin section and geochemistry all suggest limited pedogenesis occurred of underlying sediments – black organic clay – when exposed at, or near, the surface (Figure 7.4; Denham, Sniderman et al. 2009; Denham and Grono 2017). The palaeosol resulting from soil formation is an A/C profile, characteristic of immature soils in alluvial settings. Limited admixture of feature fills plausibly reflects

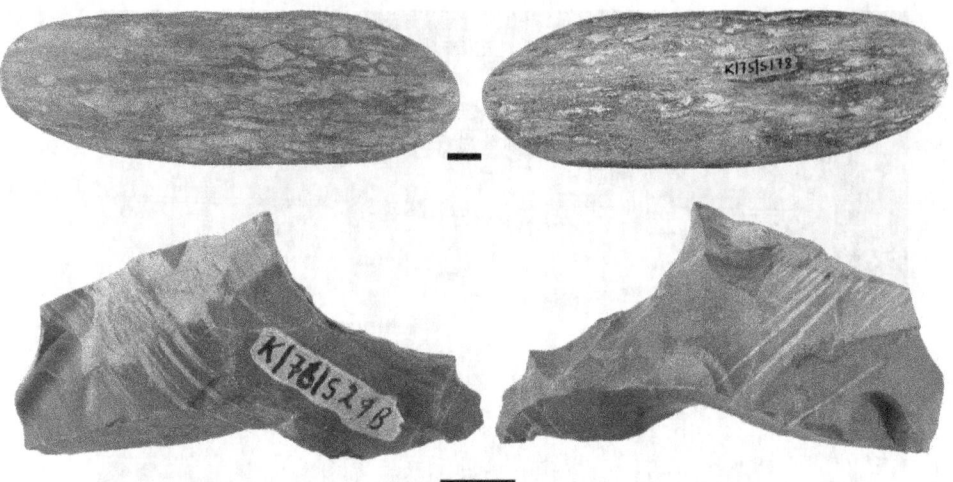

Figure 7.3 Photographs of artefacts dating to c. 10,000 years ago at Kuk Swamp

Note: (upper) possible pestle, or grinding/pounding implement (K75/S178) from palaeochannel, yielded yam (*Dioscorea* sp.) starch residues; (lower) retouched chert flake (K75/S29B) from edge of a palaeosurface feature, yielded possible taro *(Colocasia esculenta)* starch residues

Source: Fullagar et al. 2006

minimal tillage techniques, such as dibbling during planting or digging during harvesting (Denham et al. 2017a: 195). Continuous diatom and pollen records through one feature fill suggest locally drier conditions for a short time that enabled human use of the wetland margin, and increased levels of burning and limited transformation of the local vegetation (Figure 7.5; Denham, Sniderman et al. 2009).

The nature of human activities on the palaeosurface is corroborated by the fills of the adjacent palaeochannel. The admixture of soil aggregates, leaves, tree roots and woody debris in the basal fill – as well as associated palynological and microcharcoal records – are all consistent with localised clearance adjacent to the palaeochannel. Soil aggregates are not stable in watercourses and those in the fills of the palaeochannel are locally derived. The erosion of soil aggregates accords with the micromorphology of fills on the palaeosurface, which are suggestive of limited mechanical admixture (Denham and Grono 2017). Taken together with the large woody debris, an impression emerges of localised human disturbance of the vegetation, plausibly representing the clearance of a patch or plot within the rainforest with attendant soil erosion on the exposed surface. Soil erosion was plausibly exacerbated by human digging and other activities.

The interpretation of localised clearance on the wetland margin is characteristic of the watershed at this time. Pollen, phytolith and microcharcoal records all indicate sustained disturbance of lower montane rainforests on the valley floor using fire from c. 10,000 years ago (Figure 7.6; Denham et al. 2003; Denham, Haberle et al. 2004, Denham, Sniderman et al. 2009; Haberle et al. 2012). Peaks in charcoal are accompanied by increases in disturbance and grassland taxa, with slight decreases in primary forest taxa. Although disturbance to the rainforest in the Kuk vicinity was characteristic of the Terminal Pleistocene (Powell 1984; Denham, Haberle et al. 2004), it was more sustained during the early Holocene and led to the creation of a mosaic landscape, including patches of grassland and secondary forest, reflecting more persistent disturbance (Denham and Haberle 2008: 488).

Activities similar to those occurring on the wetland margin at Kuk were potentially occurring across the Upper Wahgi Valley landscape. Although the palaeosurface is unlikely to represent a naturally occurring gap within the rainforest, it could represent either a deliberately cleared patch designed to manage wild plant resources or a cleared and prepared plot designed for planting. Whichever of these latter two scenarios is correct, these were probably not specialised wetland adaptations; rather, these activities represent the extension of practices from the valley floor to the wetland margin during a short-lived drier period. The levee may have been dry enough only to intensively utilise for a few years, with potential visitation to harvest wild plants after active management had ceased. Although palaeoecological records suggest that these practices were widespread across the floor of the Upper Wahgi Valley, they are preserved only on the wetland margin where the physical evidence has been buried. Similar traces on the dryland slopes of the valley floor would have been destroyed by subsequent soil formation, erosion and cultivation.

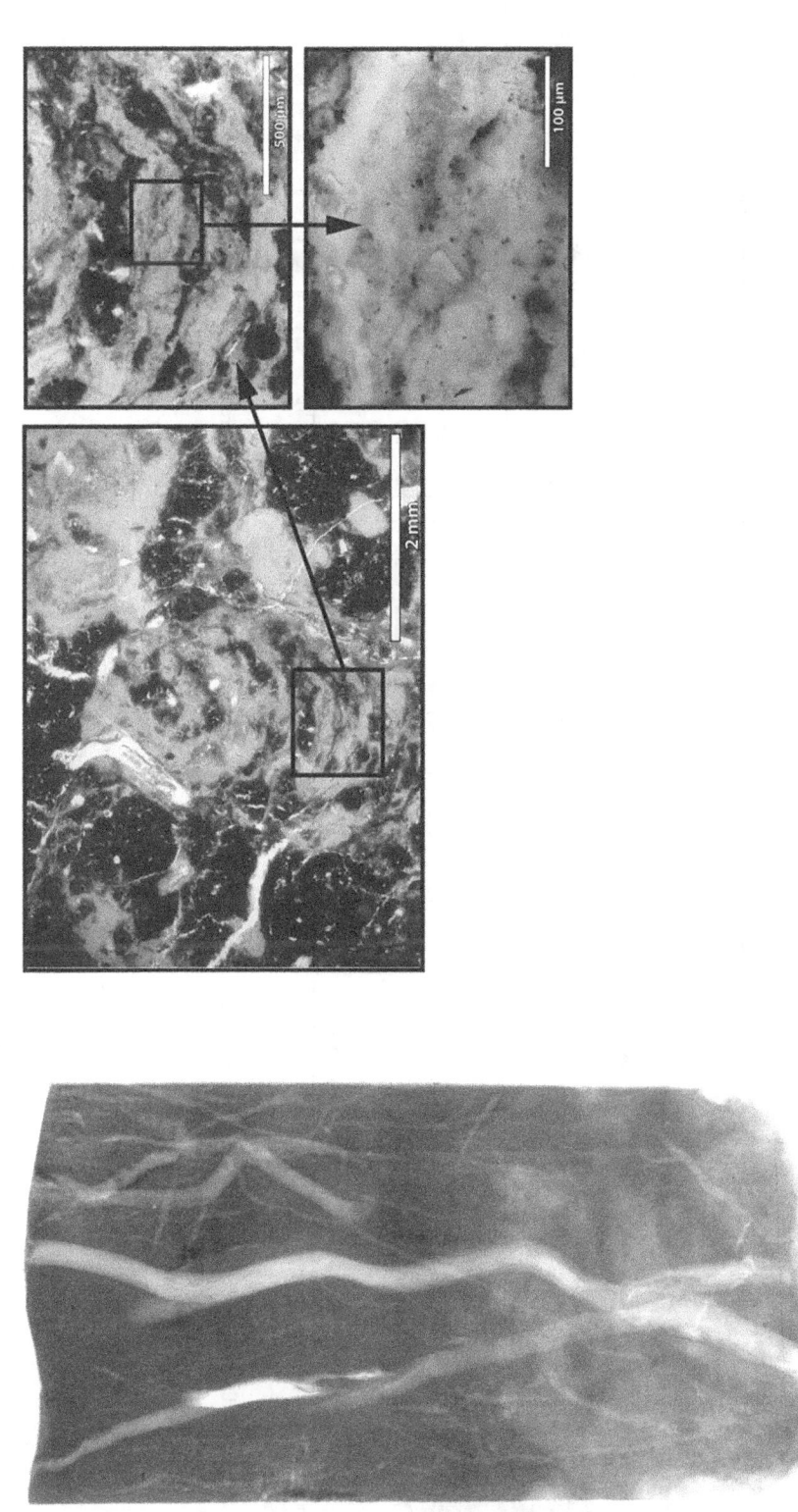

Figure 7.4 X-radiograph and photomicrographs of the fills of a feature on the c. 10,000-year-old palaeosurface

Note: The X-radiograph indicates a heterogenous feature fill suggestive of nascent soil formation in the lower third of the image, which is overlain by relatively homogenous grey clay (Source: Denham, Snider-man et al. 2009: Figure 4a). The photomicrograph panel depicts heterogeneous fabrics, including banding of organic–mineral matter and illuvial clay, which is shown in increased magnification in the panels to the right of the main image (see Denham and Grono 2017). Increased magnification shows amorphous organic–mineral matter arranged in lenticular aggregates separated by illuvial limpid and speckled clays with silt-sized particles and amorphous staining (upper image). Irregular clay illuviation with poorly sorted phytoliths, charcoal and amorphous staining (lower image)

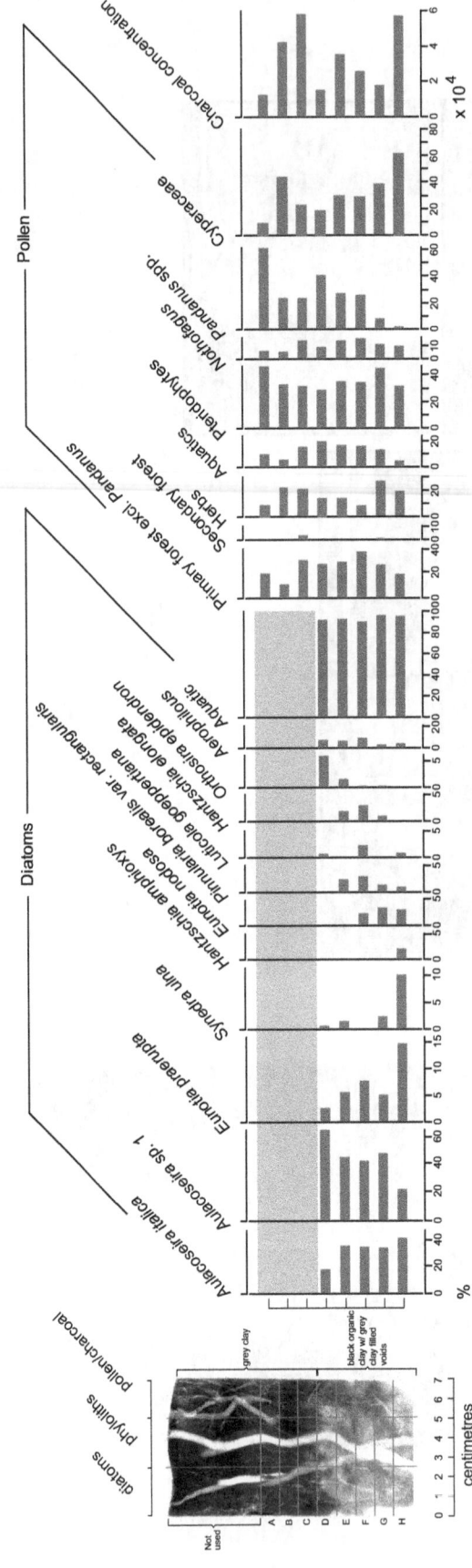

Figure 7.5 Summary diagram depicting X-radiograph slice and subsample locations, together with results of continuous and paired diatom, pollen and macrocharcoal analyses

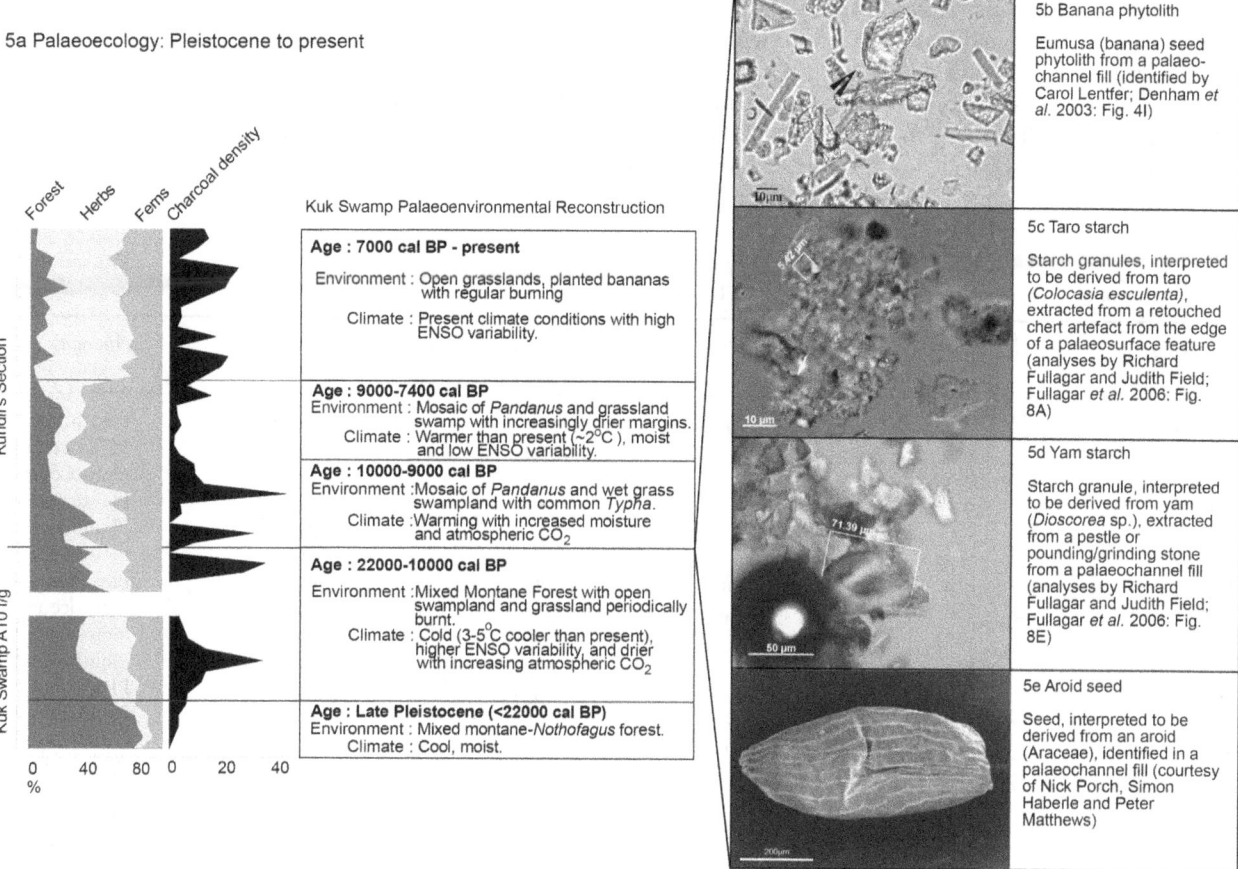

Figure 7.6 Palaeoecological (left) and archaeobotanical (right) summaries of c. 10,000-year-old evidence at Kuk Swamp

Source: Denham 2008b: 46.5

People's use of plants

A suite of archaeobotanical and palaeoecological techniques has yielded evidence for the presence and exploitation of a range of edible food plants at Kuk during the Pleistocene and early Holocene (Table 7.1; Denham et al. 2003; Denham, Haberle et al. 2004; Denham 2005c; Denham, Sniderman et al. 2009). Microfossil (phytolith, pollen, raphide and starch grain) analyses complement macrofossil (fruits, nuts, seeds and wood) techniques to yield a broad range of food plants that were present or used during these periods. The Pleistocene record has been constructed using materials and sediment samples from the stratigraphy at Kuk. The early Holocene record relies upon samples from the fills of the palaeochannel and features on the palaeosurface, as well as residues extracted from stone artefacts in associated contexts.

During the Pleistocene, palaeoecological records indicate a range of edible plants was present in the local vicinity (Lentfer and Denham 2017). None of these plants are major staples, although *Castanopsis* and *Pandanus* provide significant quantities of nuts in highland valleys today. Several other plants are presently cultivated in gardens, including an edible cane grass (cf. *Setaria palmifolia*) and vegetables, (e.g., *Abelmoschus* sp., *Oenanthe javanica*). These plants form part of the natural flora anticipated for wetlands on the valley floor and adjacent slopes (Powell 1970a). There is no evidence to indicate human exploitation of these plants, though these records provide a partial inventory of the edible plants that were available and potentially utilised by people on the floor of the Upper Wahgi Valley during the Pleistocene and early Holocene.

Around 10,000 years ago, the earliest records for several significant food plants occur at Kuk. These include the pre-Ipomoean staples of traditional highland agriculture: bananas (*Musa* spp.), taro *(Colocasia esculenta)* and yam *(Dioscorea* sp.). The lack of comparable findings for the Pleistocene reflects, in part, the nature of the archaeological record. Few palaeochannels, features and artefacts have been documented for earlier periods and none of these have been sampled and subject to the required level of appropriate archaeobotanical investigation. Climates in the Upper Wahgi Valley may have been too cold for bananas and yams during the Last Glacial Maximum (LGM; Haberle 1993; Denham and Barton 2006); but these plants would have probably grown on the valley floor throughout the Holocene, even if near their upper altitudinal limits (Denham, Haberle et al. 2004).

Indigenous vegetables that are cultivated today, include *Amaranthus* sp., *Commelina* sp., *Oenanthe javanica* and *Solanum* spp. (Powell 1970a: 200, 1981: 296, 1982a: 31–32; Powell et al. 1975: 15–32). Other species present are harvested from wild sources or as garden weeds, for example, *Cerastium* sp., *Drymaria cordata*, *Hydrocotyle javanica*, *Oxalis corniculata*, *Rubus*

Table 7.1 Archaeobotanical evidence for food plants from Pleistocene and Holocene contexts at Kuk

Species/Genus[1]	Exploited Form[2]	Edible Part(s)[3]	Archaeobotanical Evidence[4]	Kuk Phase[5]	Antiquity
Abelmoschus sp.[6]	c	l, sh	s	pre-P1	Pleistocene
Acalypha sp.	w, t	l	s, p	pre-P1	Pleistocene
Castanopsis sp.	w, t	n	w, p?	pre-P1	Pleistocene
Cerastium sp.	w	p	s	pre-P1	Pleistocene
Coleus sp.	w	l	s	pre-P1	Pleistocene
Ficus cf. *copiosa*[6]	c, w	f, l	s	pre-P1	Pleistocene
Ficus spp.	c, w	l, f	w	pre-P1	Pleistocene
Garcinia sp.	w	f, l, b	w	pre-P1	Pleistocene
Hydrocotyle sp.	w	l?	s	pre-P1	Pleistocene
Lycopodium spp.	w	sh	p	pre-P1	Pleistocene
Maesa sp.	w	f	s, w	pre-P1	Pleistocene
Musaceae	c, w	f, c	ph, st	pre-P1	Pleistocene
Oenanthe javanica	c, w	l, sh	p	pre-P1	Pleistocene
Pandanus cf. *antaresensis*	w	d	p	pre-P1	Pleistocene
Pandanus cf. *brosimos*	c, w	d	p	pre-P1	Pleistocene
Pandanus spp.	c, w	d	s, p	pre-P1	Pleistocene
Parsonsia sp.	w	n	p	pre-P1	Pleistocene
Phragmites karka	w	l, r, sh	ph	pre-P1	Pleistocene
Pouzolzia hirta	w	l, st	s	pre-P1	Pleistocene
Rubus moluccanus	w	f	s	pre-P1	Pleistocene
Rubus rosifolius	w	f	s	pre-P1	Pleistocene
cf. *Setaria palmifolia*	c, w	s	ph	pre-P1	Pleistocene
Solanum sp.	c, w	f, l, sh, t	s	pre-P1	Pleistocene
Syzygium sp.	w	f	w	pre-P1	Pleistocene
Wahlenbergia sp.	w	p	s, p	pre-P1	Pleistocene
cf. Zingiberaceae	c, w	r, l, sh	ph	pre-P1	Pleistocene
Colocasia esculenta	c	c, l	s, st	P1	Early Holocene
Dioscorea sp.	c	t	st	P1	Early Holocene
Elaeocarpaceae	w	n	p	P1	Early Holocene
Ipomoea sp.	w	sh	p	P1	Early Holocene
Musa section bananas[7]	c, w	f, c	ph	P1	Early Holocene
Ingentimusa section[7] bananas	w	f, c?	ph	P1	Early Holocene
Phragmites sp.	w	st	ph	P1	Early Holocene
Typha sp.	w	st	p	P1	Early Holocene
Commelina sp.[6]	c, w	l, sh	s	P2	Mid-Holocene
Drymaria cordata[6]	w	l?	s	P2	Mid-Holocene
Floscopa sp.[6]	c, w	l, sh	s	P2	Mid-Holocene
Viola arcuata	w	l	s	P2	Mid-Holocene
Amaranthus sp.	c	l, p	s	P2/P3	Mid-Holocene
Bidens pilosa	w	s	p	P2/P3	Mid-Holocene
Polygonum chinense	w	l	s	P4/P6	Late Holocene
Finschia sp.	w	s	s	P5	Late Holocene
Saccharum officinarum	c	st	m	P4	Late Holocene
Bambusa sp.	c, w	sh	w	P6	Late Holocene
Dioscorea esculenta[8]	c	t	m	P6	Late Holocene

(Continued)

Table 7.1 (Continued)

Species/Genus[1]	Exploited Form[2]	Edible Part(s)[3]	Archaeobotanical Evidence[4]	Kuk Phase[5]	Antiquity
Elaeocarpus sp.	w	n	w	P6	Late Holocene
Ipomoea batatas	c	t	m	P6	Late Holocene
Polyscias sp.	w	l	w	P6	Late Holocene
Oxalis corniculata	w	l, sh	s	post-P6	Late Holocene

Notes: Updated version of Denham 2005c: Table 2. Only potential food plants are included, although other uses may have been as important in the past (cf. Powell 1976, 1982b).

1 List of edible species at Kuk based on ethnographically documented use of plants in New Guinea (M. Bourke 1989 and pers. comm. 2002; French 1986; Haberle 1995; G. Hope pers. comm. 2002; Powell 1976: 108–112; Powell and Harrison 1982: 57–86; Powell et al. 1975: 15–39). Edible species have been reported from other early and mid-Holocene archaeological sites in the highlands.
2 Exploited form: c = cultivated, w = wild (used as supplementary food), t = transplanted
3 Edible part(s): b = bark, c = corm, d = drupe, f = fruit, g = gourd, l = leaf, n = nut, p = plant, r = rhizome, s = seed, sh = shoot, st = stem, t = tuber
4 Evidence: m = macrobotanical (by Tara Lewis and Jon Hather), p = pollen (by Simon Haberle), ph = phytolith (by Carol Lentfer), s = seed (by Jocelyn Powell and Laurie Lucking), st = starch (by Richard Fullagar, Judith Field and Michael Therin), w = wood (by Jocelyn Powell and Laurie Lucking). Only those cases relevant for the earliest recorded occurrence are listed.
5 Earliest records are given for Pleistocene contexts (pre-P1); Kuk Phase 1 at c. 10,000 cal BP (P1); contexts post-dating Kuk Phase 1 at 10,000 cal BP and pre-dating Kuk Phase 2 at 7000–6400 cal BP (pre-P2); respective Kuk Phase 2 (7000–6400 cal BP), Phase 3 (4350–4000 to 2700–2400 cal BP), Phase 4 (2000–1230/970 cal BP), Phase 5 (700–c.290 cal BP) and Phase 6 (250–50 cal BP) contexts; and post-dating Phase 6 (post-P6) (latest phase dates are derived from Golson 2017: Table 1.2).
6 Identification to genus or species level should be considered provisional because it is based on a single seed.
7 A review of the sections within the genus *Musa* (Wong et al. 2002), which is not universally accepted (De Langhe et al. 2009), has reclassified the five former sections into three: section Musa contains the former Eumusa and Rhodochlamys; section Callimusa contains the former Callimusa and Australimusa; and section Ingentimusa remains the same. Previous identifications of Callimusa (formerly Australimusa) phytoliths from Phase 2 and 4 contexts at Kuk (Wilson 1985: Table 3) are excluded because they were based on a single phytolith each and no such phytolith morphotypes were documented from Phase 1–3 contexts at Kuk targeted during recent research (Denham et al. 2003: 192).
8 Originally identified by Jon Hather (pers. comm. to Jack Golson), although now considered doubtful and more likely to be sweet potato *(Ipomoea batatas)* (Lewis et al. 2016).

rosifolius and *Viola* sp. Although it is possible that these resources were all natural colonisers of the wetland margin at Kuk, as at other wetlands in the Upper Wahgi Valley (Powell 1970b), their distributions may have been enhanced by clearance using fire and by localised ground disturbance.

Phytoliths of different species and sections of Musaceae are present at Kuk around 10,000 years ago (Wilson 1985; Denham et al. 2003; Denham, Golson et al. 2004; Lentfer 2009; Haberle et al. 2012). Significantly, non-diagnostic Musaceae phytoliths are present in only one sample from the base of one feature (Lentfer and Denham 2017), which may indicate that bananas were growing on-site. By contrast, the majority of Musaceae phytoliths, including diagnostic types, derive from the fills of the palaeochannel and could either derive from the watershed or from adjacent land.

Most Musaceae phytoliths cannot be differentiated, whereas some have been identified to various taxonomic levels, including the large highland banana, *Musa ingens*, and to Musa section (formerly Eumusa section). Of the Musa section bananas indigenous to New Guinea (Argent 1976; Arnaud and Horry 1997), *Musa acuminata* ssp. *banksii* is the most likely to have grown at Kuk at this time. The other species are largely confined to lower altitudes. Significantly, the *banksii* subspecies is implicated in an early stage of banana domestication, most specifically the development of parthenocarpy (De Langhe and de Maret 1999; Perrier et al. 2011), and its predomesticated form may have spread by seed. The presence of diagnostic seed phytoliths of Musa type – attributable to ssp. *banksii* – suggest that the 10,000-year-old microfossils at Kuk represent seeded forms of *Musa acuminata* ssp. *banksii* growing in the vicinity.

Starch grain and raphide residues extracted from the cutting edge of a retouched chert fragment (K/76/S29B; probably the distal end of a flake) are indicative of taro *(Colocasia esculenta*; Fullagar et al. 2006; Fullagar and Golson 2017). Processing (including usually cooking) of taro is needed to enhance taste and to liberate nutrients for digestion. Although suggestive of local exploitation of taro, the residues alone could as readily represent the exploitation of these plants in the lowlands, or the movement of taro corms into the highlands for consumption (following Denham 2007a; Summerhayes et al. 2010).

Strong corroboration that taro was growing at Kuk, or immediately inland, c. 10,000 years ago is provided by aroid seeds, characteristic of cf. *Colocasia esculenta*, in the fills of the palaeochannel. A comparison of fossil seed morphologies with modern seeds from a range of aroids is indicative of taro (Denham 2008a; Fairbairn and Matthews 2017). The identification is not definitive until a wider range of reference seed samples of aroids growing in the highlands of New

Guinea is available for study. On balance, the combination of seeds, starch grains and raphide is suggestive of the presence and exploitation of taro at Kuk, c. 10,000 years ago.

Starch grains of a yam (*Dioscorea* sp.) were extracted from a stone pestle or pounder (K/75/S178) found in the basal fill of the palaeochannel. Identification to the species level was not confirmed, either *D. alata* or *D. pentaphylla* is suggested (Fullagar et al. 2006). Use wear together with residues indicate the artefact was used to mash soft, non-siliceous raw plants, including a yam. The artefact must have been thrown or dropped into the palaeochannel locally; namely, it is too heavy and large to have been transported as bed load. It cannot be discounted that this tool was used elsewhere, or to process yams carried to Kuk from somewhere else. Such a scenario has been documented at the Ivane Valley sites, where stone artefacts have been excavated that indicate the processing of yam even though it is above the altitudinal range for yam growth (Summerhayes et al. 2010). In contrast, Kuk is within the altitudinal range of most types of yam in New Guinea and it seems more plausible that some yams were growing locally.

In sum, the most significant archaeobotanical evidence from Kuk is for the use of taro *(Colocasia esculenta)* and a yam (*Dioscorea* sp.) and the presence of *Musa* spp. from c. 10,000 years ago. These plants provide energy-rich sources of food and could have functioned as staples in the Wahgi Valley because they provide aseasonal, high-caloric resources in this part of New Guinea today (e.g., Powell 1976; Gagné 1982: 236; Dwyer and Minnegal 1991; Yen 1991, 1995). These plants need not have been cultivated during the early Holocene: wild tubers are eaten in the highlands today, particularly during famine conditions (Watson 1964; Waddell 1973; Powell et al. 1975: 15-32). Feral and wild forms of *Colocasia* taro, *Dioscorea* yams, *Pueraria* sp. and yam-like tubers (possibly *Dioscorea* of unknown species) grow in highland valleys today and are used for food (Powell et al. 1975: 21, 24-25; Yen 1991: 77, 81; Mike Bourke pers. comm. 2002).

However, bananas, taro and yams are among the most important staple crops of traditional agriculture in New Guinea, as well as the Pacific, Southeast Asia and beyond. Is the co-occurrence of these three staples at Kuk c. 10,000 years ago coincidental? Could the presence of these plants represent the spread of seeded forms as climates ameliorated at the beginning of the Holocene, or transplantation and incipient cultivation?

Natural ranges and loci of plant domestication

Several food plants were domesticated in the highlands of New Guinea, some of which are still grown only at higher altitudes (Yen 1973; Bayliss-Smith 1988). These include rungia *(Rungia klossii)*, karuka *(Pandanus julianettii* complex) and highland *pitpit (Setaria palmifolia)*. The archaeobotanical remains of three plants at Kuk dating to c. 10,000 years ago are potentially of great significance in determining the antiquity of agriculture on New Guinea. *Musa* bananas, taro and yams were originally conceived as lowland domesticates that could only have been brought to the highlands under cultivation (Yen 1995). The ecology and phytogeography of these plants is poorly understood and different distributions can be inferred for New Guinea at different times in the past.

The altitudinal ranges for the current cultivation of bananas, taro and yams suggest the potential for their cultivation in the highlands at the beginning of the Holocene (Bourke 2009; see Table 5.2). Yen (1995) has long argued that these plants were originally domesticated in the lowlands, that their current altitudinal ranges have been extended upwards through cultivation and that they were unlikely to have grown naturally in the highlands in the past. In contrast to bananas, little is known of the 'natural' ranges of wild populations of taro and various yams in New Guinea. For instance, 'wild' (potentially feral) yams of unknown taxon grow in the Upper Wahgi Valley (Powell et al. 1975). Existing ecological information is not specific enough to gauge whether these three plants grew wild in the highlands before they were cultivated there, nor when climates had ameliorated sufficiently to support their growth.

If bananas, taro and yams are of ultimate lowland derivation, then the case for 10,000-year-old agriculture in New Guinea seems to be supported. The residue evidence for the exploitation of a yam and an aroid would not be definitive on its own, as the tools could have been used to exploit plants in the lowlands by highly mobile groups who also visited the highlands (Denham and Barton 2006; Denham 2007a; cf. Summerhayes et al. 2009). The phytoliths of *Musa* banana seeds are more definitive, as are the aroid seeds, because they demonstrate plants were growing locally c. 10,000 years ago.

Until higher resolution ecological information is forthcoming, it would be unwise to infer that archaeobotanical remains of banana, taro and an unidentified yam represent the human-mediated spread of lowland domesticates to the highlands approximately 10,000 years ago. On the basis of the limited ecological information available, bananas and yams were at the limits of their natural range during the early Holocene prior to the development of more robust cultivars. In order to make these plants more productive and potentially even viable at these altitudes, people needed to intervene in the ecology of the lower montane forest through burning and clearance, as well as in the reproductive cycle of these plants.

Transitional steps to cultivation

Various models have proposed how plant exploitation practices changed in the Upper Wahgi Valley around the Holocene-Pleistocene transition (Golson 1991a, 1991b; Hope and Golson 1995; Denham and Barton 2006; Denham and Haberle 2008). In general terms, these models focus upon how people increasingly began to manage the landscape and the edible plant resources contained within it (following in a broad sense Yen 1989 and Harris 1989). Within the Upper Wahgi landscape, people most likely focussed on areas where edible resources were potentially different and more abundant than under the forest canopy. These included disturbed gaps within the forest, such as landslides and tree falls; the disturbed patches people created; as well as ecotones, riparian corridors and wetlands. Through time, people increasingly maintained these environments and created new ones within the forest using fire and stone tools (Groube 1989). The increased human intervention in the landscape was designed to increase resource density and productivity of favoured species, primarily plants, as well as to enhance hunting and the gathering of fauna attracted to the edible plant resources within those gaps and patches.

Two transitions are significant for marking the transition from resource management and intensification to cultivation. Each transition reflects transformations to the ways people interacted with their landscape and to the plants within it (amended from Denham 2015: 458–459). Initially, people began to focus on certain types of plants for their diet, including in terms of management practice, which led to deliberate planting. Subsequently, people began to create new environments within the forest; they cleared and prepared areas, or plots, for planting of favoured species.

Transition 1: At some point during the early Holocene, or potentially earlier, people began to increasingly focus upon the management of individual species: starch-rich plants such as bananas (*Musa* spp.), taro (*Colocasia esculenta*), yams (*Dioscorea* spp.) and edible grasses including sugarcanes (*Saccharum* spp.) and *pitpit* (*Setaria* spp.); oil and protein-rich palms and trees (*Pandanus* spp. and *Castanopsis* sp.) and leafy vegetables. Effectively, their dietary focus shifted from broad spectrum hunting and gathering, or foraging, to more selective exploitation and management of specific groups of plants (Denham and Barton 2006). People's relationship to the landscape changed as their relationship to the plants within it changed. They increasingly moved from using burning and clearing as ways to increase resource densities of favoured plants to the deliberate removal of reproductively viable parts from plants growing in the forest or elsewhere and replanting in managed gaps and patches. Given the dominance of vegetative propagation to traditional arboriculture and horticulture in New Guinea, planting was initially based on a pre-existing awareness of the vegetative capacity of plants (Denham 2005c, 2011; Barton and Denham 2011, 2017).

Transition 2: Subsequently, people created and prepared plots within the forest and in other environments for planting. Initially, planting occurred within managed gaps and artificially created patches in order to supplement the density and range of edible resources. With experience, people started to clear areas within the rainforest to create plots for planting. From this time, any increased resource density through vegetation clearance or disturbance (characteristic of patches) was incidental and secondary to the plant resources obtained from deliberate planting (characteristic of plots). Plots were probably small with large trees left standing after being ring-barked or pollarded due to the difficulties involved in cutting them down and to aid soil retention. Planting occurred by dibbling, namely, by making a hole in the ground using a stick, inserting a plant part and back-filling by hand.

Archaeological, chronological, palaeoecological and stratigraphic signatures of these two transitions are likely to be ambiguous and difficult to differentiate. How is it possible to differentiate resource intensification from localised, small-scale patch creation with limited planting, as well as from plots with planting using minimal disturbance, i.e., dibbling? These ambiguities are primarily methodological because fragmentary and partial evidence can be interpreted in different ways.

A novel form of plant exploitation

The multidisciplinary evidence at Kuk could represent a landscape management strategy, or it could also represent a more plant-focussed strategy. In both scenarios, people created the patch or plot on the wetland margin for plant exploitation. The multidisciplinary evidence strongly indicates localised clearance, burning and soil erosion within the rainforest on the wetland edge, mirroring practices elsewhere in the landscape (Denham, Haberle et al. 2004; Denham and Haberle 2008; Haberle et al. 2012; Denham and Grono 2017).

From a land management perspective, the short-lived manipulation of the wetland margin at Kuk represents patch creation; namely, people sought to mimic the gaps within the montane rainforest and intensify resources within the landscape. People were deliberately disturbing the rainforest in the Upper Wahgi Valley at this time, and they plausibly created this patch to intensify the density and productivity of plant resources within it. Bananas and yams favour disturbed, gap and ecotonal habitats within rainforest, rather than closed canopy; taro would favour riparian environments along the bed and banks of the palaeochannel or seeps along the wetland edge. Once created, people actively managed and exploited the essentially 'wild' resources within the patch on the wetland margin in similar ways to those on adjacent slopes.

Significantly, although there is strong evidence for taro growing on-site, it is less clear for bananas and yams. Banana phytoliths predominantly occur within the palaeochannel fill and are sparse in fills of palaeosurface features, suggesting bananas were growing within the catchment. The location of yams is even less certain, since they could be growing on-site, on the floor of the valley, or even at some distance from Kuk. Consequently, people could have exploited taro and possibly bananas within a patch on the wetland margin; while a wider range of plants, including bananas and yams, were exploited across the landscape.

Archaeological and stratigraphic interpretations suggest independently that the plot was dry only for a short period and the palaeochannel was short lived. It seems unlikely that all three plants, which do not ordinarily occur in the same types of environment, adapted in unison to this short-lived window provided by the changing hydrology of the wetland margin. Although taro could persist in the changing wetland environment and readily disperse, such an interpretation would require bananas and possibly a yam to naturally disperse from adjacent slopes to, and to become established upon, this recently appeared drier area. Such an interpretation is possible, but on balance it would seem simpler to invoke people, who brought these three food plants together and planted them in a newly emergent environment on the wetland margin.

Thus, from a plant management perspective, a plot was cleared and prepared for planting. Reproductively viable parts were collected and brought to the site from plants growing locally or growing far from the Upper Wahgi Valley. Plant sources could include essentially 'wild' or 'cultivated' plants – as occurs today throughout New Guinea. Planting of suckers (banana), corms (taro) and tubers (yam) was affected using minimal tillage techniques, namely dibbling with a stick and hand covering. Bananas and yams could be planted on better-drained ground, whereas taro could be planted in wetter locations.

Following either landscape or plant management strategies, the tending and harvesting, as well as processing prior to consumption, of plants within the patch or the plot would have been similar. Bananas, edible cane grasses and yams would have been supported using branches, posts and stakes. Drainage would have been assisted by microtopographical manipulation. Digging occurred to harvest tubers and corms, and other plant foods could have been picked, collected or cut. Even though an argument can be made that planting more parsimoniously and effectively accounts for the sudden co-occurrence of these three significant food plants on the wetland margin at Kuk, c. 10,000 years ago, the evidence is not conclusive.

Multiple lines of evidence indicate people were clearing forest and exploiting starch-rich plants in a restricted area on the wetland margin. These include a combination of archaeobotanical, archaeological and palaeoecological evidence indicative of on-site and practices, as well as those occurring in the vicinity. Practices comprise (Denham 2005c: 301; Denham 2015: 460–461):

- disturbance of montane forests using fire and stone tools to create and maintain a mosaic of habitats in the landscape;
- localised clearance of vegetation and soil disturbance on the wetland margin, as reflected in the fills of an adjacent palaeochannel;
- digging, for planting and for harvesting underground storage organs;
- possible staking of plants, such as bananas and sugarcane;
- locally growing taro and bananas, including Musa section bananas from which major domesticated varieties are ultimately derived; and, procurement and processing of starch-rich plants, including taro *(Colocasia esculenta)* and a yam *(Dioscorea* sp.).

The multidisciplinary evidence is insufficient to state that these practices together represent agriculture, that is, cultivation within a plot, as they may as equally represent incidental planting or management within a maintained patch. Three lines of evidence are lacking that would assist clarification, yet none of them are essential or definitive within the context of known cultivation practices in New Guinea (Denham 2005c: 301; 2015: 461).

First, the extent of soil preparation is not known. Immature A/C soil profiles are consistent with limited soil formation during short-lived, locally drier conditions (Denham, Sniderman et al. 2009; Denham and Grono 2017). Micromorphological evidence of limited mechanical admixture is consistent with minimal tillage techniques, as occur in many forms of swidden cultivation in New Guinea today; additional evidence for soil preparation need not be present.

Second, features and artefacts are consistent with the range of archaeological remains anticipated for planting in a patch or plot. Features indicate staking, digging and microtopographical manipulation for surface drainage, whereas artefacts demonstrate processing of starch-rich plants. Wooden artefacts are likely to have been employed for digging, planting and harvesting (Golson 1977b), yet they are unlikely to have been common or preserve.

Third, evidence for planting is equivocal. Although Musa section bananas, taro and some yams are generally considered to be of lowland derivation, and archaeobotanical evidence for all three is present at Kuk c. 10,000 years ago, the precise altitudinal range of these plants in New Guinea at that time is uncertain (Denham, Golson et al. 2004; Hope and Golson 1995). Bananas and yams were probably near the edge of their natural range and, given the vagaries of palaeoclimatic reconstruction within the highlands for the early Holocene (Brookfield 1989; Haberle 2007), a definitive conclusion regarding cultural or natural modes of dispersal to the floor of the Upper Wahgi Valley is not possible.

Conservatively, the multidisciplinary evidence for practices on the wetland margin at Kuk c. 10,000 years ago is suggestive of more intensive management of the landscape and plants. There is evidence for the creation of patches or plots and an emerging focus on starch-rich plants, which were potentially growing locally. Ultimately, though, these practices do not reflect a way of life dependent upon the cultivation of food.

8 The emergence of shifting cultivation

Denevan (2001) has argued that shifting cultivation was not a common subsistence strategy in the lowland neotropics of Amazonia before the advent of metal tools, which greatly enhanced people's ability to clear rainforest. He has further proposed that early agriculture in Amazonia was based on the intensive and relatively continuous cultivation of plots near, or adjacent to, wetlands and rivers, rather than upon extensive forms of shifting cultivation in closed canopy rainforest.

For the highlands of New Guinea the situation seems to be reversed. Early forms of plant exploitation were shifting and extensive, with a more restricted focus upon certain locations, including wetlands, through time (Golson 1977c; Denham and Haberle 2008). Early ethnographies and historical accounts document the widespread use of stone tools to clear forests in the highlands of New Guinea before the prevalence of metal implements from the 1930s onwards (Leahy 1936). As discussed for the Pleistocene and beginning of the Holocene, the disturbance of montane rainforest using stone tools and fire has a long history in the highlands. The character of these processes seems to have changed during the early-to-mid Holocene.

There is a strong palaeoenvironmental – including palaeoecological and geoarchaeological – case for the emergence of shifting cultivation in the highlands during the early Holocene, between c. 7000 and 10,000 years ago. This is largely inferred from the persistent and cumulative nature and extent of disturbance, burning and clearance of montane rainforests, as well as associated soil erosion. Direct archaeological evidence for shifting cultivation is, as yet, lacking in the highlands; the multidisciplinary evidence dating to c. 10,000 cal BP at Kuk Swamp is only indicative (reviewed in Chapter 7). A lack of evidence is not surprising given that archaeological traces of shifting cultivation in tropical rainforests are unlikely to preserve or be readily identifiable. Characteristically, plots are cultivated for 1 to 2 years, abandoned to forest regrowth, albeit with some continued harvesting of perennials, and then perhaps re-used after a period of 5 to 15 years. In tropical environments, any features and palaeosols associated with former cultivation would be relatively quickly erased, mixed and re-incorporated into the soil matrix as a result of bioturbation of tree roots and soil fauna, as well as subsequent cultivation, erosion and deep weathering profiles.

In this chapter, multiple inferential lines of evidence – palaeoecology, geomorphology, pedology and archaeology – for the emergence of shifting cultivation in the highlands of New Guinea around 7000–10,000 years ago are presented to address a series of questions. First, what lines of evidence enable shifting cultivation to be identified in past landscapes? Second, is there a neolithic signature accompanying the emergence of shifting cultivation in the highlands? Third, can shifting cultivation be inferred from palaeoenvironmental (palaeoecology and geomorphology) evidence in the highlands? Fourth, how do we conceive of the emergence of shifting cultivation in terms of the practices involved?

What are we looking for?

The possibility of identifying archaeological remains and palaeosols associated with shifting cultivation on the floors of the highland valleys during the early Holocene is unlikely, except adjacent to wetlands. Most valley slopes, as well as wetland margins, have been subject to repeated cultivation, especially over the last few thousand years, and are unlikely to preserve any traces of shifting cultivation during the early Holocene. As witnessed in the discussion of the 10,000-year-old remains at Kuk, the creation and use of a patch is difficult to distinguish from clearance and planting in a plot, even on a wetland margin where the remains are relatively well preserved.

There are several *a priori* reasons why early shifting cultivation is hard to identify. Ethnographically observed shifting cultivation practices in New Guinea vary greatly in terms of the extent of forest clearance, whether slash-and-burn or slash-and-mulch, degree of soil preparation, degree and structure of polyculture, degree of tillage for planting and harvesting, duration of plot use, degree of maintenance following abandonment, reliance on wooden tools and so on (Clarke

1971; Bourke 2001; Bourke and Harwood 2009). The earliest activities in the highlands were probably at the less intensive range of this scale and mimicked in several significant ways rainforest dynamics. Practices included:

- localised clearance mirroring that of a gap or patch in the rainforest;
- burning, especially in more seasonal locales, following cutting of undergrowth and potential ring-barking of trees;
- no or minimal preparation of the soil, because burning or surface mulching would increase soil fertility in the short term;
- minimal tillage involving planting and subterranean harvesting using a digging stick, which would leave a limited archaeological trace likely to be indistinguishable from the repeated harvesting of underground storage organs (USOs);
- potential use of a plot for a few years prior to abandonment and forest regeneration, because nutrient cycling is slower in the slightly cooler climates of the highlands;
- minimal maintenance following abandonment except for occasional visitation and tending to protect and encourage favoured species and to harvest perennials; and,
- use of stone and wooden tools to assist clearance, planting and harvesting, although digging sticks are the predominant tools of garden cultivation.

Except where remains of features are preserved, most likely in wetlands, these activities are archaeologically indistinguishable from standard rainforest dynamics in areas subject to human disturbance through burning and patch creation. On most valley slopes, any features would be re-incorporated into the soil matrix through soil formation under regenerated forest. Wooden tools are unlikely to preserve in the acidic and aerated soils of the valley slopes, and any lithic artefacts would be hard to provenance given the extent and duration of soil formation and mass movement after original archaeological deposition.

Further, there is unlikely to be any long-term or cumulative evidence of soil enrichment through the addition of domestic refuse in the highlands, such as in house gardens or the vicinity of garden shelters, because these structures are usually only occupied for a few years prior to abandonment. Many societies practising shifting cultivation in the highlands and highland fringe today do not often live in nucleated settlements; rather, they occupy dispersed hamlets throughout the rainforest. They characteristically cultivate an area and establish houses for only a few years before abandonment (Rappaport 1968; Clarke 1971; Riebe 1974). Consequently, it is unlikely that most highland or highland fringe slopes would preserve pedological evidence of long-term soil enrichment, as documented for *terra preta* and *terra mulata* soils in Amazonia (Lehmann et al. 2003).

Today, shifting cultivators tend to move within prescribed territories on a semi-rotational basis, often returning to ancestral areas after 15–25 years or so after abandonment. This may not have been the case 7000–10,000 years ago. At that time, population densities were plausibly much lower than today, as were territorial constraints on movement to new areas. Certainly, people would have returned to old patches and plots to manage and harvest favoured foods, including harvesting perennials – especially palms, pandanus and trees – and vegetatively reproducing plants, as well as for hunting and the gathering of small vertebrates and invertebrates, such as grubs from the trunks of felled trees. Without the territorial constraints imposed by neighbouring groups, people are likely to have expanded across forested landscapes in more extensive ways than in the recent past.

In sum, direct archaeological evidence of shifting cultivation in the highlands during the early Holocene would be hard to identify on valley slopes, most of which have been subject to repeated cultivation, soil formation and erosion for millennia. These practices are unlikely to have any archaeological or pedological visibility away from the wetlands. Furthermore, nascent forms of cultivation are potentially indistinguishable, at least archaeologically, from a range of other practices and processes that plausibly occurred in domesticated landscapes across the island (Yen 1989; Terrell et al. 2003).

Is there a neolithic signature?

Arguably, there is no clear lithic signature accompanying the emergence of agriculture in the highlands of New Guinea. The absence of an unequivocal lithic signature of early agriculture may indicate that the stone tools used to facilitate plot preparation were not greatly different from those used during pre-agricultural periods. Golson originally suggested that the tool technology used by agriculturalists was not necessarily complex and the majority of agricultural implements were wooden, for example, digging sticks and spades (Golson 1977b). Despite these reservations, three stone artefact types warrant special consideration in terms of potential agricultural associations: the ground stone axe/adze; tanged implements, or hoes, and the mortar and pestle complex. These artefacts were plausibly used for felling trees, working the soil and food processing, respectively.

Ground stone axe-adzes

Ground stone axe-adzes were 'an important component of most tool kits in the Western Highlands' by 5000–6000 years ago and would greatly have increased the efficiency of forest clearance (Christensen 1975a: 33). Christensen based this interpretation on the results of his excavation of Manim 2, and Sue Bulmer's excavations at Kiowa and Yuku (see Bulmer and Bulmer 1964: 66). Tools with edge grinding dated to the early Holocene or Pleistocene at Kafiavana (White 1972: 195), Kosipe (White et al. 1970: 162), Yuku (Bulmer 1975: 31; cf. Bulmer 1977b: 57) and Nombe (Golson 2001: 196). Christensen (1975a: 32; after White 1967) noted that the Kafiavana collection provided evidence of early Holocene grinding on the faces, sides and cutting edge.

The most significant site for understanding the agricultural associations of ground stone technology is Manim 2, which is located less than 20 km from Kuk in the Upper Wahgi Valley. There appear to be correlations between the local production and widespread adoption of ground stone axe-adzes at Manim 2, forest clearance on the valley floor and the agricultural chronology at Kuk. The earliest ground stone axe-adzes at Manim 2 come from Level 21 in Test Unit I (Burton 1984: 227–228; Mangi 1984: 106). They are described as 'two miniature axes of ovoid section made from local stone' (Figure 8.1; Burton 1984: 227; Figure 10.14).

A new AMS dating programme based on plant macroremains from Test Unit I at Manim 2 provides minimum ages for the wider adoption of ground stone axe-adzes, as well as of ground stone technology, in the Upper Wahgi Valley region. The oldest ground stone axe-adzes conservatively date to 6570–6410 cal BP and date to 7430–7290 cal BP based on a more literal reading of archaeostratigraphic association. Taken together with rough outs and axe adze flakes, the antiquity of ground stone axe-adze production most likely pre-dates 8340–8180 cal BP. The more accurate and precise dating of ground stone tool technology at Manim enables robust correlations with the well-dated palaeoecological record of environmental transformation on the floor of the Upper Wahgi Valley (see ahead).

In the highlands, the widespread adoption of ground stone technology during the Holocene is more significant than its original innovation in the Pleistocene. The emergence of better ground and polished axe-adzes would have increased the efficiency of clearing and disturbing forest (Figure 8.2), thereby assisting the expansion of agriculture and the associated social and palaeoecological transformations in the interior of New Guinea during the mid and late Holocene (from S. Bulmer 1966 to Christensen 1975a).

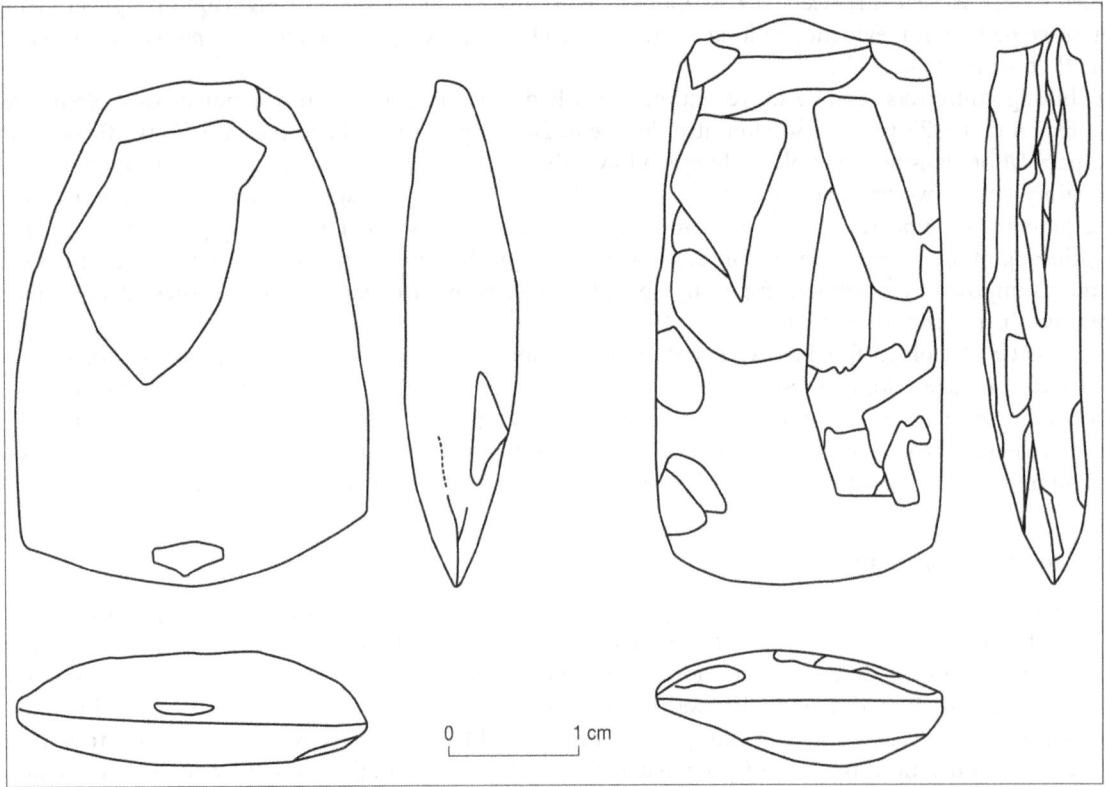

0   1 cm

Figure 8.1 The oldest ground axe-adzes at Manim 2 rockshelter

Note: (left) artefact 6280A; (right) artefact 6280 (Redraft of artefacts depicted in Burton 1984: Figure 10.14)

Figure 8.2 Hewa man felling a tree with a ground stone adze

Notice: Note that although the tool is being used as an adze, it has a symmetrical cutting edge.

Source: Peter White, 1967

Tanged blades

Tanged blades have been reported for several sites in the New Guinea highlands including Kuk (Allen 1970: 180, 1972: 187) and Wañelek (S. Bulmer 1973: 14–15, 1977a: 66–67) and those in the Wurup Valley (Figure 8.3; Christensen 1975b) and the eastern highlands (in S. Bulmer 1977a: 67). There are reports of similar tools within and beyond New Guinea (in S. Bulmer 1977a: 66–67). Although an exact function is uncertain, these implements have been interpreted to be hoes or digging tools (S. Bulmer 1977a: 67; Golson 1977b: 159–160; Steensberg 1980: 110). Except for Wañelek, these tanged blades and other cutting implements of potential agricultural function, for example, stone knives (Golson 1977b: 155–156), are chance finds for which an original provenance is unknown.

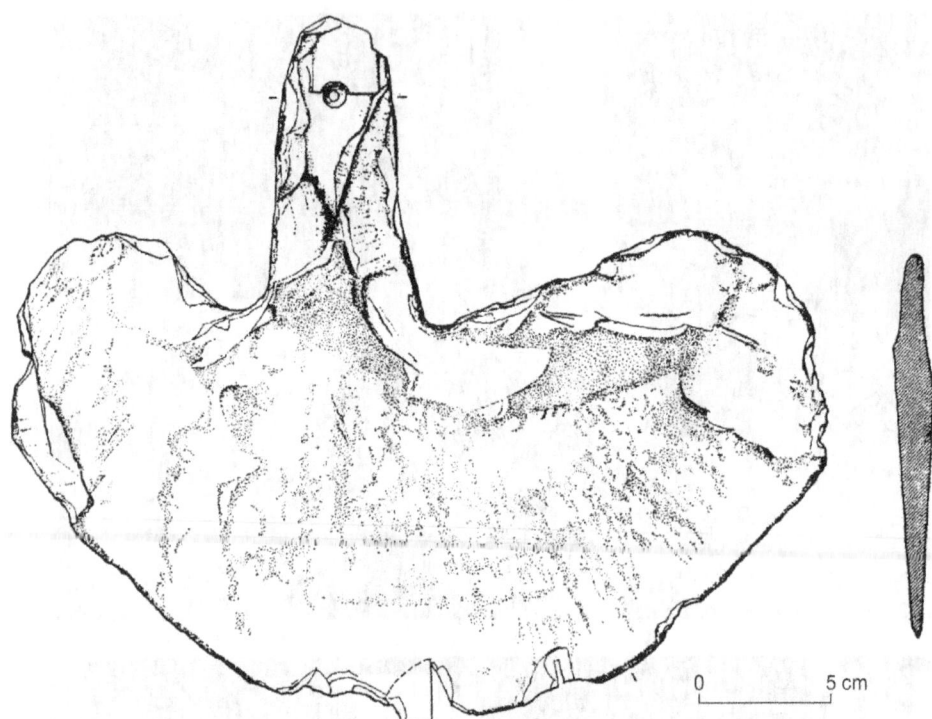

Figure 8.3 A tanged blade, or 'hoe-like object', from the Wurup Valley

Source: Jack Golson, reproduction of original image in Christensen 1975b: Figure 1

Stone mortars, pestles and figurines

Stone mortars and pestles have been collected across eastern New Guinea, though they are not ubiquitous and few are reported from the western half of the island (Pretty 1965; Golson 2000; Swadling and Hide 2005). Mortars and pestles are distributed across New Guinea and the Bismarck Archipelago, and discrete interaction spheres have been inferred based on forms and decorative motifs (Torrence and Swadling 2008).

The antiquity, function and cultural associations of mortars and pestles are enigmatic (Denham 2014: 585-587). Early dates for mortars and pestles range from approximately 8000–7500 cal BP at Kuk to c. 3500 cal BP at NFB (Golson 2000; Swadling and Hide 2005). Most contexts are poorly dated and have limited chronostratigraphic control. The exception is NFB, where radiocarbon dates on charcoal from contexts above and below a stone mortar suggest an age of c. 3500 cal BP (Watson and Cole 1977). Such a date accords with Swadling and Hide's (2005) assertion that mortars and pestles were largely abandoned with the advent of Lapita pottery in the Bismarck Archipelago around c. 3470–3250 cal BP (Denham et al. 2012).

A geoarchaeological interpretation for the formation of the infilling of a former inland sea in the Sepik-Ramu basin corroborates a pre-3500-year-old antiquity for the distribution of mortars and pestles (Swadling and Hide 2005). An inland sea existed in the current location of the Sepik-Ramu Basin from c. 8000 years ago until its infilling c. 4000 years ago (Swadling et al. 1989; Chappell 2005). While mortars and pestles occur on mid-Holocene floodplains and deltas, as well as on slightly higher ground and in adjacent uplands, they do not occur in the location of the former inland sea (Swadling and Hide 2005). This distribution suggests the exchange networks through which mortars and pestles were distributed, as well as their manufacture (for the most part), had potentially ceased by c. 3500–4000 years ago.

The processing function of mortars and pestles is unclear. The range of proposed uses includes the processing of acorns of *Castanopsis acuminatissima* (R. Bulmer 1964), mashing of taro (Swadling and Hide 2005), grinding of seeds (Telban 1988), preparation of kava (*Piper methysticum*, Ambrose 1991) and cracking of *Pandanus antaresensis* drupes (Golson 2000). Ancient stone mortars and pestles have attained new magical and symbolic properties, such as a cure for venereal disease (Kasprus 1973: 157–159), or are re-incorporated into contemporary practices, such as garden magic, male rituals and so on (Ballard 1995). Among some communities, these objects are not considered to have been made by people and have little significance. In recent times, mortars and pestles were still used by communities on Buka and in the Solomon Islands for food preparation (Swadling and Hide 2005).

The broader cultural associations of stone mortars and pestles in the New Guinea region are unclear. They may represent part of a cultural complex with other artefact types that were made around the same time and sometimes ornamented in similar ways, for example, club heads and figurines (e.g., Bulmer and Bulmer 1964). Plant material preserved in a crack within a foetus-shaped stone figurine collected in the Ambum Valley, Enga Province was radiocarbon dated to c. 3500 cal BP (Tworek-Matuszkiewicz 2001; cf. Egloff 2008). The antiquity of this figurine broadly accords with that of the well-dated stone mortar at NFB, and provides a chronological foundation for the stylistically-based stone mortar, pestle and figurine complex (Denham 2014: 587).

Contrary to earlier views (Bulmer and Bulmer 1964), the ambiguity and distribution of mortars and pestles overlap with that of horticultural practices. On the basis of these associations, Swadling and Hide (2005) proposed that mortars and pestles are primarily found in areas of taro horticulture in the hinterlands of coastal embayments. It seems more plausible that these implements were used adaptively based upon social and subsistence requirements. Namely, people used stone mortars and pestles of different form and size for practices ranging from the sacred to the profane.

Palaeoecology, geomorphology and landscape change

Geomorphological evidence, mainly sedimentation type and rates, has been used independently and in conjunction with palynological studies to infer the extent of vegetation clearance, and hence horticultural activities, in catchments. Accelerated deposition rates at rockshelters in the highlands have been interpreted to reflect soil erosion following clearance. For example, Gillieson et al. attributed the marked increase in erosion rates at Nombe after c. 6500 cal BP to horticultural intensification (Gillieson et al. 1986: 315, 1987). Similarly, Hughes interpreted increased erosion rates from c. 2300 cal BP at Manim 2 to represent the commencement of grassland tillage on the valley slopes (Hughes 1985: 400).

Rates of deposition within wetlands have been used to estimate the changing impact of people on their environment (Oldfield et al. 1985: 390) and as a surrogate measure of dryland agricultural activities. For instance, at Kuk, accelerated rates of deposition of grey clay within an alluvial fan occurred between 10,000–6400 cal BP (Golson and Hughes 1980: 296–298; Hughes 1985; Hughes et al. 1991). The 10,000-year-old palaeochannel and palaeosurface were filled and sealed by this massively structured grey clay. The grey clay formed a fan emerging from the low-lying hills and drainage basin to the south of the wetland. Similar deposits, superimposed on Pleistocene fans, extended south onto the northern margins of the wetland from Ep Ridge. The grey clay was interpreted to represent accelerated erosion (Hughes 1985; Hughes et al. 1991) and to be a product of agricultural clearance for swidden-type, dryland cultivation in the southern catchment (Golson 1991b: 485).

Despite the convincing character of these interpretations, there are a number of problems in using crude deposition rates to infer anthropogenic disturbance in wetland environments. First, and as with the palynological evidence, potential anthropogenic influences need to be considered in terms of climatic and other known effects, for example, landslides, earthquakes and pyroclastic events. Second, there is a need to take account of erosional hiatuses in the sequence, such as those documented at Lake Ambra (Powell 1982c), Telefomin (Hope 1983) and Lake Ipea (Oldfield et al. 1980). Third, regional contexts (Hope et al. 1988: 617) and changing depositional environments undermine the comparability of sedimentation rates. Landscapes and landforms change through time in response to numerous stimuli, only some of which may be known. For example, the deposition of catchment sediment in an alluvial fan on a wetland margin will vary spatially in response to fan drainage patterns, wetland hydrology and changes in base level. Fourth, sedimentation type and rates in the Highlands vary in response, often synergetic, to detrital, pyroclastic and limnic inputs (Hope et al. 1988). Consequently, inter- and intra-site comparisons of deposition rates are problematic proxies of anthropogenic clearance.

Ferrimagnetic and chemical analyses of sediment columns have been conducted to complement palynological and geomorphological analyses at archaeological and wetland sites in the highlands (e.g., Oldfield et al. 1980; O'Garra 1981; Gillieson et al. 1986, 1989; Worsley and Oldfield 1988). Given problems of interpretation in the New Guinean context, ferrimagnetic analyses have largely only aided in the identification of tephras, many of which were already identified by other means (Haberle 1994: 194–195). Ideally, elemental analyses could differentiate the degree of weathering for different sediment sources within the catchment (O'Garra 1981: 27; Thompson and Oldfield 1986); in practice, most distributions tend to reflect standard weathering profiles and the incorporation of nutrient rich tephra (Oldfield 1988: 532–534). Ideally, elemental analyses should be interpreted in conjunction with pollen and charcoal frequency records.

Pollen, charcoal and gross sediment records are more robust when used in combination even though problems persist in eliciting the influence of humans from other factors in the observed signals (Oldfield 1988). Primarily, issues arise in seeking to interpret these long-term records in terms of archaeologically meaningful periods and practices. Is there a simple correspondence between level of forest disturbance and intensity of subsistence practice that can be read from the palynological records? Do early Holocene disturbances of the rainforest represent more intensive subsistence strategies, such as shifting cultivation, than previous types of forest disturbance during the Pleistocene?

In contrast to other highland landscapes, early Holocene warming and increased precipitation did not lead to forest encroachment and the replacement of grasslands in the Upper Wahgi Valley. Rather forest advance was 'muted' and a mosaic of grassland and forest subject to episodic burning persisted (Haberle et al. 2012). Haberle et al. (2012: 136) ask: 'What were people doing at Kuk Swamp to restrict the advance of forest due to rapid and significant climate change during the Terminal Pleistocene [and early Holocene]?'

The palynological record from Kuk is somewhat unique in the highlands because it shows major changes in vegetation history in a relatively continuous form from c. 10,000 to 7000 years ago (Figure 8.4). The record is primarily derived from a series of samples collected through the massively structured grey clay unit, which comprises the deposition of an alluvial fan from the southern catchment on the valley floor (Hughes 1985; Hughes et al. 1991, Denham and Grono 2017; Hughes et al. 2017). The 10,000-year-old palaeosurface at Kuk is buried beneath this grey clay unit, and the earliest archaeological evidence of mound cultivation occurs stratigraphically above it. The sedimentological characteristics of grey clay, which is composed of phytoliths in a highly weathered clay fabric, suggest both colluvial and alluvial processes of deposition (Denham 2003a). The pedogenic characteristics reflect residual and *in situ* soil formation processes; the latter includes extensive micromorphological evidence for clay illuviation (Denham 2003a; Denham and Grono 2017). Taken together, sedimentological and pedogenic investigations of grey clay are suggestive of increased erosion within the catchment, with soil formation and weathering on the alluvial fan during periods of lower water table.

Haberle et al. (2012: 131) split the palaeoecological record within the grey clay unit at Kuk into two discrete periods, or zones, based on the pollen, microcharcoal and phytolith assemblages:

> Zone Kuk-2 (10,200 cal BP- mid grey clay unit) is characterized by substantially higher values for rainforest taxa often associated with swamp forest habitats including *Pandanus*, *Schefflera*, Myrtaceae and *Macaranga/Mallotus*. Woody non-forest taxa remain poorly represented; the grasses and Melastomataceae show increased yet variable values. Ferns also increase and the aquatic reed, *Typha*, has its peak representation for the diagram in the zone. Charcoal values are highly variable and peaks tend to coincide with grass pollen % increases. Grass (including the swamp grass *Phragmites*) and Arecaceae/Zingiberaceae phytoliths increase moderately. [Musa] Eumusa and *Musa ingens* phytoliths are present in this zone. Palynological richness values are highest in this zone.
>
> Zone Kuk-3 (mid grey clay unit – 7000 cal BP) observes a slight reduction in a number of rainforest taxa such as *Nothofagus*, *Castanopsis/Lithocarpus* and the gymnosperm taxa. *Pandanus* reaches its maximum values in this zone as do a number of minor sub-canopy rainforest taxa including Palmae and *Psychotria*. Woody non-forest taxa values are at a minimum. There is a clear increase in Asteraceae (Tubuliflorae type) and *Ipomoea* sp. along with a decrease in grasses and sedges. Ferns reach their maximum representation, though there is a sharp reduction in *Typha* towards the top of the zone and *Myriophyllum* is absent. Grass phytolith percentages remain stable and Palmae/Zingiberaceae phytoliths continue to increase moderately. [Musa] Eumusa phytoliths are present in this zone. Palynological richness decreases gradually and charcoal is low.

Something different was occurring across the Upper Wahgi Valley landscape during the early Holocene, from approximately c. 10,000 years ago. Disturbance of the montane rainforest using fire and stone tools continued, if not increased, on the valley floor to create a mosaic of expanding grasslands and disturbed montane forest, primarily composed of sub-canopy taxa (Haberle et al. 2012). For a period during the early Holocene, the arboreal signal becomes dominated by *Pandanus*, suggesting a swamp forest locally at Kuk. The contribution of other arboreal taxa within the catchment indicates degraded lower montane rainforest on the valley floor. At the same time, mid to upper montane forest had retreated to the ridge-lines, peaks and higher valley walls above 2000 m.

Phytolith frequencies and intact phytolith chains indicate bananas were common and growing, and were most probably planted, locally in the Kuk vicinity (Figure 8.5; Denham et al. 2003; Haberle et al. 2012; Lentfer and Denham 2017). People continued to exploit starch-rich plants, as evidenced by the taro *(Colocasia esculenta)* and yam *(Dioscorea* spp.) residues on a unifacially flaked, volcanic cobble core at Kuk (K/75/S179; Fullagar et al. 2006; Fullagar and Golson 2017). The range of plants and practices are similar to those at c. 10,000 years ago on the wetland margin, but the environmental context changed considerably over the next 3000 years. At c. 10,000 years ago, the lower montane forest on the valley floor was largely intact with evidence of occasional localised, anthropic disturbance. By 7000 years ago, the lower montane forest was severely degraded and effectively being maintained in that state. People inhabited and sought to obtain food within this increasingly modified landscape.

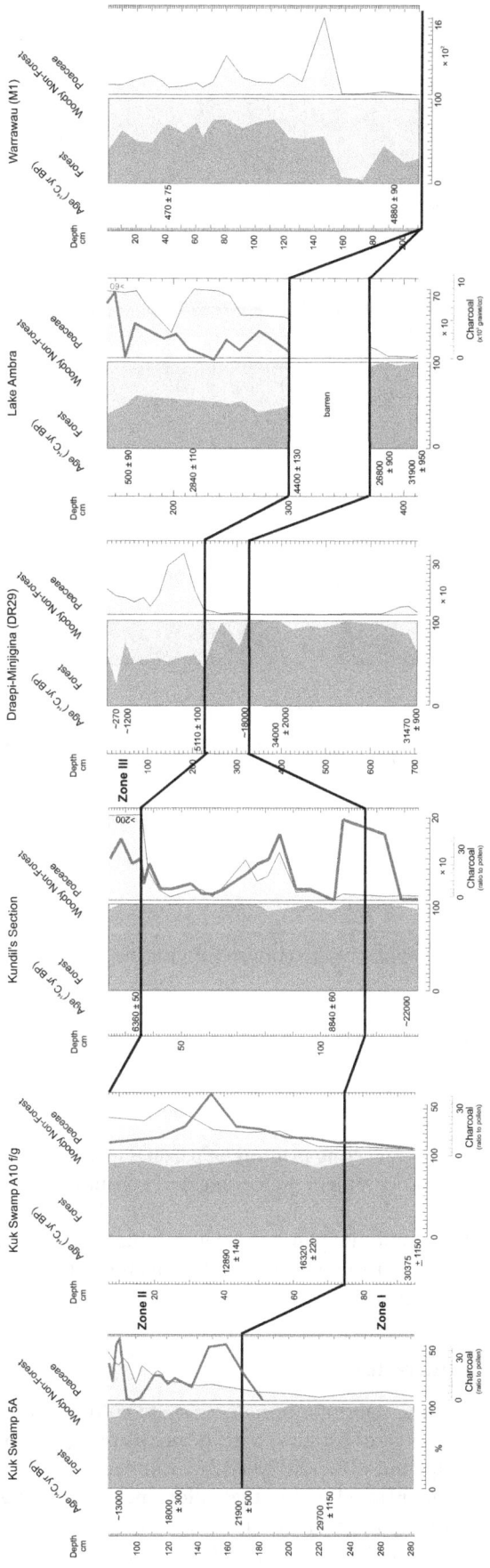

Figure 8.4 Diagram of summary arboreal pollen with Poaceae (pollen sum for all taxa based on total forest and woody non-forest taxa), charcoal particle counts (where available) and radiocarbon chronology (~ = inferred age) from four swamp sites in the Upper Wahgi Valley (32,000 uncal BP to present).

Note: Pollen data derived from Powell 1984 unpublished manuscript (Kuk 5A and Kuk A10 f/g), Powell 1982a: 219 (Draepi DR 29), Denham, Haberle et al. 2004: Figure 8.5 (Kundil's Section), Powell 1982a: 221 (Lake Ambra) and Powell 1970a: Figure 8.7 (Warrawau M1). Solid lines mark comparable pollen assemblages at the six sites based on pollen assemblages and correspond approximately to inferred ages given in the original texts: Zone I (before 21,000 cal BP), Zone II (c. 21,000–7000 cal BP) and Zone III (c. 7000 cal BP –present). Note changing scale for % Poaceae and charcoal counts

Source: Denham and Haberle 2008: Figure 4

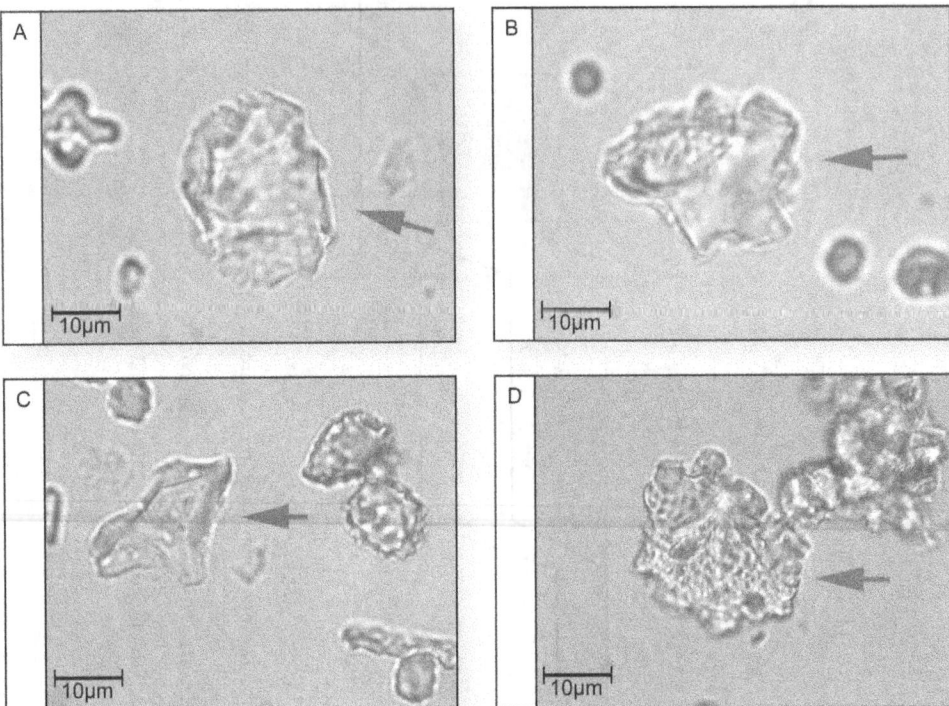

Figure 8.5 Ancient banana phytoliths recovered from archaeological excavations at Kuk Swamp

Note: (SEM images, A–D). A and B dorsal and lateral view of Eumusa seed phytolith recovered from sample 5, Kuk dated to 6,990 to 6,440 cal BP; C lateral view of another Eumusa seed phytolith from sample 5; and D dorsal view of Eumusa seed phytolith recovered from sample 28, Kuk pre-dating 3,000 cal BP. Morphotypes shown in A–C are specific to *Musa acuminata*. The morphotype with lobate margins shown in D occurs in *M. acuminata* and a similar morphotype occurs in *Musa schizocarpa*

Source: Perrier et al. 2011: Figure 3

Although not definitive, these multiple lines of evidence are consistent with the signals anticipated for shifting cultivation (Denham 2013b: 110):

- a vegetation mosaic of primary rainforest, secondary forest and grasslands on the valley floor, representing progressive disturbance, clearance and burning in the catchment;
- increased erosion rates within the catchment, reflected in increased rates of deposition on an alluvial fan on the wetland margin at Kuk (Hughes et al. 1991);
- periodic peaks in banana phytolith frequencies, reflecting planting of bananas locally (Lentfer and Denham 2017); and
- residues of taro and yam, reflecting targeted procurement, processing and consumption of starch-rich plants.

The exact nature of activities is uncertain, yet the multidisciplinary evidence is highly suggestive of some form of shifting cultivation on the floor of the Upper Wahgi Valley during the early Holocene. Indeed, it is hard to envisage another set of practices that could account for the observed phenomena.

From patch to plot: the transition to cultivation

Within the montane rainforests of New Guinea, the transition from foraging to cultivation is marked by the transformation from the creation, exploitation and maintenance of patches to deliberate planting in plots. Previously planting may have occurred adventitiously within patches, gaps and other disturbed environments, yet there is a qualitative difference to planting in specially prepared plots. The species planted within these plots include parts taken from wild or feral species in the rainforest, as well as from old loci of exploitation (whether gap, patch or plot). Such practices mirror those occurring in the interior of New Guinea today and, through time, will have led to the development of 'domestication' characteristics in some plants – those repeatedly reproduced from cultivated stock – whereas the same degree of pheno-typic or genotypic transformation would not be expected in plants repeatedly cultivated from wild or feral stock.

From this perspective, there is no necessary relationship between planting, cultivation and domestication. Planting pre-dated or coincided with cultivation, hereby defined as planting within prepared plots. By contrast, plants may have become domesticated in different ways through long and variable associations with people. Once taken beyond the natural range, a sub-population of plants may rapidly acquire phenotypic and genotypic traits resulting from being grown in a new environment; however, this can occur with or without cultivation. Conversely, wild populations of plants repeatedly cultivated in plots may never generate any phenotypic or genotypic traits representative of domestication if they are continually interbreeding with wild populations.

The transition to shifting cultivation has been characterised in terms of land management practices (Denham and Haberle 2008: 488, 490):

> Hypothetically, there is a relatively smooth set of transformations from gap exploitation to gap maintenance to patch creation (all characteristic of foraging) to plot preparation (characteristic of swidden-type agriculture). In the highlands, such transformations may have occurred during the early Holocene, if not before. Patch creation and shifting cultivation mimic gap exploitation and maintenance in many ways, including: spatially focussed but temporally transient exploitation; short-duration plant exploitation prior to revegetation; juxtaposition of plants with different structural properties, i.e., trees, vines, herbs, shrubs and grasses; and use of fire and tools (stone and wooden) to foster favoured plant growth (Denham and Barton 2006: 258–259).
>
> The transformation from gap maintenance to patch creation is marked by the deliberate clearing of a space within the forest, and plausibly within grasslands, to promote the growth of favoured plants. Groube (1989) has argued that people mimicked gaps formed by landslides, tree fall and fluvial processes through the artificial creation of patches using stone tools and fire from initial settlement over 40,000 years ago. Certainly palaeoecological records in the highlands, including the Upper Wahgi Valley, signal anthropic disturbance of forests using fire during the Pleistocene (Haberle 2007). Through time, and following the digging up of useful plant parts – perhaps with the careful replacement of viable plant parts to ensure resource continuity – the range of activities occurring in patches increased and became more systematic. The systematisation of burning, clearing and digging in turn marks the difference between patch creation and plot preparation.

The transition to shifting cultivation has also been characterised in terms of plant management (Denham and Haberle 2008: 490):

> Another major difference between patch-based foraging and plot-based shifting cultivation in New Guinea is an increasing dependence upon the vegetative propagation of starch, fat and protein-rich plants, as well as leafy vegetables, and the transplantation of wild and feral seedlings, in conjunction with the continued tending of pre-existing stands. Planting is claimed to have a Pleistocene antiquity in Melanesia (Yen 1996) and has been suggested as a means through which some bananas (of Musa section), taro and some yams moved from the lowlands to the highlands of New Guinea in the early Holocene (Yen 1995; cf. Denham, Haberle et al. 2004: 852).

Thus, although currently indeterminate, the palaeoenvironmental record of shifting cultivation in the Upper Wahgi Valley provides an indirect line of evidence for planting. Transformations in the social sphere would have accompanied the initiation of shifting cultivation. People's mobility may have decreased as plant exploitation became more focussed on maintained patches and shifting plots. Even though not sedentary, general mobility may have decreased as people invested more time and labour in the maintenance of spatially fixed resources that were becoming more and more significant to their diets. People continued to engage in extensive foraging activities, as evidenced by the *Pandanus antaresensis*-dominated early Holocene archaeobotanical assemblage at Manim 2 (Donoghue 1988), with long-distance and long-duration hunting and foraging expeditions extending to higher and lower altitudes. However, a greater percentage of time was spent managing, harvesting and processing resources within demarcated territories around maintained patches and shifting plots on the valley floors. In the absence of clear archaeological evidence, though, arguments for shifting cultivation practices in the highlands during the early Holocene are inferential.

9 The adoption of mound cultivation during the mid-Holocene

Contemporary and ethnographic accounts of cultivation using mounds are widespread in the highlands and lowlands of New Guinea (Waddell 1972; Bourke and Allen 2009; Hitchcock 2010). In the highlands today, mounds are associated with sweet potato cultivation (Waddell 1972: 42–56). In this chapter, mound cultivation will be shown to have originally been developed for the cultivation of other crops, because the practice pre-dates the introduction of the sweet potato to New Guinea by several millennia. Mounding represents more intensive land use in terms of labour inputs, duration of use and ground preparation than previous shifting cultivation, whether slash-and-burn or slash-and-mulch.

Following his exploration of the Upper Wahgi Valley from 1933 onwards, Mick Leahy (1936: 242) speculated that the technology of mound cultivation originated in the wetlands and subsequently spread to dryland slopes. Even though the earliest archaeological evidence for mound cultivation derives from wetlands, it is argued here that the adoption of mound cultivation was a response to environmental degradation and the establishment of grasslands on the valley floor. From this perspective, mound technology was not a wetland adaptation, even though early archaeological evidence for it is only preserved in wetlands.

Dramatic deforestation in the Upper Wahgi Valley

A dramatic change in palaeoecology of the Kuk catchment occurs around 7000 years ago. The rainforest and swamp forest signals drop precipitously with a corresponding increase in ferns and herbaceous taxa, primarily grasses and sedges (Denham, Haberle et al. 2004; Denham and Haberle 2008; Haberle et al. 2012). Effectively, the valley floor within the Kuk catchment was degraded to grassland and other disturbed vegetation communities at this time. These grasslands were maintained by periodic burning and, presumably, cultivation. Such degraded environmental conditions are characteristic of previously cultivated areas across the highlands today (Gillison 1972).

No other palaeoecological records in the Upper Wahgi Valley are continuous as far back in time as that from Kuk. Several other records show the existence of grasslands and heavily disturbed environments, albeit with localised periods of forest recovery from at least 5000 years ago (Powell 1970a, 1982b: 218; Golson 1977c: 617–618; Denham, Haberle et al. 2004; Sniderman et al. 2009). This anthropic landscape dominated by grasslands persisted on the floor of the Upper Wahgi Valley until the arrival of Australian-based gold prospectors and a government officer in 1933 (Leahy 1936).

The palaeoecological transition occurred rapidly, because it corresponds to a well-dated stratigraphic boundary between grey clay and overlying black clay at Kuk. The Bayesian calibration of multiple radiocarbon ages has enabled the age of this transition to be determined accurately (Denham 2003a; Denham et al. 2003). Uppermost grey clay yields a radiocarbon date of 7420–7210 cal BP and the features associated with mound cultivation cut into the upper surface of this grey clay date to 6950–6440 cal BP (Denham 2003a: Table 9.3).

The only other highland valley with a comparable record of vegetation history is the Upper Baliem Valley, where 'widespread forest disturbance, burning and myrtaceous swamp forest clearance' occurred around 7800 years ago (Haberle 2003: 153). Extensive grasslands were not established there until around 3000 years ago (Haberle et al. 1991). Haberle (2003) has used these types of palaeoecological indicators to characterise the emergence of agricultural landscapes in the highlands.

Deforestation on the floor of the Upper Wahgi Valley around 7000 years ago was relatively rapid, occurring within a few hundred years. The people, whose ancestors had practised forms of shifting cultivation within a mosaic landscape – comprising stands of primary and secondary forest, grasslands and other disturbed habitats – were now inhabiting a very different type of environment. Just as the forest was largely gone from the valley floor, having been replaced with extensive grasslands, so too the variety of wild resources available within that mosaic landscape was gone. People had to adapt relatively quickly to living within resource-poor grasslands and accessing wild resources higher up the valley walls.

People would have struggled to practise shifting cultivation in grassland environments. Today, cultivators in the highlands and highland fringes express an aversion to making new plots in grasslands; the dense root mats entail considerable work to physically fragment, till and remove, even after burning. Ordinarily, people inhabiting grasslands use more intensive forms of tillage and cultivate their plots for longer than shifting cultivators in rainforest.

The options available to the inhabitants of the Upper Wahgi Valley around 7000 years ago were to leave the area or adopt new forms of cultivation. To migrate to new lands may not have been so readily achieved. People develop an attachment to their ancestral territories. Places are named after the dead and the living, and landscapes are inhabited by spirits of the dead who need to be looked after by their descendents (e.g., Rappaport 1968). So the landscape not only accumulates the environmental signatures of past human activities, but it also accumulates multiple layers of cultural meanings. Further, people may not have been free to move; they could have been restricted by neighbouring groups or a shortage of suitable land. Whatever the reason, people continued to inhabit the valley floor, which archaeological evidence suggests was enabled by the adoption of new forms of cultivation.

The archaeological evidence for mound cultivation at Kuk

The bases of former mounds used for cultivation around 6950–6440 cal BP have been excavated on the wetland margin at Kuk. These finds are the earliest, unequivocal (Hope and Golson 1995: 823) and fully acceptable (Golson 1989: 679), archaeological evidence for cultivation in the New Guinea highlands (Golson 1977d: 48–49, 1991b: 484, 489; Denham 2003a, 2003b, 2006, 2007a; Denham et al. 2003; Denham, Golson et al. 2004, 2017b; Denham, Haberle et al. 2004). Two types of archaeological remains have been identified; an integrated web of features that clearly define the bases of mounds and a more discrete type that seems to lack an overall pattern (Golson 1977c: 616–617; Denham 2003a, 2003b). Golson (1976b: 3) inferred the integrated form to occur on lower ground, whereas the discrete form was restricted to higher and drier ground.

The integrated palaeosurface is comprised of the truncated bases of former mounds, defined as sub-circular-to-irregular 'islands' in grey clay by a microtopographical network of features (Denham 2003b). Archaeological remains of these former mounds were exposed in multiple excavations, with slightly varying form and regularity (Figures 9.1 and 9.2). Golson (1977c: 616) characterised the archaeological finds as 'a web of short channels, so disposed as to define roughly

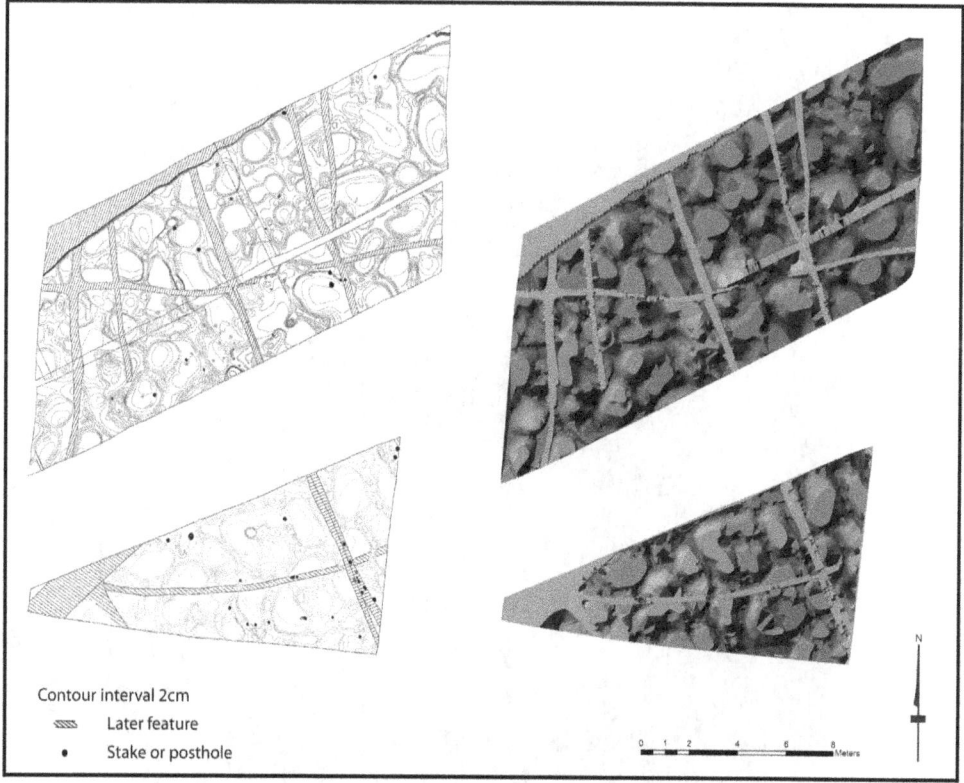

Figure 9.1 Palaeosurface of former cultivation mounds exposed in excavation at Kuk Swamp

Note: (left) contour map of trench bases (2 cm contour interval, originally planned by Art and Cherie Rohn, 1977); (right) digitised and coloured palaeosurface (prepared by Rob Patat and Uri Gilad)

Figure 9.2 Photographs showing archaeological excavations of palaeosurface at Kuk (left), Mugumamp (middle) and Warrawau (right)

Source: Jack Golson, taken by Ed Harris, 1977

circular clay islands of about a metre diameter' (Figure 9.3). The cut features around the mound bases consist of deeper basins, shallower hollows and interconnecting runnels. Numerous stakeholes and postholes are found in association with the mounds, often being associated with deeper basin-like features.

At Kuk, many of the features are shallow and the fills are not clearly differentiated from overlaying black clay, except where lenses of a phytolith-rich deposit, called 'R+W', are found near the base. Stratigraphically, grey clay defines the former mounds, which originally would have been higher and composed of overlying soil as well as the spoil heaped up from digging the surrounding features. With time, the soils on the mounds were either eroded to become the fills of the adjacent features, or reworked and incorporated into the overlaying black clay stratigraphic unit.

The discrete type has been inferred to represent 'denudation and truncation of more integrated forms' (Denham et al. 2017b: 207), whereas scattered features in other areas are suggestive of different and less intensive practices taking place between integrated plots. Characterisations include 'saucer-shaped' depressions with stakeholes (Golson and Hughes 1980: 299) and a less organised pattern of inter-cut features (Golson 1977c: 617). The exact function of these discrete features is uncertain; on the basis of similar morphological characteristics to the integrated palaeosurface, they are considered to represent a form of cultivation.

Artificially constructed palaeochannels were originally interpreted to articulate with and drain this palaeosurface (Golson 1977c: 615–616; Golson and Hughes 1980: 298); however, the artificial character of contemporaneous palaeochannels and any association with them are doubtful (Denham, Golson et al. 2014b, 2017b). The slightly sinuous planforms, cross-sectional morphologies and course gradients, are unusual, yet consistent with natural watercourses. Two palaeochannels are broadly contemporaneous with the palaeosurface; they are probably natural watercourses that flow over 300 m west of the closest evidence for mounds (Denham 2003a: 160–176, 2003b: 163–164; Denham, Golson et al. 2004: 279–280). Any association between palaeochannels and palaeosurface is thus indirect; palaeochannels allowed human use of the southern portion of the wetland margin because they aided local drainage and limited flooding by incident water from the southern catchment.

Palaeoenvironments within cultivated plots

Although the palaeoenvironments at Kuk have been characterised throughout the Holocene (Denham et al. 2003; Denham, Haberle et al. 2004), methodological questions remained regarding the archaeological associations of samples from the fills of features on the mounded palaeosurface (Denham, Sniderman et al. 2009). In order to resolve uncertainties, two features (3003 and 3004) associated with mound cultivation at Kuk were sampled and subject to contiguous multi-proxy analysis (X-radiography, diatom, pollen and microcharcoal; Denham, Sniderman et al. 2009).

The two features contained similar stratigraphy (Figure 9.4): an upper lens of 'R+W', a black clay fill and the underlying grey clay into which the feature was cut. 'R+W' is a phytolith-rich deposit dating to 6440–5990 cal BP, which was formerly considered to be a tephra (Denham et al. 2003). Thin section microscopy, radiography and XRD indicate that this deposit was not a tephra, but was composed almost exclusively of phytoliths (Denham 2003a; Coulter et al. 2009). The exact conditions under which this deposit formed are unknown. The predominance of grass phytoliths suggests that grasses, which dominated the local vegetation at this time, were cut, cleared and burned or mulched; detrital materials were washed into the deeper features where they became concentrated. The higher frequencies of charcoal visible on thin sections within these fills support this interpretation, with grass ash being noted as containing high percentages of biogenic silica (Matthews 1995).

In both monoliths, the lower boundary of 'R+W' is diffuse, representing some intermixing with black clay fills. The underlying grey clay is marbled with black clay, suggesting the downward movement of black clay into voids within the grey clay. The latter voids represent bioturbation resulting from faunal burrowing and root activity.

The diatom compositions reflect a transition from wetter to drier conditions from grey clay to the feature fills. Upward through the black clay fill and 'R+W' lens above grey clay, aerophilous taxa become dominant. Further, '[t]he increasing proportion of *Diadesmis contenta* (Grunow ex van Heurck) Mann, absent from older sediments at Kuk, suggests drier conditions, shallowing and, perhaps, shading by vegetation, because optimal conditions are at the air-water interface' (Denham, Sniderman et al. 2009: 730). The grey clay below the features fills is characterised by poor diatom preservation, but one sample with sufficient diatoms for interpretation is suggestive of much wetter conditions, probably standing water. The wetter conditions characteristic of grey clay contrast with the moist and drier conditions within the fills of both features and suggest damp soil conditions around the bases of the mounds when they were in use.

The pollen signal in both features, as with others associated with this palaeosurface (Denham, Haberle et al. 2009; Haberle et al. 2012), is similar: grey clay samples are relatively rich in *Pandanus* spp. and primary forest taxa, whereas feature fills are dominated by herbs with some secondary forest taxa and higher charcoal frequencies (Denham, Sniderman et al. 2009: 731). Transitions in pollen and charcoal frequencies that suggest sediment mixing and/or pollen translocation

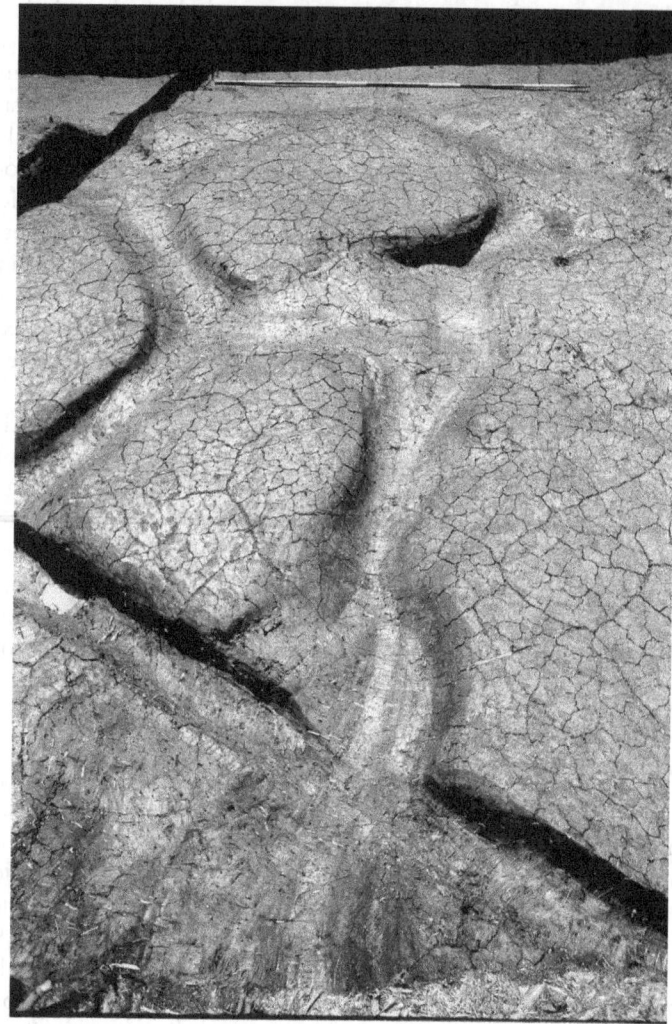

Figure 9.3 Photographs of mounded palaeosurface: prior to excavation with R+W deposit in-fill (upper) and after excavation (lower)

Source: Jack Golson, taken by Alistair Marshall, 1976

at the base of the features. For feature 3003, the transition is reflected in increasing Poaceae and decreasing *Pandanus* spp. values; whereas for feature 3004, admixture along the boundary is greater and is reflected in intermediate Poaceae and charcoal values, as well as gradually decreasing *Pandanus* spp. values. These admixed palaeoecological signals at the base of features are consistent with the X-radiographs indicating bioturbation.

The multi-proxy analyses of these two features indicate several major transformations to the local and extra-local environment occurred at 6950–6440 cal BP: a damp, soil environments (diatoms) characterised the areas between mounds (archaeology), and the wetland was transformed to grassland maintained by periodic burning (pollen and microcharcoal).

Multidisciplinary consilience

On the basis of the archaeological findings, several feature types are clearly artificial, including the mounds and associated stakeholes and postholes. The regularity of mound form within and between excavated areas reflects their construction in accordance with a rough, practical template for agricultural mounds (Golson 1991b: 489). Stakes and posts were seemingly used to support and tether bananas, sugarcane and other edible cane grasses within cultivated plots. Stone artefacts, manuports and elevated charcoal frequencies are associated with the palaeosurface.

The recurrent associations between feature types reflect deliberate design. Stakeholes and postholes are repeatedly associated with deeper basins, namely the deeper features within the microtopographical network. Although the precise function of this association is unclear, it represents a deliberate and repeated activity.

The formation factors associated with several forms of comparable, natural microrelief have been investigated and discounted for Kuk (Golson and Hughes 1980; Denham 2003a; see Sullivan and Hughes 1991). Non-agricultural modes of formation were considered for both palaeosurface types given morphological similarities to known microtopographic forms. For example, the integrated and discrete palaeosurface types are morphologically comparable to network and 'melon-hole' gilgai, respectively (Hallsworth et al. 1955: 3 and 9, respectively). Although there is a possibility that the kaolinitic grey clay represents the weathered end-product of formerly smectitic clays, which are associated with gilgai formation, this seems unlikely. The fundamental properties of this deposit, including its mineralogical composition, are thought to be inherited, that is, pre-date erosion from the southern catchment. Gilgais tend to form where there is high base cation, sodium and carbonate concentrations; smectitic clays; and pronounced seasonality. None of these conditions pertain to prehistoric or recent environmental conditions along the wetland margin at Kuk, which has amorphous or kaolinitic clay fractions, acidic profiles and relatively aseasonal climates. Additionally, no micromorphological features associated with gilgai formation or vertisol development were present, such as coarse particle alignments, characteristic birefringence fabrics or slickensides (Edelman and Brinkman 1962; see Denham and Grono 2017).

Feature fills indicate heterogeneity suggestive of mechanical disturbance and are often associated with elevated charcoal levels in thin section. In the field, the fills are massively structured and undifferentiable from black clay in the field. Meso and micro-scale analyses (X-radiography and thin section description, respectively) reveal several pedofeatures potentially corroborating archaeological interpretations of palaeosurface function and clearly differentiate early feature fills from black clay (Denham 2003a; Denham and Grono 2017). Micromorphological examination indicates extensive heterogeneity within feature fills, especially in contrast to adjacent stratigraphic units. Admixture is represented by intercalations and displaced granules of grey clay (Figure 9.4). The heterogeneity of these fills is interpreted to represent a primary fill composed of inwashed, rolled and slumped material from the edge of newly constructed, unvegetated mounds.

Some of the stone artefacts collected from the palaeosurface show evidence of plant processing, including a heavily weathered basalt cobble (K/77/S17) with potential taro residues (cf. *Colocasia esculenta*; Fullagar et al. 2006). The exploitation of this aroid indicates continuity with previous periods. Taro was originally proposed as the primary staple of early agriculture in the highlands (Golson 1977c: 616). The paucity of stone artefacts is not surprising, as most tools traditionally used in gardening are wooden (Golson 1977b). Stone artefacts would be used to work wood and process foods prior to consumption (Fullagar and Golson 2017). Only wooden digging sticks would be required to turn the soil to construct mounds. Indeed, some of the putative stakeholes and some undercut areas along the edges of mounds plausibly represent working of the soil using digging sticks.

The fills of one early palaeosurface feature exhibit elevated *Musa* banana phytolith frequencies suggestive of planting in the immediate vicinity (Denham et al. 2003; Denham, Haberle et al. 2004; Haberle et al. 2012; see Wilson 1985). Some Musaceae types are fire tolerant and can rapidly recolonise grasslands after burning, that is, *Ensete* spp. (Carol Lentfer pers. comm. 2002). *Musa* spp. would be expected to colonise disturbed patches from adjacent disturbed or forested habitats, but these no longer existed at Kuk at this time. The locally elevated *Musa* spp. frequencies are interpreted to represent deliberate planting within grasslands maintained by periodic burning. Lower values in other feature fills represent spatial variability in preservation and cultivated plant densities, characteristic of polyculture, or multi-cropped, plots.

Mounds at other sites

Comparable archaeological evidence to that at Kuk has been documented at two other wetlands in the Upper Wahgi Valley (Denham 2003b: 169–173; Denham. Golson et al. 2017b: 215–220): Mugumamp (Harris and Hughes 1978) and Warrawau (Golson 2002). Finds at two other sites, Kana in the Middle Wahgi Valley (Muke and Mandui 2003) and Ruti Flats in the Lower Jimi Valley (Gillieson et al. 1985), do not provide clear evidence of mounds, are contentious in terms of form and function, and are probably more recent; they are not considered here (Denham 2003b).

In 1977, Harris and Hughes (1978) conducted small-scale open area excavations and test pitting adjacent to the south slopes of Mugumamp Ridge in that part of the North Wahgi Swamp referred to as South Swamp. A palaeosurface comprised of features directly comparable to the basins, runnels, islands and postholes at Kuk was exposed and interpreted to have functioned in an analogous way (Figure 9.2; Harris and Hughes 1978; after Golson 1977a: 616). A ditch running in one trench was originally interpreted to be contemporary with, or earlier than, the palaeosurface, although this interpretation is questionable and the association discounted (Denham 2003b: 170–171). The palaeosurface was infilled with Kim tephra (also known as R ash), which has been radiocarbon dated to 3980–3630 cal BP (Coulter et al. 2009). Consequently, 'the Mugumamp gardens' provide later evidence for mound cultivation than Kuk (Harris and Hughes 1978: 442).

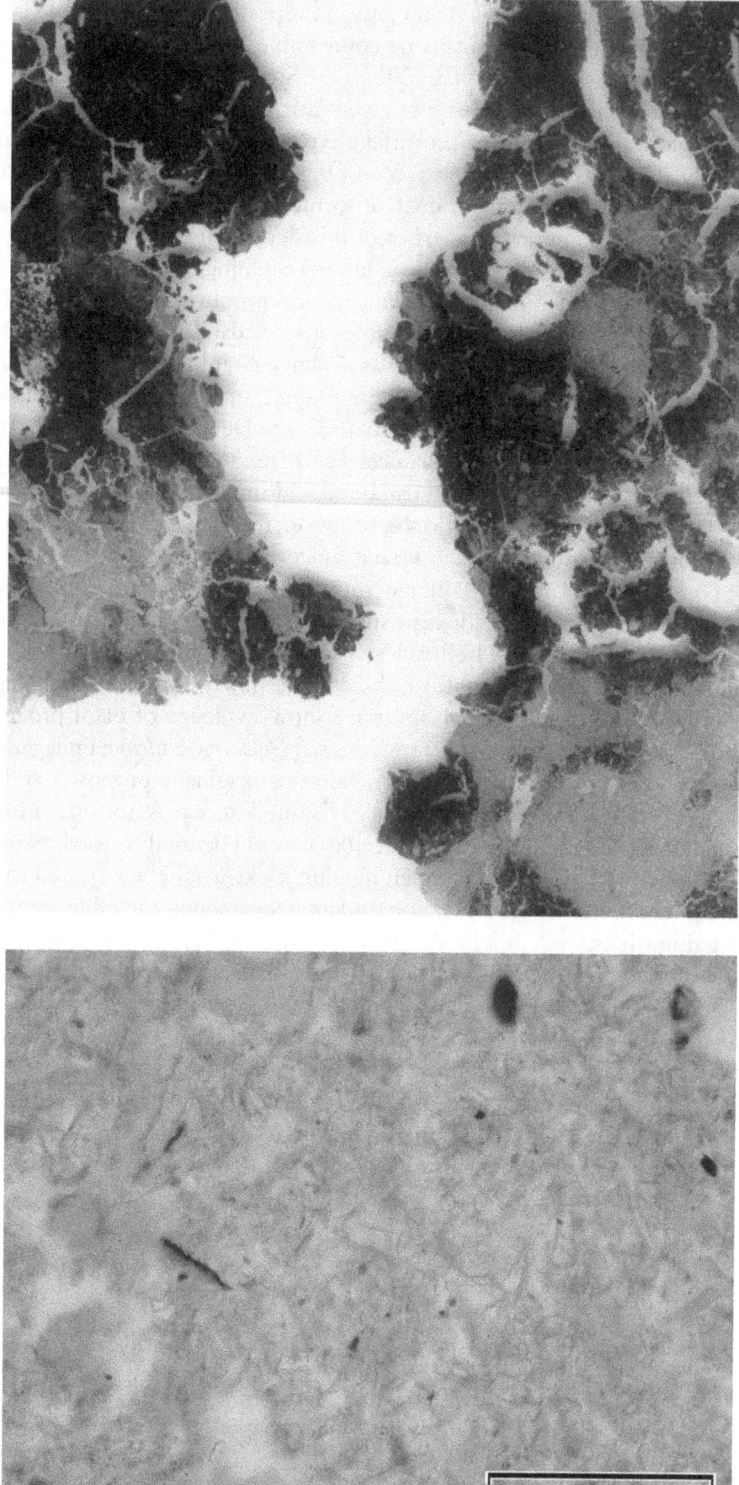

Figure 9.4 Thin section slide of heterogeneous fill within a feature on the mounded palaeosurface feature containing 'R+W' deposit (upper, image width 3 cm) and detailed microphotograph of 'R+W' showing dominant phytolith composition (lower, image scale bar is 100 μm)

In 1977, a number of archaeological excavations originally undertaken at Warrawau in 1966 (Golson et al. 1967; Lampert 1967) were re-excavated with the intention of re-examining the ditch networks in the light of the Kuk sequence (Golson 1982a: 121). A palaeosurface similar to that at Kuk was exposed in the base of one re-excavated trench (Figure 9.2); it had not been recognised in 1966 when the trench floor was wet, muddy and obscured (Jack Golson pers. comm. 2000). The palaeosurface was composed of 'a series of basins with R ash [Kim tephra], much larger and deeper than at Kuk, but similarly linked by interfluves and defining islands' (Jack Golson site diary 1977, page 74, cited in Denham 2003b: 173). Tephrochronology, radiocarbon dating and stratigraphy indicate that the Warrawau palaeosurface is later than that at Kuk and older than that at Mugumamp.

Another two trenches at Warrawau were re-excavated to expose a probable palaeochannel. Radiocarbon dates on wood from the basal fill accord with a stratigraphic position beneath R ash, and date to c. 5800–5300 cal BP (Denham 2003b: 173). During re-excavation, fragments of gourd exocarp were noted and collected from towards the base of this feature, at approximately 40–60 cm below R ash. These fragments were not identified to genus or species and unfortunately are not available for analysis. Golson (2002: 70–71) has suggested that these gourd fragments, which are older than c. 5000 cal BP, are likely to be wax gourd *(Benincasa hispida)*, following its occurrence at Kana around 3000–2000 years ago (Matthews 2003), rather than the original identification of bottle gourd *(Lagenaria siceraria*; Powell 1970a: 144–145, Powell 1970b: 199).

A sediment core (M1) collected through the fills of the palaeochannel at Warrawau provides a c. 6000 year sequence of land use change (Powell 1970a: 155–159, 1982b: 221; Powell et al. 1975: 43–44, 46–48). The palaeoecological record documented clearance of the primary forest and establishment of secondary forest on the dryland slopes within the catchment from the beginning of the diagram at around 5800–5300 cal BP. In contrast to other sites in the Wahgi Valley at this time – Kuk (Haberle et al. 2012), Lake Ambra (Powell 1981: 306) and Draepi-Minjigina (Powell 1970a: 165–186) – the decline in forest taxa was not matched by a corresponding increase in non-forest taxa. Regeneration of the primary forest occurred locally during the late Holocene.

Taken together, the archaeological evidence for mound cultivation at Kuk, Warrawau and Mugumamp spans almost 3000 years. These wetland archaeological sites show the persistence of mound cultivation within the Upper Wahgi Valley, which continues to the present day in parts of the valley, as well as elsewhere in the highlands. While the antiquity of this innovation is at least 6950–6440 cal BP, its widespread adoption was necessitated by the dramatic decrease in montane rainforest at that time. As with other innovations, mounding broadened the technological repertoire that cultivators in the highlands could draw on, depending upon circumstance.

The purpose of mounds

Following previous interpretations by Golson and co-workers, the integrated palaeosurfaces at Kuk, Mugumamp and Warrawau represent mounded cultivation along a wetland margin. As in contemporary gardens in the highlands, a mound is formed by digging soil from the surrounding ground and piling it in the centre. Multiple interpretations of the function of these mounds are possible.

Golson (1977c) originally proposed that mounds were an innovation designed for water control. Water was retained in the deeper basins during periods of low runoff, whereas the microtopographical network enabled the drainage of water away from the mounds during periods of high runoff (Golson 1977c: 616, 1977d: 49). In a poorly drained environment, such as along a wetland margin, mounds enable the cultivation of plants with different edaphic requirements (following Golson 1977c). Water-tolerant plants, for example, taro *(Colocasia esculenta)*, were potentially planted along the edges and in the bases of the runnels, and water-intolerant plants, for example, sugarcanes *(Saccharum* spp.), bananas *(Musa* spp.), yams *(Dioscorea* spp.), *Setaria palmifolia* and mixed vegetables, were planted and staked on the raised 'beds' (after Golson 1977c: 616, 1981: 57–58; Powell et al. 1975: 42). Archaeobotanical remains from feature fills, associated contexts and stone tool residues provide evidence consistent with this type of horticulture. From this perspective, mound cultivation was originally restricted to the wetland margin, from whence it became a widespread practice on the valley floors and slopes of some highland valleys (following Leahy 1936).

Multidisciplinary investigations of the integrated, early palaeosurface at Kuk suggest that the fills of the cut features between the mounds were not deposited under water. Diatom assemblages indicate that the conditions at the base of these features were moist, rather than containing standing water. Micromorphological and X-radiographic analyses of fills do not show any bedding structures or particle alignments suggestive of deposition in water, even within the phytolith-rich 'R+W' deposit. These lines of evidence suggest that standing water was not the dominant depositional environment within the features surrounding the mounds. The relatively low porosity and infiltration capacity of the underlying clay, twinned with a relatively high water table along the wetland margin, suggest that water could have collected on the surface periodically and, in such a low gradient and vegetated environment, would not have freely drained as surface flow.

Mounds provided slightly drier areas on the wetland margin, but this may not be the rationale for the initiation of the practice in the highlands.

In the highlands today, mound cultivation serves to reduce the risk of frost damage, enhance fertility through the incorporation of mineral and organic inputs, and maintain productivity. Mounds enable prolonged cultivation within a plot, especially in grasslands, because new organic matter (i.e., ash and mulch) can be incorporated into the mound to act as a fertiliser, as well as to raise the internal temperature of the soil surrounding roots and underground storage organs (such as corms, rhizomes and tubers). Further, mounds enable cold air drainage on a microtopographical scale, thereby reducing frost damage to plants growing on them.

From this perspective, mounding was an agricultural innovation to enhance and, in some places, enable cultivation in highland climates. It need not have been restricted to wetlands, because it occurs widely across valley floors and slopes today (Brookfield and Brown 1963). Mounds created microclimates that made viable and increased the productivity of plants near, at or beyond the usual altitudinal range of cultivation, including bananas and yams (see Table 5.2).

Around 7000 years ago, within the degraded landscape of the Upper Wahgi Valley, mounds represent a way to maintain fertility and to prolong cultivation within a cleared plot. Rather than enabling cultivation for only a year or two before abandonment, as is typical under shifting cultivation, mounds enable continuous cultivation of plots for prolonged periods. Consequently, as well as representing an adaptation to highland climates, mounds represent an adaptation to the denuded grasslands carpeting the valley floor.

At Kuk, the discrete palaeosurface was originally interpreted to represent an *ad hoc* system of cultivation, possibly including water control on better-drained, slightly-sloping ground. Surface water was considered to have been less of a problem on sloping ground, despite the low infiltration capacity of the underlying strata and probable impediments to overland flow, for example, vegetation. A less integrated system of cut features would be required to manage the water sufficiently to enable cultivation of the desired crops. However, the different form of the discrete palaeosurface type could also represent the relocation of mounds within plots during successive plantings.

The integrated and discrete palaeosurface forms represent two types of mound construction, opening-out and shifting-mounds, respectively, which are still practised in the Kuk vicinity today (Powell et al. 1975: 11, 12; see Figure 4.7). Within a shifting-mounds scenario, only the preserved bases of deeper features would survive successive plantings, as reflected in the discrete feature morphologies. Thus, the discrete palaeosurface solely represents the differential preservation of a once more integrated palaeosurface depending upon elevation, slope, intensity of more recent cultivation, as well as other factors.

To conclude, mound cultivation was plausibly devised as an adaptation to highland climates. The use of mounds had probably occurred on a limited scale in forms of shifting cultivation before 7000 years ago; the archaeological evidence was not preserved on the wetland margin because it was too wet to cultivate at that time. With the deforestation of the Upper Wahgi Valley floor, the use of mounds became more widespread because they enabled more prolonged cultivation of cleared plots within grasslands. Along the wetland margin, mounds served an additional function; they enabled the cultivation of plants with different edaphic requirements. Even though the bases of mounds are well preserved only in a few places along wetland margins, this was probably not the primary reason for the widespread adoption of the practice.

Significantly, the construction of mounds marks the earliest form of soil tillage in the highlands. Whereas shifting cultivation requires only dibbling, effectively a zero tillage strategy, the construction of mounds requires turning of the sod and piling of soil. The admixture of soil layers in the creation of mounds represents tillage, which can be effected using relatively simple tools (Golson 1977b).

A later development

There are numerous features recorded in plantation drain walls and occasionally in archaeological excavations at Kuk that post-date the mounded palaeosurface and pre-date the drainage of wetland margins using ditches. These features had similar morphologies to older features, but they were often isolated and interrelationships were not usually elucidated or explored through excavation. As a result, the function of most features is not clear.

One composite, curvilinear feature (504) has been investigated in detail (Denham et al. 2017b: 213–214). It consisted of inter-cut, rising and falling basins and channel sections, which were individually similar to older features documented on the wetland margin, that is, concave sides and bases (Figure 9.5). Two radiocarbon dates on different organic fractions of fill indicate that the feature formed by at least 4840–4440 cal BP (Denham et al. 2003). The general direction of flow within this feature is unclear from the contoured plan. It was not considered to solely function as a drain, because water would have collected within the deeper basins. Given the relatively flat local topography, the feature is unlikely to

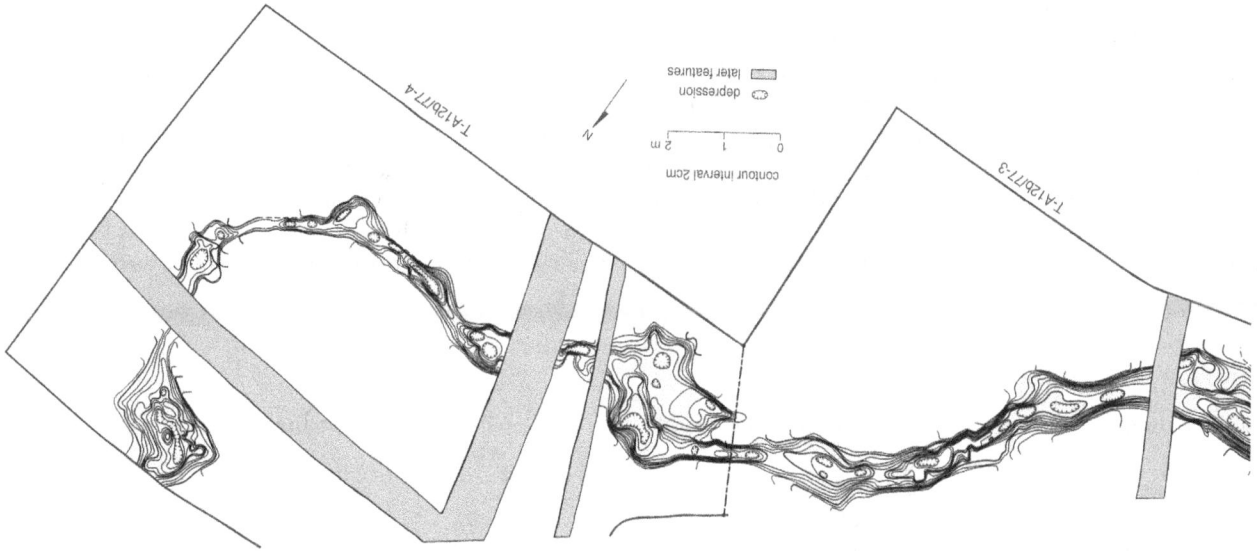

Figure 9.5 Plan of curvilinear feature (upper; feature 504; Source: Denham 2005a: Figure 9.3) and accompanying photograph (lower; view to southwest, Source: Jack Golson, taken by Ed Harris, 1977)

Note: The original field plan drawn by Art and Cherie Rohn in 1977 was digitised by Rob Patat and Uri Gilad

represent natural rill formation through the concentration of surface runoff (cf. Morgan 1995: 17–19). The sinuous runnel was interpreted to have been dug to enable localised drainage within or adjacent to a cultivated plot.

The fills of this feature comprised admixed and poorly sorted aggregates of soil, secondary-deposited tephra and dense macrocharcoal fragments (Lin 2016). Aggregates fell, rolled or washed into the feature, probably relatively rapidly. The highly admixed fills are suggestive of localised erosion, probably following tillage of land adjacent to the feature.

The fills also contain the highest Musaceae phytolith frequencies documented at Kuk, representing up to almost 20% of all phytoliths (Denham et al. 2003). The abundance of banana phytoliths is remarkable, given that the local environment was grassland at this time. Grasses produce plentiful phytoliths in comparison to bananas and the high Musaceae frequencies suggest intensive banana cultivation in the vicinity.

Even though the morphology of the sinuous runnel is unusual in comparison with earlier features, it could have functioned in similar ways. The low surface gradients in the vicinity and lack of a clear drainage direction within the feature suggest it was dug to aid water table control for *Musa* bananas and potentially other crops growing on adjacent land, rather than representing rill formation on an exposed ground surface. Irrespective of its mode of formation, the feature rapidly filled with aggregates, charcoal and vegetation debris associated with earth moving, burning and cultivation locally. As well as facilitating drainage for cultivation, the feature could have been used to grow water-tolerant crops, such as taro. The feature would provide a slightly wetter environment for taro to grow, especially after it had filled with nutrient rich soil and other debris. It is unclear whether this curvilinear is a precursor of more formalised drainage of wetlands using ditches.

Summary of agronomic innovation

Several agronomic practices become archaeologically visible from 6950–6440 cal BP in the highlands:

1   plant cultivation using mounds;
2   tillage (implicit in the construction of mounds); and
3   deliberate planting of bananas (*Musa* spp.).

These practices probably occurred before they become visible in the multidisciplinary record at Kuk. Even though the practices did not probably originate at this time, or necessarily within the Upper Wahgi Valley, they were more widely adopted after 7000 years ago. Plant cultivation occurred within, and contributed to the formation of, a grass-dominated landscape, largely denuded of forest.

The adoption of these practices represents a continuing elaboration of plant exploitation practices from the early Holocene. Even though they do not necessarily represent something new; these practices were 'bundled', or deployed, in fundamentally new ways. The adoption of mound cultivation expanded the plant exploitation repertoire people could draw on and provided a means of subsistence that enabled communities to permanently occupy the grasslands of the Upper Wahgi Valley and, presumably, other locales.

10 The digging of drainage ditches during the late Holocene

The use of ditches to drain wetlands is a widespread agricultural practice among horticulturalists in the highlands and lowlands of New Guinea (Ballard 1995, 2017). The advent of ditches to drain wetlands for cultivation marks a major transformation in the nature and organisation of cultivation practices in the highlands. This technological innovation also precipitated, or was associated with, changes to people's relationship with land and with each other.

Drainage ditches in the highlands

Today, ditches are constructed in the highlands for a variety of purposes: to drain wetlands for cultivation; to delineate land boundaries across slopes; for defence around settlements; to keep pigs away from gardens and settlements; and to demarcate special places, such as burial grounds (Ballard 1995, 2001). Here the focus is on the early (pre-2000 cal BP) use of ditches, usually in interconnected networks, to drain wetlands for cultivation.

Modern ditches and ditch networks are of various alignment, planform and cross-sectional morphology (Ballard et al. 2013; Ballard 2017). Comparable variability is reflected in the archaeological record (Ballard 1995, 2001; Golson 1977c; Bayliss-Smith and Golson 1992a, 1992b, 1999; Denham 2005a; Bayliss-Smith 2007). Ditches are usually dug to create tiered drainage networks that are laid out in curvilinear, dendritic, trellis, rectilinear or gridded forms.

Although ditch networks require relatively high labour inputs to construct (e.g., Steensberg 1980: 87–95), the enhanced fertility of soils in wetlands enables greater yields and continuity of cultivation. Moreover, the availability of soil water for crops in wetlands ensures more reliable yields during times of climatic stress, such as the increased frequency of droughts in the highlands following intensification of El Niño Southern Oscillation (ENSO) events during the mid-to-late Holocene (Moy et al. 2002). Due to the combined benefits of higher yields and risk reduction, drained wetlands have become increasingly important and reliable foci of agricultural production.

The Tambul spade

The oldest archaeological evidence of ditch digging in the highlands is the Tambul spade and the ancient ditch from which it was collected (Figure 10.1). Golson and co-workers (Golson 1997a) describe a wooden, hastate-type spade recovered during modern drainage of Tambul High Altitude Experimental Station (HAES) in the Upper Kaugel Valley (see Figure 3.2). The term 'hastate' is a name given to a sub-group of paddle-shaped spades in the highlands (Powell 1974). Similar types of spade have been ethnographically documented among the Kapakau of the Paniai Lakes in western New Guinea (Ishige 1977: 103), where they were used for the maintenance, as opposed to the digging of ditches until the recent past. Powell et al. (1975: 13–14, also see Steensberg 1980: 95–97) document a range of other possible uses for this implement, including 'marking out garden plots, cutting surface turf and shaping and lifting soil out of planting holes for bananas, sugarcane and taro' (Golson 1997a: 162). The shallow feature from which the artefact was recovered suggests that it was used to work soil in a wetland context.

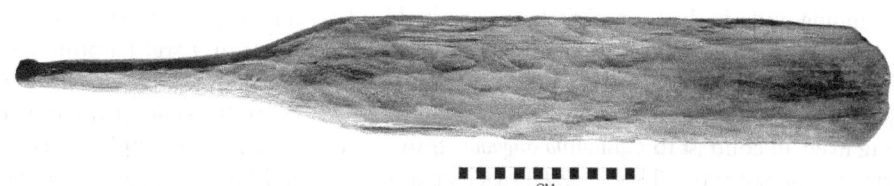

CM

Figure 10.1 Photograph of the Tambul spade

Note: Used with permission of Jack Golson, taken by Bob Cooper and Darren Boyd

Table 10.1 Radiocarbon dates for early ditches at wetland sites in the highlands

ANU #	Location	Context	Material	Radiocarbon Age (BP)	Calibrated Date (cal BP)[1]	%
Tambul (Golson 1997a: 155)						
ANU-2282	Ditch	basal fill	wood, hastate-type spade	3930 ± 80	4570–4150	.985
					4120–4090	.015
Kuk (Denham et al. 2003)						
OZF-240	Ditch 353	primary fill	pollen/charcoal, < 80 μm	3780 ± 50	4350–4330	.022
					4300–4060	.848
					4050–3980	.130
OZF-239	Ditch 353	primary fill	pollen/charcoal, < 80 μm	4000 ± 30	4530–4410	1.00
ANU-8055	Ditch 309	secondary fill, below Y ash	peaty clay < 600 μm	2650 ± 80	2950–2700	.800
					2670–2480	.200
ANU-8056	Ditch 309	secondary fill, below Y ash	organic clay < 500 μm	2480 ± 80	2730–2360	1.00
ANU-11185	Ditch 350	primary fill	diffuse charcoal	2890 ± 80	3260–2840	.981
					2820–2800	.019
Warrawau (Manton's) (Golson et al. 1967: 370)						
ANU-43	Ditch, Cutting M	basal fill	wood, digging stick	2300 ± 120	2710–2040	1.00
Minjigina (Powell 1970a: 174)[2]						
ANU-255	Cooking pit	fill	charcoal	2310 ± 90	2710–2630	.089
					2620–2560	.039
					2540–2110	.873
ANU-277	Ditch	fill	charcoal	2280 ± 90	2010–2020	.005
					2040–2500	.949
					2590–2610	.011
					2640–2700	.035
Haeapugua (Ballard 1995: C40)						
ANU-7800	Ditch, LOJ/a	basal fill	charcoal	2390 ± 230	2950–1880	1.00
Kana (Muke and Mandui 2003)						
ANU-9382	Feature I, Drain WD5	basal fill	bulk sediment	2970 ± 70	3340–2950	1.00
ANU-9487	Feature 1, Drain WD3	basal fill	gourd exocarp (*Benincasa hispida*)	2450 ± 200	2950–2000	1.00

Source: Denham 2005a: Tables 4 and 5

Notes:
1 All calibrations were undertaken to two sigma, Method B, Calib 4.1, IntCal98 atmospheric curve (Stuiver and Reimer 1993).
2 A date of 2280 ± 90 BP (ANU-277) obtained on the basal fill of a ditch at Minjigina is not considered to be representative of Phase 3, but has been interpreted to derive from a Phase 4 feature (Golson 1982a: 121; Golson and Steensberg 1985: 376).

A radiocarbon age on the spade indicates the spade dates to 4570–4090 cal BP (ANU 2282; Table 10.1). The age is problematic because the spade had been subject to freeze-drying and impregnation with polyethylene glycol 400 (Ambrose 1975, 1990; Golson 1997a: 157–159). Radiocarbon dating occurred after removal of hydrocarbons using a technique that had enabled complete decontamination of similarly treated Australian boomerangs (Ambrose 1975: 9–10). Remnant contamination must be considered a possibility given that the spade is c. 2000 years older than the next oldest wooden digging implement in New Guinea.

Additionally, the age of the spade is indistinguishable from a date obtained on a buried timberline in Tambul Swamp (ANU 1059A, 3950 ± 90 BP; Golson 1997a: 155). The near-identical ages might suggest that old wood had been dug up from the swamp and used to make the implement. Even though this possibility should not be fully discounted, ancient wood is unlikely to be strong enough for use as an agricultural implement. In the ethnographic past, people stored wooden digging sticks and other implements in standing water in the base of ditches in the highlands to aid conservation (Steensberg 1980: 99).

If the date of the spade is taken as an accurate estimate of its manufacture, it may still harbour an 'old wood' effect (following Schiffer 1987). The exact species of tree used to manufacture the spade is unknown, although it was noted as being possibly *Dacrydium* sp., *Podocarpus* sp., or *Phyllocladus* sp. (Golson 1997a: 160). Some of these species are relatively slow growing and long lived, in contrast to *Casuarina oligodon*, from which many digging implements in the Wahgi Valley were made in the more recent past (Powell 1982a: 29; Powell et al. 1975: 13, although a range of tree species are reported in Gorecki 1978; Powell 1974: 28, 1982a: 29). An old wood effect may slightly increase the antiquity of the Tambul spade relative to the Wahgi sites, although this is unlikely to be more than a few hundred years.

Taken at face value, and despite uncertainties, the Tambul evidence is highly suggestive of anthropic drainage of the wetland approximately at c. 4600–4100 years ago. The drainage of the wetland was likely a short-lived phenomenon, and later evidence for drainage has not been reported. The altitude of Tambul at 2170 m is beyond the range of many cultivars, especially bananas and yams (see Table 5.2; Bayliss-Smith 1985, 1988: 155; Golson 1997a: 145–146).

The findings from Tambul accord with Walker and Flenley's interpretation of anthropic forest clearance at 2500 m at Sirunki in Enga Province from c. 4500–3000 cal BP (inferred ages; Walker and Flenley 1979: 339–340). The degradation of primary forest to secondary forest, disturbance and open land taxa was inferred to reflect the altitudinal expansion of settlement based upon the exploitation of tree crops, hunting and the management of other useful plants.

The archaeological sequence of early ditches at Kuk

Extensive archaeological excavations were undertaken at Kuk in order to map and investigate early ditch networks (Figures 10.2 and 10.3). Ditches were also recorded in, and interpolated between, modern drain walls. The drainage network complexes, associated chronologies and artificial morphologies of these networks are summarised here (for detail see Denham 2003a: 216–262; Denham 2005a; Denham, Golson et al. 2004, 2017c). The investigations at Kuk are the most extensive and intensive of any archaeological site in the New Guinea region.

Chronology

Three broad chronological sub-phases have been established (Figure 10.4): early with a distinctive 'new grey clay' fill, mid-late with level or slightly dipping Y ash and black clay fills (< 10 cm) and late with steeply dipping Y ash and black clay fills (> 10 cm). The antiquity of early ditch networks at Kuk is based largely on tephrochronology; materials in the fills of only a handful of ditches have been directly radiocarbon dated. Most unarticulated features had a black clay primary fill with an overlying lens of Baglaga tephra (Y ash), which fell at c. 2650–1950 cal BP (see Table 3.4). The degree of dip of the ash was the only distinguishing characteristic. Even though degree of dip is known to be an unreliable indicator of age, in many cases there are no other criteria for chronostratigraphic grouping.

Direct radiocarbon dating of palaeochannels and ditches has proven problematic and hinders the disaggregation of the major sub-phases (see Table 10.1; Denham 2003a; Denham et al. 2003; Denham 2005a). However, an intensive dating project has been undertaken on the associated tephras – Baglaga (Y), Niupela (Np) and Kim (R) (see Table 3.4), which provides a framework for determining the antiquity of early drainage networks (Coulter et al. 2009; Sniderman et al. 2009). At Kuk, multiple early ditch networks and ditches post-date Kim tephra (R ash; 3980–3630 cal BP) and pre-date Baglaga tephra (Y ash; 2650–1950 cal BP). Based on radiocarbon dating (Denham 2005a), the earliest sub-phase ditches date to c. 4350–3980 cal BP and pre-date Kim tephra.

Ditch complexes

Five relatively early ditch complexes pre-date Baglaga deposition at Kuk. The extent, ages and articulations of these five complexes, as well as numerous isolated or 'floating' features, were not always determined, and it is likely that parts of the wetland margin were periodically drained throughout this period. The five ditch complexes are (Figure 10.2):

1   107 complex filled with a 'new grey clay' deposit;
2   353 complex with some features filled with 'new grey clay';
3   585 complex of ditches, some of which were filled with 'new grey clay';
4   203 complex of ditches filled with black clay and overlain by Baglaga (slightly dipping); and,
5   350/393 complexes filled with black clay and overlain by Baglaga (dipping).

Most palaeochannels at Kuk that are contemporary or articulated with early ditch networks have very different planforms and cross-sectional morphologies to older palaeochannels at Kuk (Figures 10.5 and 10.6). Palaeochannels 102, 103, 106 and 107 have relatively straight courses, sometimes with marked angular changes in direction between straight sections (102 and 106). The cross-sectional morphologies of these palaeochannels are more characteristic of the steep and straight sides found in ditches. By comparison, older palaeochannels that are interpreted to be natural, as well as one roughly contemporaneous one (channel 108), are shallower and wider, exhibit wavy or slightly sinuous courses, and have undergone extensive subaerial erosion (e.g., slumping) along the banks.

The earliest integrated ditch networks were associated with new grey clay deposits or channel 107 and date to c. 4350–3980 cal BP. All three complexes had rectilinear alignments (Figure 10.2): ditches 242 and 340 joined directly with

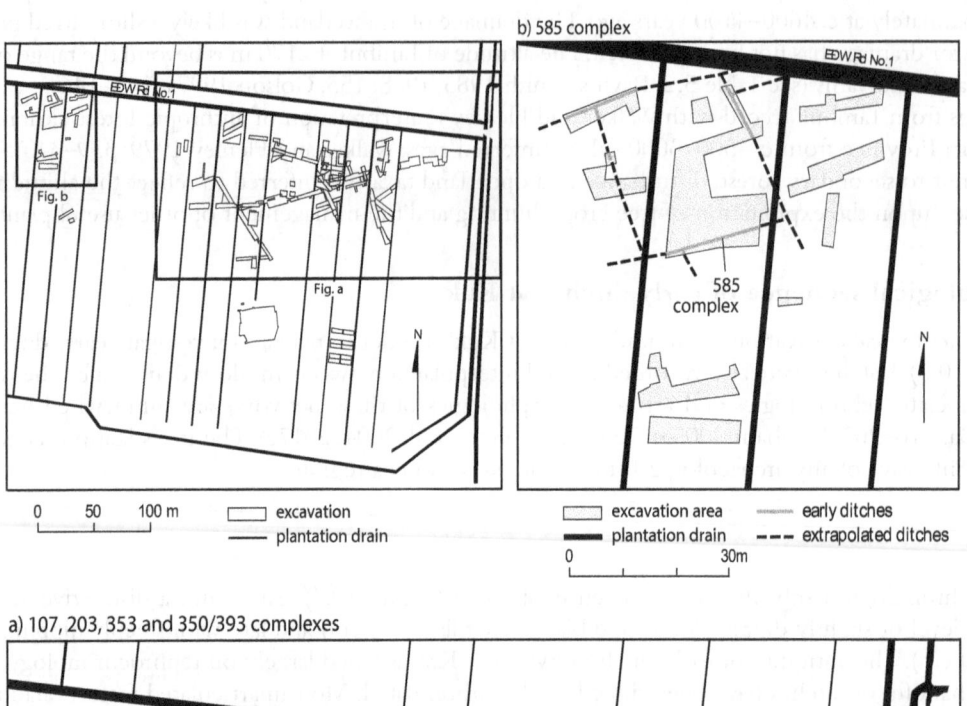

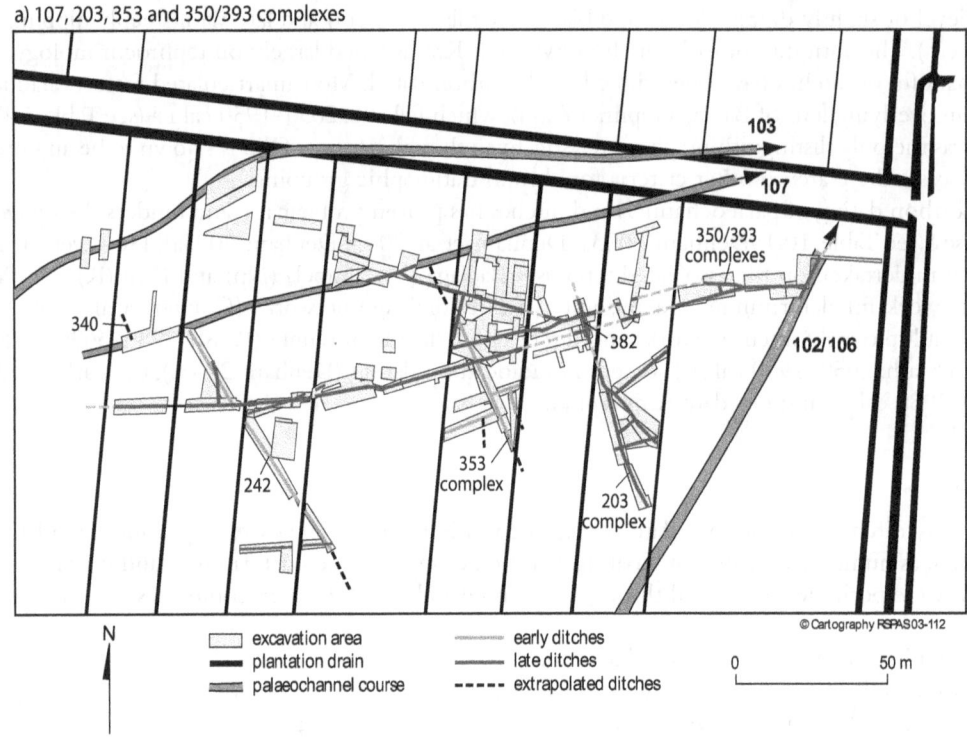

Figure 10.2 Early ditch networks exposed in archaeological excavations at Kuk Swamp

Source: Denham 2005a: Figure 4

channel 107, an integrated network of ditches articulated with 353, and an integrated network of ditches (585) delimited a possible enclosure. The three ditch complexes were inferred to be broadly contemporaneous on the basis of shared orientations, fill types and rectilinear design.

The network design of the 353 and 585 complexes differed. The 353 complex had a drainage hierarchy with three distinct levels and corresponding functions (Denham et al. 2017c: 228):

1   Ditch 353 was a major water conduit to the palaeochannel (presumed to be channel 107).
2   Ditches 225 and 513 were perpendicular tributaries to ditch 353, which served as major plot-dividing drains; and,

Figure 10.3 Photographs of archaeological excavations at Kuk Swamp exposing early ditches cross-cut by later ditch networks of varying antiquity

Note: Upper image showing excavations in 1975 (Source: Jack Golson, taken by Klim Gollan, 1975) and lower image showing excavations in 1999 (Source: Tim Denham, 1999)

3 Ditches 516 and 519 were slightly oblique tributaries to ditches 513 and 225, respectively, which served as minor plot-dividing drains.

In contrast, the 585 complex did not seem to have a defined drainage hierarchy, although only a limited area was exposed in excavation. Ditches of different sizes articulated to define an enclosed area. The enclosed area measured approximately 29 m northwest-southeast by 24 m southwest-northeast. This enclosure is the earliest drained plot identified in the New Guinea region.

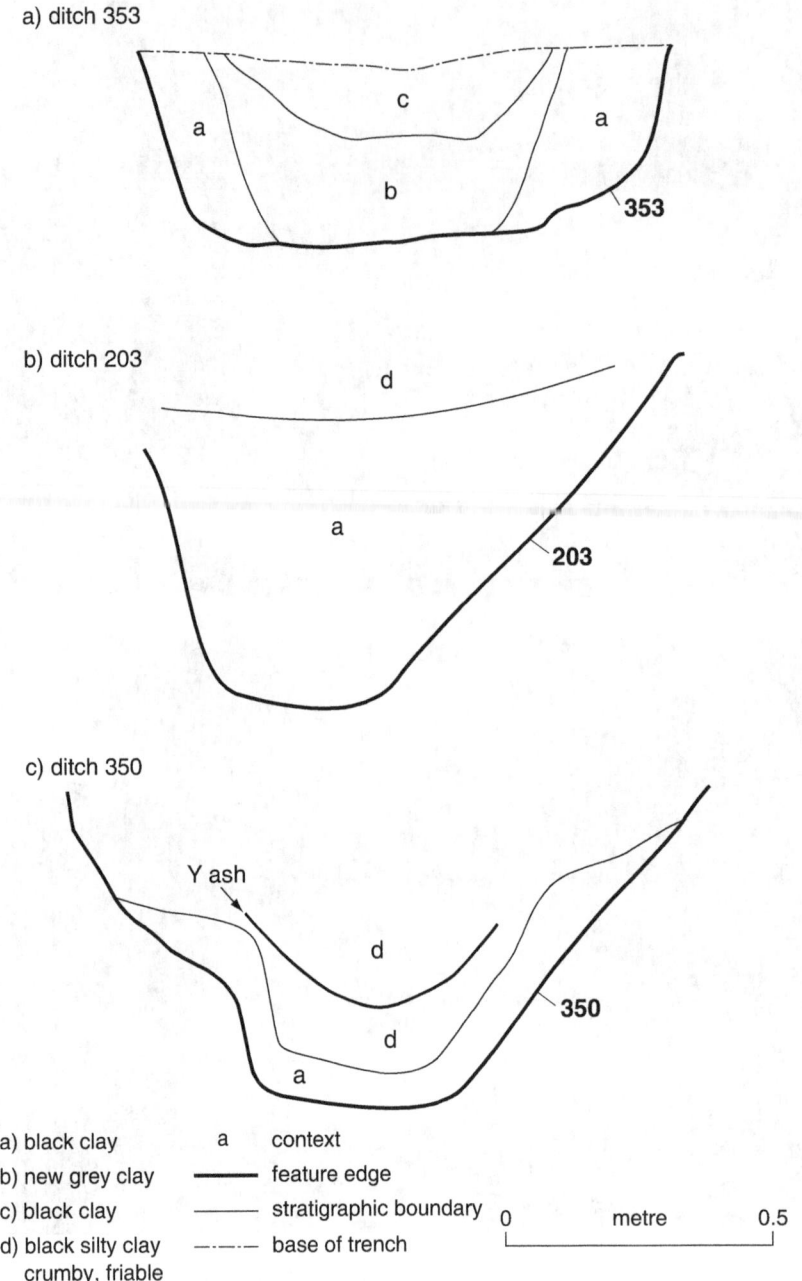

Figure 10.4 Section drawings illustrating different fill types for early ditches at Kuk Swamp

Note: early (ditch 353), mid-late (ditch 203) and late (ditch 350)

Source: Denham, Golson et al. 2004: Figure 17

The late sub-phase, characterised by the 203 and 350/393 complexes, dates to pre-2730–2360 cal BP. These ditches are overlain by Baglaga tephra. The spatial design of these ditch networks included offset and oblique (dendritic), perpendicular (rectilinear) and A-frame (triangular) junctions (see Figure 10.2). The fills were a black clay primary fill with overlying and steeply dipping Y ash.

The 350/393 complex was inferred to articulate with channel 106 and a remnant, post-abandonment depression of channel 107. An association between the 203 complex and a palaeochannel was not established. The forms of the 203 and 350/393 complex ditches were slightly different. The 203 complex ditches tended to be narrower, deeper and more

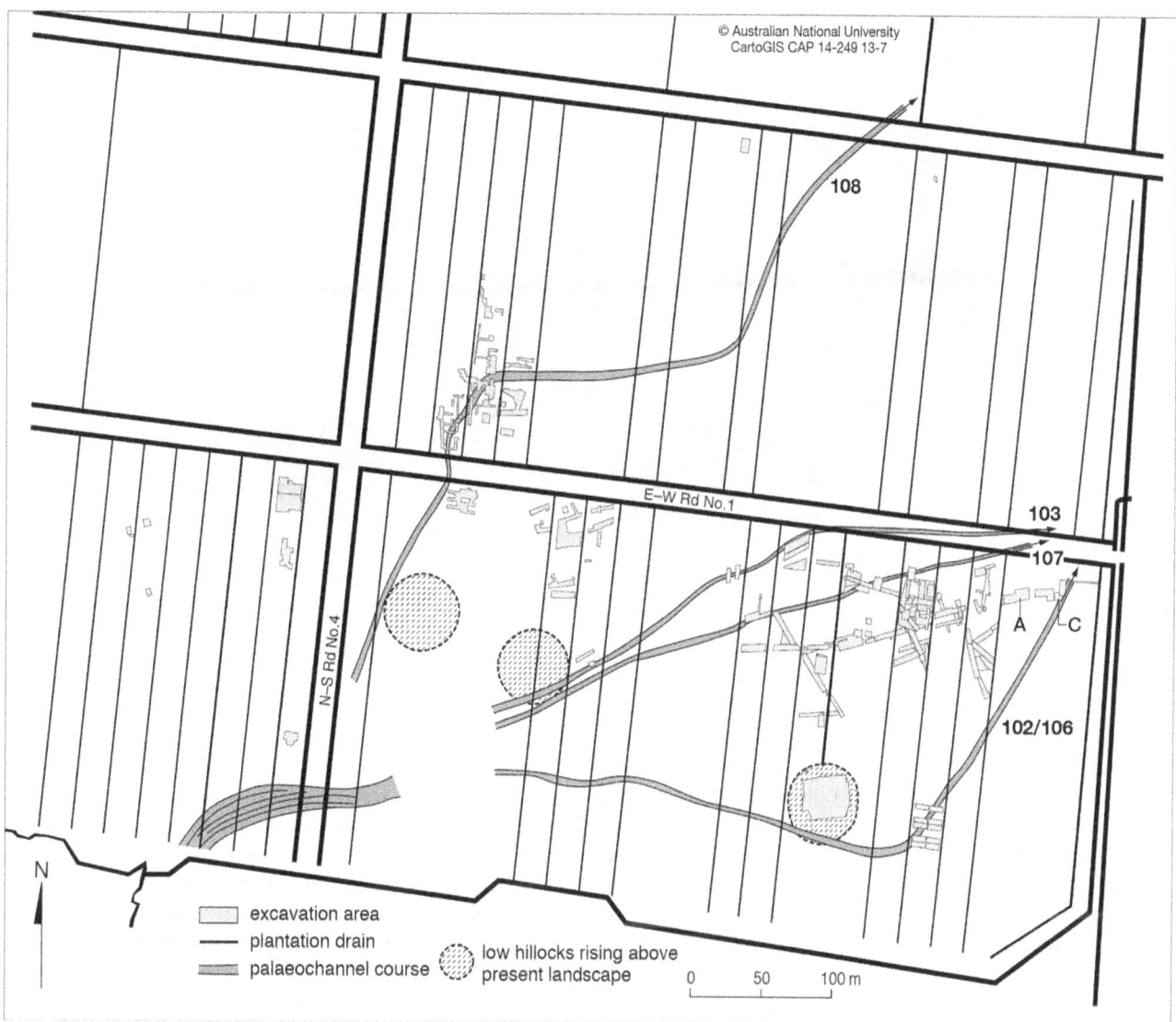

Figure 10.5 Plan of c. 4000–2500-year-old palaeochannels on the wetland margin at Kuk Swamp.

Note: Notice relatively straight courses and channels 102/106 and 107 articulate with ditch networks

Source: Denham et al. 2017c: Figure 13.7

angular, although not in all sections. The 350/393 complex ditches were generally wider with more rounded sides and bases (Denham et al. 2017c: 228). The differences in form may reflect the amount of time the ditches were in use prior to abandonment. A ditch open for only a short time would preserve signs of its original digging, that is, steep/vertical sides and angularity. In contrast, a ditch open for an extended period prior to abandonment would be expected to widen and become rounder as a result of subaerial erosion. Alternatively, the differences in form may reflect different digging techniques, styles and idiosyncratic preferences of individuals and communities.

The 203 complex is earlier and does not articulate with the 350/393 complex. Recutting and modification for the 203 and 350/393 complexes are indicated by the recutting of ditch 203 for a later ditch section (382), and the recutting of the 393 complex for ditch 350. Temporal and cultural continuity is implied by the persistence of the spatial structure between both complexes. The structural continuity reflects a similar form of network design, despite some variability in individual ditch form. The continuity of practices through time suggests that the 203 complex only slightly pre-dated the 350/393 complex, and that groups with a shared cultural tradition made both networks.

Figure 10.6 Photograph of Jack Golson recording the section of channel 107

Source: Jack Golson, taken by Klim Gollan, 1975

Artificiality

Cross-sectional morphologies and planform indicate that these linear features are artificial ditches. In cross-section, the morphologies of most ditches and palaeochannels were continuous across space, maintained smooth parallel edges in plan, were U-shaped or sub-rectangular with steep and smooth sides in cross-section and exhibited only the occasional evidence of within-channel recutting. The alignment of at least one ditch (393) did not respect the topography, suggesting it was dug for purposes other than drainage. In planform, individual features were predominantly straight with no evidence of lateral migration and were arranged in rectilinear and composite drainage networks comprised of T-junctions, A-frames and other angular junctions. Artificial palaeochannels functioned as carrier ditches within drainage networks that discharged normal and peak flows, provided outlets for tributary ditches and potentially acted as a reservoir during times of drought (after Castle et al. 1984: 77–79).

The early sub-phase, associated with the 107, 353 and 585 complexes, as well as mid-late interpolated drainage networks exhibited rectilinear patterns in which the ditches articulated at perpendicular or near perpendicular junctions. Trellis and rectangular drainage patterns usually form due to lithological control, and such a pattern would not be expected across the low-gradient margins of a wetland (Gregory and Walling 1973: 52). These ditch networks constitute a coaxial field system that exhibits a prevailing or dominant axis of orientation: '[M]ost of the field boundaries either follow this axis (axial boundaries) or run at right angles to it (transverse boundaries)' (Fleming 1987: 188; cf. 1989: 63).

The late sub-phase patterns, associated with the 350/393 and 203 complexes, exhibit characteristics of artificial and natural drainage networks. Several elements of both complexes are clearly artificial. First, the rectilinear network was composed of terminal T-junctions and straight ditches that traversed slightly undulating topography without deviation. Second, the triangular form of the 'A-frame' junctions would not be expected in non-anthropogenic drainage networks. The 'A-frames' in the 203 and 350/393 complexes are of regular form and composed of straight ditch sections that form isosceles triangles in which the internal angles of juncture are nearly equal. Such a form minimises the distance to be dug between two diverging ditches. Third, and taking into account the likely extent of differential shrinkage between organic and inorganic deposits at Kuk, the basal levels of ditch 393 do not indicate a unilinear direction of flow. Therefore, some ditches were not dug solely for drainage. Ditch 393 may have facilitated drainage across some lower-lying areas, whereas in other areas it served an alternative function, potentially as a boundary marker or as an area to plant water-tolerant crops, such as taro. Multiple functions have been noted for ethnographic ditches, or *gana*, at Lake Haeapugua that followed the same alignment across wetlands and adjacent dryland slopes (Ballard 1995: 95–98, 180–184, Table B13).

Other elements of the late sub-phase complexes suggest adaptations and innovations based on natural drainage networks and forms. First, the general spatial arrangement of the 350/393 complex resembles a dendritic drainage pattern with some modifications, for example, A-frame and T-junctions (after Gregory and Walling 1973: 52). Given the artificiality of this drainage system, the offset junction, for example, between 290 and 393, may have been an adaptation to serve a utilitarian purpose, perhaps to minimise erosion, scouring and backing up of water at a confluence (after Knighton 1998: 59). Second, the outfall of ditch 393 into ditch 327 over channel 107 indicates that a pre-existing natural depression on the ground surface was utilised for drainage.

The emergence of rectilinear field systems

> If nothing else, field systems are monuments to the confidence of their creators.
>
> (Fleming 1989: 74)

By c. 4350–3980 cal BP, rectilinear ditch networks were dug along the southern wetland margin at Kuk. The earliest complexes (107, 353 and 585) share alignments and represent a coaxial field network, with at least one enclosed plot being present and other areas forming perpendicularly aligned drainage networks. These ditches clearly differ in form from earlier horticultural practice in the highlands. Previous practices, especially those associated with wetlands, represent the generalised and piecemeal use of plots at different points in time; in a generative sense, they are irregular and potentially 'aggregate' (after Bradley 1978). In contrast, the early ditch networks at Kuk have defined directions of flow and demarcate whole areas of the landscape at a spatial scale greater than individual plots; they are regular, planned and 'cohesive' (after Bradley 1978). As Fleming (1989) noted for the reaves of Dartmoor, these ditch complexes are social constructions, which need not be read solely in terms of edaphic requirements, demographic causation and hydrological efficiency (see Ballard 1995; ; Bayliss-Smith and Golson 1999; Bayliss-Smith 2007).

The advent of ditches at Kuk is presaged by discrete curvilinear features, such as feature 504, which were probably dug to lower water tables locally and enable cultivation of adjacent land. These sinuous runnels are possible precursors to linear, water table control ditches. Subsequent drainage and the demarcation of space using ditch networks occurred shortly, perhaps only decades or a few hundred years, later than feature 504. The reasons for the adoption or innovation of ditching are unknown. They could include intractable water control problems that could not be adequately mitigated by carrier ditches alone, that is, palaeochannels, or an increased reliance of people on the wetlands within an increasingly resource-poor grassland and drier climate, or a combination of fluctuating water tables and the agronomic requirements of more intensive banana cultivation.

The reasons for field drain depths and varying network designs at Kuk are also unclear, namely were they to lower or maintain water table levels (after Trafford 1975: 7–8)? Ditches at different levels within the network hierarchy had different functions: shallow field ditches to intercept surface and subsurface runoff (interceptor ditches), deeper ditches between fields to manage the water table (water table control ditches) and larger ditches, or palaeochannels, to carry the water away from the wetland margin (carrier ditches) (after Castle et al. 1984: 77–79). For early drainage networks at Kuk, most ditches are considered to have functioned as water control ditches and the networks are assumed to articulate with palaeochannels, even though in many cases such articulations have not been demonstrated. The physical and biological requirements of prehistoric drainage were determined by numerous factors including slope, drainage basin hydrology, sediment load, soil characteristics and crop requirements; the social requirements are unknown.

The dispersal of ditch digging, 2750–2150 cal BP

Several other wetlands in the highlands exhibit ancient ditch networks of variable antiquity (Golson 1982a; Denham 2007a). These have only been archaeologically investigated in the Tari Basin and Wahgi Valley. Wooden digging implements have also been recovered and ancient ditches recorded at several other wetlands across the highlands, including the eastern highlands, yet these have not been systematically investigated. Where radiocarbon dating has occurred, the earliest finds post-date those from Kuk and Tambul. However, the synchronous occurrence of the first evidence of wetland drainage at multiple sites suggests a discrete period at c. 2750–2150 cal BP during which the drainage of wetlands using ditches spread, presumably together with various types of cultivation practice (Denham 2005a: 347–348; Denham et al. 2017c: 230–235).

Warrawau

Archaeological excavations by Jack Golson, Ron Lampert and Wallace Ambrose at the Manton's site on Warrawau Plantation, Upper Wahgi Valley, in 1966 were the first to determine the pre-Ipomoean antiquity of agriculture in the highlands. A pointed wooden digging stick associated with an ancient ditch was radiocarbon dated to 2710–2040 cal BP (ANU-43; Golson et al. 1967; Lampert 1967; Powell 1970a: 142–146; 1982b: 216; see Table 10.1). The digging stick was collected from the basal fill of a ditch at 115 cm below surface (Powell 1970a: 146). Although no additional information is presented, the provenance of this digging stick and its associations with an artificial ditch are secure. Powell interpreted this artefact and ditch – together with other digging implements, wooden stakes, ditches and stone axe-adzes of uncertain age – to be comparable to contemporary agricultural practices in the region (Powell 1970a: 145).

Minjigina

At Minjigina in the Upper Wahgi Valley, limited archaeological excavations by Ron Lampert in 1967 (Lampert 1970) and recording in plantation drains by Jocelyn Powell and others revealed 'digging sticks, ditches and an associated cooking pit' (Powell 1970a: 172–174, 1970b: 199). Radiocarbon dates obtained on charcoal from the cooking pit (ANU 255) and on the basal fill of a ditch (ANU-277) were statistically contemporaneous to each other and to the Warrawau date (Table 10.1).

The basal fill of a ditch was dated to 2700–2000 cal BP (ANU-277; Table 10.1); however, the dated material was interpreted to represent slumped material from the ditch walls as opposed to a primary deposit (Golson and Steensberg 1985: 376). The presence of tephra, inferred to be Olgaboli (Q ash) dipping into the ditch above a buried digging stick, suggested that the feature was more recent, namely, c. 2000–1200 years old (Golson 1982a: 121; Powell 1970a: Plate 37B). The wooden implement was noted as being 'a similar small diameter, pointed digging stick' (Powell 1974: 22) to that recovered and dated at Warrawau.

Charcoal collected from an adjacent cooking pit yielded a radiocarbon date of c. 2710–2110 cal BP (ANU-255; Table 10.1). Although not indicative of cultivation or drainage, the cooking pit suggests that the wetland was dry enough for people to dig a cooking pit there at this time. Today, people often make small cooking pits while cultivating their plots, as well as around settlements and at ritual and ceremonial sites. Similar types of activity could well be represented by the cooking pit at Minjigina. The definite presence of people and drier conditions locally at this time circumstantially corroborate the antiquity of the ditch.

Haeapugua

Chris Ballard undertook extensive archaeological and ethnohistorical investigations of wetland field systems in the Tari Basin during the early 1990s (Ballard 1995, 2001; Ballard et al. 2013). His research focussed on linking archaeological excavations to ethnography for the last few centuries, but he also uncovered earlier evidence of linear ditches at site LOJ in Haeapugua (Ballard 1995: C40–46). Although there were major problems with the interpretation of the lower stratigraphy at Ballard's sites (Ballard 1995: 195–200), one feature (LOJ/A) was clearly cut into light grey clay and filled with dark grey clay (Ballard 1995: C10: 40). The feature was linear in plan and was cut by a later ditch.

A radiocarbon date of 2950–1880 cal BP (ANU-7800; Table 10.1) obtained on dispersed charcoal collected from the base of the feature corresponded to a period of ditch digging at Kuk, Warrawau and Kana, as well as potentially at Minjigina. Ballard considered this to represent 'the earliest evidence for agriculture in the [Tari] region' and 'cultural continuity with the current Huli-speaking populations' (Ballard 1995: 213). He cross-correlated his archaeological findings to a nearby palynological record of vegetation history: 'a transitional zone towards the end of which the carbonised particle count and grass and sedge values increase[d] dramatically' (Ballard 1995: 213; cf. Haberle 1993: 214–215, 1998). Early ditched drainage networks at Haeapugua remain to be properly characterised because only one early feature has thus far been identified and dated.

Kana

In 1993 and 1994, John Muke and Herman Mandui (2003) undertook fieldwork at Kana in the Middle Wahgi Valley. The wetland had formed on an alluvial terrace and was being drained to create a coffee garden. Several anthropic features, including ditches, were recorded and sampled in the walls of the modern plantation drains. The antiquity of these archaeological features has been inferred on the basis of radiometric and tephrochronological data. Of most significance for an inter-site characterisation of early ditch networks in the highlands are two radiocarbon dates, one on a sediment sample from a feature fill and another on exocarp fragments of wax gourd (*Benincasa hispida*; Matthews 2003; see Table 10.1). The dates on both samples potentially correspond to the period during which ditches were first constructed in different regions of the highlands.

Muke and Mandui (2003) consider one of the sampled features (feature I) to be older than the radiocarbon date of 3340–2950 cal BP (ANU-9382, Table 10.1) obtained on a bulk sediment sample. Their interpretation of greater antiquity is based on the feature's curvilinear morphology, potentially being analogous to the inter-mound features at Kuk, as well as its inferred stratigraphic position beneath a tephra (Komun, or R) dating to 3980–3630 cal BP (Denham et al. 2003; Coulter et al. 2009). However, the morphology of the feature should not be used as a guide of antiquity, because curvilinear features are constructed between mounds and for various purposes in gardens today. Furthermore, the interpretation of the tephra overlying this feature, as well as others at Kana, is uncertain because the radiocarbon date is younger than the putative pre-Komun context from which it was obtained. The tephrochronological interpretations of Komun at Kana require reassessment and, consequently, the antiquity of feature I is uncertain.

In contrast, a radiocarbon date on wax gourd *(Benincasa hispida)* exocarp fragments is considered to be more secure, as it was obtained on a discrete sample of organic matter. The gourd exocarp was collected from the basal fill of a seemingly linear feature (as best could be determined in the drain walls; Muke and Mandui 2003) and dated to 2950–2000 cal BP (ANU-9487, Table 10.1). The sample provides a reliable, albeit imprecise indicator for the antiquity of wetland cultivation at Kana.

The gourd was identified on the basis of macrobotanical examination of exocarp fragments and seeds (Matthews 2003). The antiquity of wax gourd in New Guinea is uncertain; significantly, it is thought to originate and have been domesticated in Asia (Bates and Robinson 1995). Although water dispersal east of Wallace's Line should not be discounted, it was probably brought to New Guinea by people. Although of only minor significance as a food crop today (following Golson 2002), Matthews states (2003: 190):

> By broad ethnographic analogy, across many countries, we can suggest multiple possible uses for the wax gourd at Kana, as a food, a medicinal plant, or container. All parts of wax gourd, except the roots, have been reported as edible: mature and immature fruit, seeds, leaves, flowers, and vine tips.

It seems unlikely that the wetland field systems at Kana were constructed solely for the cultivation of wax gourd. Rather, these plots, like those at other wetlands, would have been polycultural and have incorporated a range of major and minor crops, as well as plants for medicinal, ritual and other purposes.

Additional technological considerations

On the basis of current archaeological evidence from Island Southeast Asia and Near Oceania, the use of ditches to drain wetlands in the highlands of New Guinea appears to be an indigenous innovation (Golson 1985; Denham 2005a). The technology of ditch digging could have originated outside the highlands, even though there is a current lack of comparably aged archaeological evidence for ditches and wooden digging implements elsewhere in the New Guinea vicinity. Irrespective of whether the innovation is of highlands derivation or was introduced from elsewhere, there are several technological aspects that require consideration.

The design, planning and organisation of drainage

The advent of ditch digging not only marks a technological innovation in terms of the practices involved, but it also represents a watershed in terms of how people were able to project technology onto the landscape. The design of a drainage network is not just about the mechanics of digging ditches, it also requires an ability to design and organise the construction of a multi-tiered drainage network. The scale of forethought required to envisage the design of such complex engineering projects exceeds anything undertaken previously by highland societies. Although the degree of political centralisation required to undertake irrigation and drainage projects is context specific and often overemphasised (Wittfogel 1957; Earle 1978) – given that many relatively egalitarian societies undertake such projects in New Guinea – high degrees of organisation and cooperation are required to coordinate labour during construction and maintenance, as well as to allocate land for cultivation. The construction of ancient and modern drainage networks in New Guinea clearly demonstrates that the organisation of labour for large-scale drainage projects can be undertaken through consensus, persuasion and mutual aid by relatively decentralised societies (Kropotkin 1902).

Wooden tools

Just as the construction of ditches marks a new technological innovation, so too does the invention of wooden tools to enable their construction and maintenance (Steensberg 1980; Golson and Steensberg 1985). Relatively simple wooden digging sticks were almost certainly used for millennia to forage for tubers, to dig and dibble, to construct mounds and so on. The advent of ditch digging required the adoption of a new tool kit to mark out alignments in the turf, dig ditches in wetlands, turn the soil, as well as to dredge and maintain ditch bases and walls, respectively. It is uncertain whether the precursor of wooden spades in the highlands was the digging stick, the canoe paddle or the tanged stone blade, or 'hoe' (see Figure 8.3; Christensen 1975b); an origin is open to speculation.

A range of different types of wooden implement have been described ethnographically and from archaeological contexts in the highlands. Various typologies have been proposed, with associated functions (Figure 10.7; also see Steensberg 1980: 95–97; Golson 1977b):

> At least five types of implement can be distinguished, namely the large spatulate blade (with blade tapering to handle or angled to it), the smaller paddle-shaped spade, the short handled hastate spade, the large and the small diameter pointed digging sticks.
>
> (Powell 1974: 21)

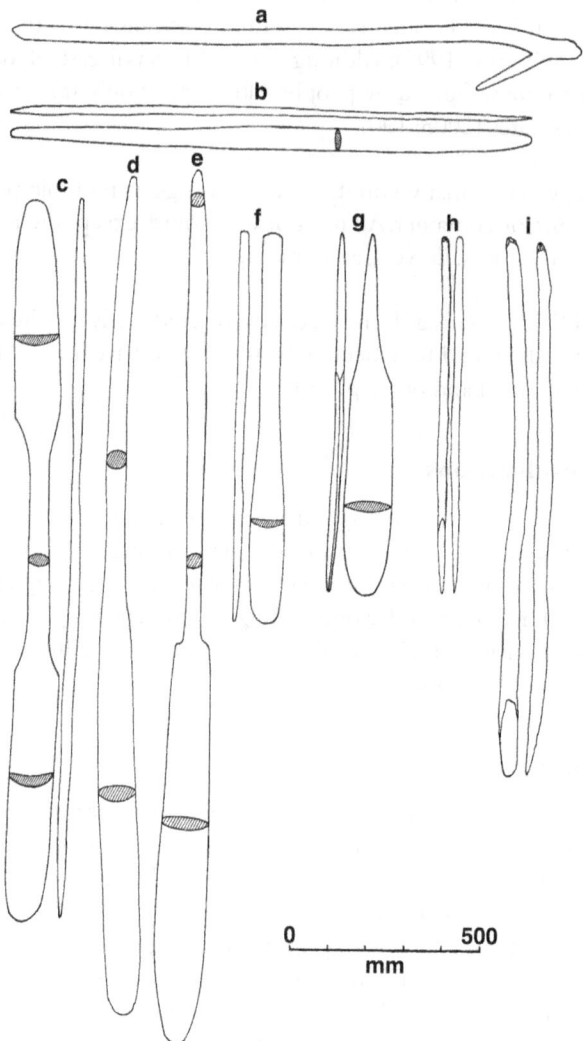

Figure 10.7 Major types of wooden agricultural implements in the highlands

Note: (a) long-handled hook, (b) 'bush-knife', (c) double-headed spade, (d–g) single-headed spade, (h–i) digging sticks

Source: Golson 2017b: Figure 19.2

Although originally proposed for the Mount Hagen region, Gorecki (1978: 185) suggested that this typology could be expanded slightly, with the addition of the double-ended spade, and extended to the whole of the highlands.

Only a few of the implements collected from across the highlands come from secure archaeological contexts or have been directly radiocarbon dated (Powell 1974; Powell et al. 1975; Gorecki 1978; Steensberg 1980: 95–100; Golson and Steensberg 1985). As discussed earlier, the Tambul spade is the oldest wooden digging implement in New Guinea. A digging stick radiocarbon dated at Warrawau is the next oldest; a similar implement was collected from near the base of a ditch at Minjigina, but is of uncertain age. These are the only wooden digging implements with relatively secure chronostratigraphic contexts that potentially pre-date 2000 years in the highlands. Numerous other wooden artefacts have been recovered from archaeological ditches at wetlands in the highlands; most are undated or date to within the last few hundred years (e.g., Powell 1970a, 1974, 1982a; Gorecki 1978; Golson 2017b).

The wooden digging implements recovered from archaeological contexts at Kuk and Warrawau (Powell 1982a: Table 10.2) have been identified to a wide variety of different species. Wooden digging sticks from these sites comprise single, undated specimens of *Acalypha, Dodonaea, Phyllanthus*, Podocarpaceae and Rutaceae, as well as five examples made of *Casuarina* approximately dated to 100–2500 years ago. Wooden spades comprise single, undated specimens of black palm, *Casuarina, Dodonaea, Hydrocarpus, Neonauclea* and *Nothofagus*. In another table of wooden digging tools from wetland archaeological sites in the Mount Hagen region, which are all undated, the majority of spades (13 of 17) and digging sticks (5 of 8) are *Casuarina* sp. (Powell 1974: Table 10.3). On the basis of the antiquity of *Casuarina* tree fallowing in the highlands, most wooden digging implements of this genus are probably less than 1200 years old.

It is unfortunate that only a small sample of wooden digging implements from highland contexts has been radiocarbon dated. As Powell (1974: 22) noted over 40 years ago:

> While further direct datings of implements are required, the data outlined above may suggest early specialisation of gardening implements followed by long retention with little change.

Cultivation in ditches

Drainage ditches can often be perceived as 'negative' features, in which ditches are not considered to be productive, but rather they enhance the productivity of adjacent land. In the highlands' context, ditches can also be considered 'positive', or highly productive features that expand the diversity of environments, habitats and niches for planting. Ditches plausibly increase the amount of available land for wetland and dryland planting. Before the introduction of the sweet potato, ditches may have been the foci of cultivation in some areas of the highlands, because ditches could have enabled the wet cultivation of taro, with a diversity of other crops being grown on the intervening drained plots. At this time, taro together with bananas and some yams are inferred to have been the primary staples of highland cultivation.

Planting seed in a vegetative world

No significant food plants enter the archaeobotanical record with the advent of ditches. Traditional crop plants were already present in highland valleys. Additionally, only limited archaeobotanical investigations have occurred within contexts dating to the last 4000 years.

The presence of wax gourd at Kana potentially marks the earliest definitive and securely dated evidence for planting from seed in the highlands (although consider Whistler 1990). Earlier domesticated forms of some crops may have been reproduced from seed, such as bananas, sugarcane, taro and yams. These crops are usually vegetatively propagated in the New Guinea context today and genetic studies suggest long periods of clonal reproduction in the past, albeit with some sexual reproduction (Grivet et al. 2004; Lebot et al. 2004; Malapa et al. 2005; Perrier et al. 2011). Although non-specific gourd remains have been reported at Warrawau from comparably aged or potentially much older contexts (Golson 2002), these have not been securely identified or directly dated, making any interpretation problematic. By contrast, the wax gourd at Kana has been securely identified and directly dated. Today, bottle gourd and wax gourd are both planted from seed in the highlands (French 1986).

The potential planting of wax gourd from seed around c. 3000–2000 years ago represents another addition to the repertoire of cultivation practices in the highlands. Prior to this, plants were most likely cultivated vegetatively. The wax gourd find in an agricultural context at Kana indicates seed-based cultivation in a largely vegecultural world.

Agricultural innovation and transfer of practical knowledge

As discussed with respect to mound cultivation, the initial innovation of ditch digging is much earlier than its widespread adoption. The earliest evidence for the digging of linear ditches is broadly synchronous at Tambul and Kuk, being

4570–4090 cal BP (ANU 2282) and c. 4530–3980 cal BP (OZF-239 and OZF-240), respectively. Although multiple periods of subsequent ditched drainage have been documented at Kuk, the practice does not become more widespread until 2750–2150 cal BP (Table 10.1). The first traces of ditch digging occur at Haeapugua, Kana and Warrawau during this period, as well as potentially at Minjigina, where radiocarbon dates derived from a cooking pit and a ditch could suggest contemporaneous cultivation of the wetland. The synchronicity of these finds is more than coincidental and marks a major shift in the nature of cultivation across a broad region of the highlands.

Before the 2750–2150 cal BP period, the use of ditches to drain wetlands was a highly localised phenomenon, perhaps practised only by a few groups inhabiting the proximal Upper Kaugel and Upper Waghi Valleys, although it potentially occurred in adjacent valleys. These neighbouring groups likely shared various cultural traditions and probably engaged in exchange relationships, such as marriage. If marriage relationships were similar to contemporary ones in this part of the highlands, namely they were patrilocal and group identity was putatively patrilineal (e.g., Strathern 1972), then women were exchanged between groups. Through intermarriage, as well as through the temporary movement of men to visit or stay with relatives and the more permanent incorporation of non-agnates following famine, migration or warfare, adjacent groups would have come to share knowledge of how to dig ditches in order to drain wetlands for cultivation.

Around 2750–2150 cal BP, social relationships and exchanges in this part of the highlands changed. Although climatic, demographic and environmental contexts for the expansion of ditched drainage are significant (consider Bowers 1968), the relatively sudden expansion and transfer of practical knowledge, or the 'know-how', regarding the drainage of wetlands for cultivation was socially mediated; it did not occur in a social vacuum. Whereas previously only groups living within a restricted region drained wetlands for cultivation, now groups inhabiting a much broader area, minimally encompassing the Tari Basin and the Upper and Middle Wahgi Valley, as well as further afield, acquired that practical knowledge.

Whereas generalised cultivation practices, such as mounding, were widespread and occurred in wetland and 'dryland' (adjacent slopes) environments, the inception of ditch digging was a more specialised practice that enabled drainage and cultivation of wetlands. Although the ultimate cause for the dispersal of ditch digging and the expansion of social interaction is open to speculation, the transfer of this practical knowledge was experiential; namely, it could have been affected only by people who had undertaken or witnessed the practices involved – whether men or women. Only people who were familiar with the organisation of labour, digging of ditches in wetlands, design of drainage networks, manufacture of wooden tools and cultivation of wetland soils would have directly, or personally, introduced these practices to another group inhabiting an adjacent valley or region. These practices could not have been transferred between groups inhabiting wetlands as 'ideas' or 'things'; rather, they would have been transferred by people with practical experience of digging them.

As opposed to generalised cultivation practices, such as mounding and plot preparation, which could readily be adopted by neighbouring groups, the transfer of ditched drainage would have occurred between groups inhabiting wetlands. The transfer of practical knowledge to drain wetlands along the floor of the Upper and Middle Wahgi Valley – from Kana to Minjigina – is readily envisaged because there is a string of almost inter-visible wetlands that were plausibly interconnected socially. Another mechanism, other than nearest neighbour analysis, is needed to account for the relatively synchronous transfer of practical knowledge to the Tari Basin and potentially other areas further afield.

The expansion of the social world at this time could represent the realignment or dissolution of long-held political alliances and enmities. Only localised expansion in the transfer of practical, or experiential, knowledge had occurred previously. To affect the transfer of practical knowledge over a much larger area suggests that the scale and geographical connectivity of social interaction expanded greatly at this time. Plausibly as a result of demographic growth, larger groups came into more regular contact. Other than warfare, these groups would have needed political and social practices in order to regulate increasing contact. It seems plausible that the mechanisms developed to moderate inter-group interaction would include the formalisation of various types of exchange relationships, including marriage, that brought disparate groups into more regular contact.

People from different social, cultural and linguistic groups have probably always come together for extended periods in order to exploit various resources, whether for hunting or the gathering of wild plant resources. Today, groups from across the southern flanks of the highlands still come together for several weeks to exploit abundant *karuka* (*Pandanus brosimos/ iwen/julianettii* complex) on Mount Au (fieldwork observation, Karimui, 2008). However, the nature of social life may have changed around 2750–2150 cal BP. As populations grew, more formal social events may have emerged that enabled groups to come together for exchange, such as precursors to the Hagen *moka* (Strathern 1971); for shared raiding and warfare; and ritual activities and ceremonies, including funerals, *singsing*, brideprice payments and marriage ceremonies. These types of social event entailed extended networks of alliance and enmity that brought formerly disparate groups together. Social relationships fostered and enabled the transfer of ideas, things and people – together with their practical knowledge.

Social transformations in the highlands

The agrarian changes detailed earlier, in terms of ditched drainage to create wetland plots and the transfer of associated practical knowledge, are contemporary with major transformations in the archaeology of highland societies. As discussed with respect to the emergence of institutions to regulate inter-group interaction, these transformations had a range of social implications.

Ditches and sedentism

Leaving claims of Pleistocene house structures to one side, several open sites in the highlands date to c. 4800–2500 cal BP and correspond in age, albeit broadly, to the earliest ditch networks (Table 3.1; Figure 10.8). The open sites of NFB, NGG and NGH consist of hearths and artefact scatters (Watson and Cole 1977), whereas Wañelek exhibits a variety of structural elements potentially representing repeated occupation (S. Bulmer 1977a, 1991; Gaffney, Summerhayes et al. 2015; Denham 2016a). On the basis of current archaeological evidence, and excluding Pleistocene-aged claims for NFX and Wañelek, early house structures do not pre-date c. 4800 years ago in the highlands and none of these are located close to drained wetlands. Only more recent habitations have been excavated or documented adjacent to several cultivated wetlands in different parts of the highlands (Harris 1977; Gorecki 1982; Ballard 1995; Lewis et al. 2016; Golson et al. 2017).

Despite limited archaeological evidence, there is a broad synchronicity between the antiquities of early house building and early ditched drainage of wetlands in highland New Guinea. The rise of group territoriality and increased sedentism are likely to accompany the advent of more permanent ditched field systems in wetlands, as well as more intensive forms of dryland cultivation. The investments of labour fixed in drainage networks and the requirement of repeated maintenance suggest that people would have needed to live nearby on a more permanent basis. The potential shift to more sedentary living is implied by the decreased use of caves and rockshelters in the highlands from c. 4000 years ago (Sutton et al. 2009).

Group identity, territoriality and inscription

Ditches reflected and inscribed in the landscape, or potentially precipitated, a new relationship between people and land. In turn, this relationship fostered novel group and individual identities, or relationships between people, manifest spatially through group territoriality.

Figure 10.8 Photograph of traditional turtle-backed house, Lower Jimi Valley

Source: Tim Denham, 1990

Claims to land across the interior of New Guinea today are to varying degrees based on a combination of inheritance and use. Groups lay claim to traditional territories, even though these have usually changed and shifted through time. Within groups, an individual is rarely able to claim absolute ownership; rather, any claim to land is based on use demonstrated through clearing, cultivating and gathering.

As forms of cultivation changed in the highlands in the past – from swidden to mound to ditch – more fluid socio-spatial relationships would have become increasingly 'fixed' in space and time. People cultivated the same plot for longer. The extended temporal duration of cultivation would have fostered an increasing connection between groups and the places they inhabited. At the same time, certain types of land, such as wetlands, became more important foci for cultivation and presumably settlement.

Under forms of swidden cultivation, plots in the rainforest are often cultivated for only a few years, even though perennials can be revisited, tended and harvested for prolonged periods after plot abandonment. The adoption of mound cultivation, which seems to have been an adaptation to grasslands, enabled more prolonged cultivation of the same plot for several years prior to abandonment. As people cultivated plots for longer, their attachment to that land would have become stronger. The advent of drainage networks takes the temporal duration of cultivation and attachment to place to a different socio-spatial level. Drainage networks can have an inter-generational lifespan.

Individual plots can be cultivated for extended periods before being left fallow; a mosaic of plots in different stages of cultivation and fallow can exist within a drained wetland at any given time (Ballard 1995, 2017). People maintain, rejuvenate and realign drainage networks through time. Ditches and drainage networks are clearly marked on the land and they have a physical persistence, or temporal duration, that is much greater than under swidden cultivation and mound cultivation. As such, drainage networks inscribe and demarcate social relationships on the land. Through the acts of digging, cultivation and maintenance, named individuals and groups become associated with specific ditches, plots and areas of a wetland. This physical connection to the land for food production has a socio–spatial correlate; it creates an increasing attachment to place and sense of territoriality.

It can be envisaged that the ways in which ditches and drainage networks demarcated social space in wetlands were eventually transferred to the drylands, or adjacent valley slopes. The necessity was not agronomic, but was stimulated by a need for groups to inscribe their identity in the landscape. Ballard (1995) has described how *gana* of the Huli in the Tari Basin literally run out of the wetlands and up the adjacent valley slopes without deviation.

Although the advent of ditches can be entwined with the development of group identity and territoriality, these were probably localised phenomena in the highland interior of New Guinea before at least 2000 years ago. Even in the recent ethnographic past, groups in the interior had a range of different relationships to the land, both in terms of their cultivation practices and in terms of social space. Additionally, the widespread occurrence and intensification of these social transformations is more characteristic of the last few hundred years following the adoption of pig husbandry and intensive sweet potato horticulture (following Modjeska 1982; Golson and Gardner 1990).

Some thoughts on timing and causation

The advent of ditches in the highlands of New Guinea, around 4500–4000 cal BP fits chronologically within the framework of the c. 4200 cal BP aridity event and the mid-Holocene intensification of El Niño Southern Oscillation (ENSO) events. The 4200 cal BP event has been traced across Africa, Arabia, India, Southeast Asia and East Asia (Ruan et al. 2016), as well as more globally (Booth et al. 2005); it has been implicated in numerous cultural impacts, including the demise of the Harappan civilisation in the Indus Valley and the Bronze Age in the eastern Mediterranean (Staubwasser et al. 2003; Berkelhammer et al. 2013). However, the extent and severity of this aridity event in the southern hemisphere is unclear (Haberle and David 2004).

ENSO events are considered to have increased in magnitude and frequency over the last several thousand years, although precise dates vary (Moy et al. 2002). Major El Niño events are associated with drier climates and periodic droughts in the highlands of New Guinea, as documented in recent decades (e.g., Allen 2000). Within this climatic context, people may have increasingly focussed upon wetlands for cultivation, given that they provide reliable environments for crop growth during droughts. This in turn may have facilitated the initiation of more intensive cultivation of the wetlands, such as ditched drainage.

At the same time as these climatic events, the lower montane forest environments in several inter-montane valleys were increasingly degraded to disturbed and secondary forest, and grassland from at least 4500–4000 years ago (Flenley 1979; Walker and Flenley 1979; Powell 1982b: 224; Corlett 1984; Haberle 1993; Hope and Haberle 2005). In some places within the Upper Wahgi Valley there is limited forest recovery, such as at Warrawau after 4000 cal BP (Powell et al. 1975: 43; Powell 1982b: 218), whereas at others there is localised clearance, such as at Kindeng between 3300–3000 cal BP (Powell 1982b: 223). Forest recovery in the Upper Wahgi Valley was not dramatic (Denham and Haberle 2008); the following c.4000–2000 cal BP period is best referred to as stabilisation of a disturbed environment based on evidence at Ambra Crater (Sniderman et al. 2009), Lake Ambra (Powell 1981), Kuk (Haberle et al. 2012) and Draepi-Minjigina (Powell 1970a: 165–186).

The degradation of these highland environments plausibly reflects a combination of human and climatic causes. Certainly, the vegetation communities are characteristic of agricultural landscapes in the highlands today (Haberle 2003). However, other ways in which people used these climatically sensitised montane landscapes after 4500 cal BP, especially burning, may have had an increased effect on vegetation – both in terms of disturbing forests and inhibiting forest recovery. The cumulative effects of these transformations could readily account for the extension and maintenance of grasslands, which together with drier climates, would have added to a local reliance on wetlands as centres of food production in deforested landscapes. Significantly, ditched drainage occurred within degraded landscapes in the Tambul vicinity (Walker and Flenley 1979), at Kuk (Denham et al. 2003; Denham, Haberle et al. 2004; Denham, Sniderman et al. 2009), at Warrawau and Minjigina (Powell 1970a) and at Haeapagua (Haberle 1993). There are no suitable palaeoecological records from the Kana vicinity to enable comparison.

Climate change and environmental transformation may provide contexts for the initial drainage of wetlands around 4500–4000 years ago. However, these contexts do not account for why and how ditch digging arose as an innovation. The archaeological finds from the highlands of New Guinea have no contemporaries or precursors in lowland New Guinea (Swadling and Hope 1992), in Island Melanesia (cf. Leavesley and Troitzsch 2007), or to the west in the Indo-Malaysian Archipelago (Golson 1985), although few investigations have been conducted in suitable or sufficiently stable environments to detect them. Certainly the advent of ditches pre-dates Lapita pottery in the Bismarck Archipelago at 3470–3250 cal BP (Denham et al. 2012), which has long been associated with the arrival of new agricultural practices there (Kirch 1997; Spriggs 1997; cf. Specht et al. 2014), as well as the introduction of pottery to the highlands around 3000 cal BP (Gaffney, Ford et al. 2015; Huff 2016). On the basis of current knowledge, ditch digging and the construction of rectilinear field systems are indigenous innovations on New Guinea.

11 Later innovations, introductions and adoptions

The sequential history of agricultural emergence and transformation for the highlands of New Guinea presented thus far has considered major technological innovations associated with early agricultural development. Agricultural history is viewed as an expanding repertoire of practices that coexist to varying degrees and in varying combinations across social space and through time. This perspective is justified because constituent practices and forms of plant exploitation do not occur just within a set period, only to then be abandoned; rather, they persist through time and are still practised by people living in the highlands today.

In this chapter, the multidisciplinary evidence for a series of later technological innovations and introductions associated with the continuing development of agriculture in the highlands is discussed. These transformations are discussed in approximate chronological order, although many overlap. As time depth decreases towards the present, the reliance on archaeological and palaeoecological data gives way to a greater reliance on ethnography and ethnohistory. Importantly, the timing of an innovation or introduction may be much earlier and, potentially, much less significant than the timing of its widespread adoption.

Tillage

Some contemporary cultivation practices in New Guinea involve minimal disruption of the soil, such as dibbling and digging for tubers (Chapter 4; Bourke 2001). Similar types of practice are implicated in the earliest forms of plant exploitation and cultivation in the highlands. Digging for tuberous plants during the late Pleistocene and early Holocene resulted in limited turning and admixture of the soil. Planting using dibbling sticks is likely under the earliest forms of swidden cultivation during the early Holocene. Subsequent practices imply a form of tillage for the construction of mounds and for the digging of ditches, which entails the spreading of excavated material on adjacent land surfaces (Steensberg 1980).

Golson (1977c: 620–621) suggested that the early ditch networks at Kuk and elsewhere were abandoned around 2000 years ago following the widespread adoption of soil tillage practices on slopes around cultivated wetlands. The resultant increased efficiency of dryland agriculture was considered to have removed the need to continue labour-intensive drainage of the wetland. Tillage enhanced the fertility of the dryland soils by aeration and the physical comminution of grass root mats. The adoption of tillage was inferred from a shift in sedimentary regime at Kuk from massively structured, black clay to spheroidal soil aggregates (Bayliss-Smith and Golson 1992b: 3; Golson 1977c: 620–621, 1977d: 49–50, 1982a: 121–122). Similarly, Hughes interpreted increased erosion rates from 2300 years ago at Manim 2 in the Wurup Valley to represent the commencement of grassland tillage on adjacent slopes (Hughes 1985: 400).

An alternative interpretation of the stratigraphy, in particular the formation of the spheroidal aggregates (crumb and granular structures) at Kuk, is advocated here. Golson interpreted the occurrence of a buried soil aggregate layer at Kuk to represent deposition on the wetland following soil erosion in the catchment, which in turn resulted from the adoption of tillage there (Golson 1977c: 620–621). Such an interpretation is unlikely because soil aggregates are not generally well preserved or transported any distance in flowing water (Tharp 1984); namely, they tend to disaggregate when eroded and are deposited across floodplains and wetlands as sediments not aggregates (Stoops 1989: 99; Denham 2003a: 282–283). Consequently, the layer of spheroidal aggregates at Kuk represents *in situ* soil formation and is associated with high biological activity in the A horizon of a palaeosol (Denham and Grono 2017).

Tillage occurred with the construction of mounds for planting, as well as during ditch construction. Subsequently, tillage was variably employed in dryland and wetland cultivation practices. Tillage was a generalised practice that was not implicated in shifts between dryland and wetland cultivation through time.

Formalisation of ditch networks

Although the earliest ditch networks date to c. 4400–4000 cal BP, as witnessed through finds at Tambul and Kuk, ditched drainage for the next 1500 years or so is limited to Kuk. Around 2750–2150 cal BP ditched drainage expanded to several other wetlands in the Upper Wahgi Valley and to other valleys in the central highlands. The geographical distribution of these drainage networks almost certainly reflects sampling biases, because ditch networks pre-dating 2000 years ago have been found in most wetlands subject to archaeological excavation. Only cursory archaeological investigations have occurred in wetlands in western New Guinea or in the eastern highlands (Golson 1982a; Denham 2007a; Ballard et al. 2013; Wright et al. 2013).

Although rectilinear components have been documented in early drainage networks at Kuk, the majority of networks identified through archaeological excavation in the central highlands that post-date 2000 cal BP are rectilinear, or gridded. In successive, reconstructed drainage networks at Kuk, the alignments, extent and size of enclosed plots vary, but all are based on a grid design (Figure 11.1; Bayliss-Smith and Golson 1992a, 1992b, 1999; Bayliss-Smith 2007). Having said

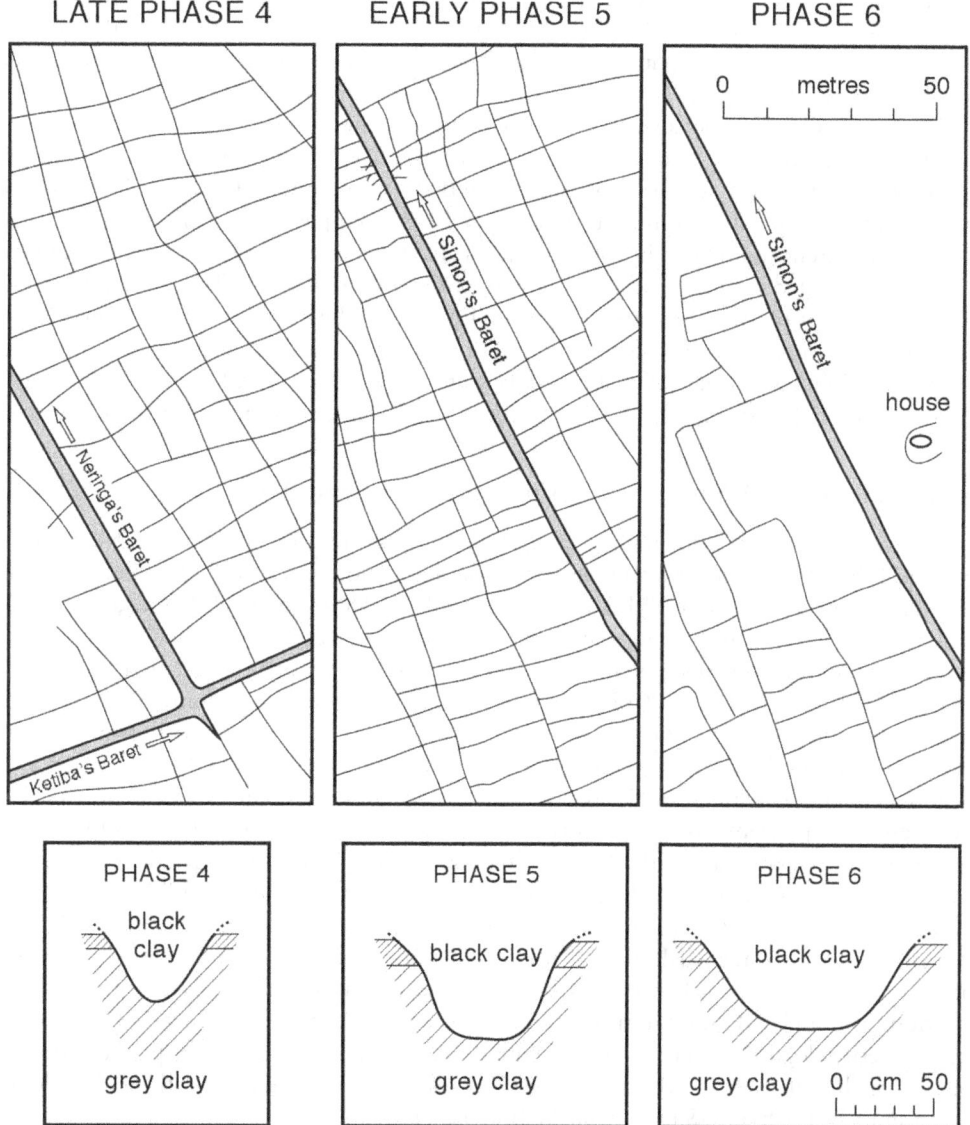

Figure 11.1 Plans and characteristic sections of gridded ditch network at Kuk Swamp

Note: Phase 4 = 2000–1230/970 cal BP. Phase 5 = 700–c. 290 cal BP and Phase 6 = 250–50 cal BP

Source: Bayliss-Smith 2007: Figure 2

this, many networks elsewhere in the New Guinea highlands and lowlands have retained curvilinear components into the present (Ballard 2017).

Various rationales for the systematisation of drainage networks through time have been proposed (Bayliss-Smith 2007: 132; also see Bayliss-Smith et al. 2017a-c):

- ditches as a reflection of economic rationality;
- ditches as the outcome of social inequality;
- ditches as a response to population crisis;
- ditches as an investment in future security;
- ditches as the symbol of property in the landscape; and
- ditches as an adaptation to climate change.

Different rationales reflect the currency of different interpretative frameworks at different points in time, rather than anything fundamental in the archaeological evidence.

Within these ancient field systems, ditches demarcate quadrangular plots used for food production. Fence alignments adjacent to ditches at Kuk date to the last 800 years (Sutton et al. 2009; Bayliss-Smith et al. 2017c). These fences would have become necessary only after the introduction of pigs *(Sus scrofa)*, namely, to keep them out of cultivated plots and to potentially keep them inside fallow plots.

The emphasis and reliance upon wetland plots have shifted over the last 4500 years in response to various factors, including droughts triggered by more frequent and intense ENSO events (Moy et al. 2002) and longer-term climatic trends (Haberle and David 2004), crises in food production in the dryland sphere (Golson 1977c) and episodic abandonment following warfare, as recorded in ethnohistories (Ketan 1998). The periodic abandonment and redigging of ditch networks may only be apparent and reflect the intensity and spatial variability of usage of wetlands and areas within them, relative to the areas investigated archaeologically (following Gorecki 1979 and Ballard 2001). Where present, wetlands are likely to have formed part of the agricultural repertoire for extended periods in conjunction with the cultivation of dryland plots; people maintained plots on valley slopes and in wetlands to cultivate plants with different edaphic requirements, as well as a risk minimisation strategy (following Bowers 1968; Ballard 1995). Rather than wetlands and drylands being viewed as different or alternating spheres of agricultural production, they are complementary and co-occurring.

The introduction of animal domesticates from Island Southeast Asia

The animal domesticates associated with traditional agricultural and hunting practices in New Guinea are domestic chicken *(Gallus gallus)*, dog *(Canis familiaris)* and pig *(Sus scrofa)*. These domesticates all ultimately originated on mainland Eurasia and were dispersed variably across ISEA between ca. 4000–3000 cal BP (e.g., Thomson et al. 2014, Larson et al. 2012 and Larson et al. 2007, respectively). None are implicated in the emergence of agriculture in the highlands of New Guinea. These three domesticates are associated with, albeit variably, early Lapita sites dating from 3470–3250 cal BP in the Bismarck Archipelago (Denham et al. 2012; Specht et al. 2014) and more commonly co-occur on newly colonised islands of the Western Pacific from c. 3250–3100 cal BP (Matisoo-Smith 2009). Animal domesticates did not always disperse together; for example, dogs and pigs did not accompany the early settlers to New Caledonia (Anderson 2009), and sometimes animals were deliberately eradicated, such as the pig on Tikopia around 1600 AD (Kirch and Yen 1982).

The oldest domesticated pig remains in ISEA date to c. 4000 cal BP at Nagsabaran in northern Luzon (Piper et al. 2009) and likely reflect introduction across the Batanes Strait from Taiwan. However, genetic research (Larson et al. 2007) suggests these were not the pigs that eventually spread across ISEA to Near Oceania and Remote Oceania (Dobney et al. 2008). Claims for mid-Holocene pigs on New Guinea (White 1972; S. Bulmer 1975; Gorecki et al. 1991) have been challenged and dismissed following direct AMS dating of key faunal remains (Hedges et al. 1995) and re-excavation of key sites (O'Connor et al. 2011). Additionally, the dates for ISEA and the Bismarck Archipelago imply that animal domesticates were not introduced to New Guinea until after 3500 years ago. In terms of the New Guinea highlands, Sutton et al. (2009: 53) state 'archaeozoological evidence is inadequate to address questions of when pigs and dogs were introduced to New Guinea, although both were clearly present by 1500 years ago'. Chickens enter the archaeological record of the highlands within a similar time frame (Sutton et al. 2009: Table 11.3).

Before the introduction of domesticates, people on New Guinea had no pre-existing experience with domestic animals. Although human-mediated translocations of marsupials occurred from New Guinea to Wallacea and to the Bismarck Archipelago (Heinsohn 2010), these constituted anthropogenic expansion of the wild resource base rather than domestication. Similarly, people in the interior of New Guinea are known to raise wild cassowary *(Casuaris* spp.) chicks, but there is no evidence for breeding.

In this context, the rearing of domestic animals must have radically changed people's relationships to their environment and socio-spatial practices. Indeed, the rearing of chickens is still foreign to some groups, who often eat them rather than keep them for eggs or breeding (fieldwork observation, Lower Jimi Valley, 1990). Despite their significance to ethnographic portrayals of highlander societies, pigs were not present, or present in any number, in the highlands before approximately 1500 years ago and most likely only after 800 years ago.

Plant introductions from Island Southeast Asia

Presumably there were many pre-Magellan plant introductions to New Guinea from Island Southeast Asia throughout the Holocene (Table 11.1). One plant with an elusive history is *kudzu* (*Pueraria montana* var. *lobata*).

The antiquity of *Pueraria montana* var. *lobata*, ordinarily to be considered an Asian domesticate, in the highlands is uncertain. *Kudzu* is cultivated in the highlands for its tuber, which is slow maturing and considered to be a lesser food plant, or minor crop, that was important as a ceremonial, reserve and famine food in the highlands (Watson 1964, 1968; Strathern 1969). French (1986: 18) reports that '[w]ild forms grow between 30 and 1860 m. Cultivated forms are more common in higher altitude areas up to 2700 m'. The extended altitudinal zonation of cultivated varieties relative to wild forms suggests forcing of the altitudinal range of domesticated forms under cultivation. A long time depth for *kudzu* cultivation is implied by its use as a food in ceremonies, which is usually associated with plants that people view as being ancestral or traditional.

Archaeological investigations of house sites at Kuk in the early 1970s have uncovered evidence for other plants potentially introduced to New Guinea. Archaeobotanical finds of sweet potato *(Ipomoea batatas)* and sugarcane *(Saccharum officinarum)*, as well as other crop plant remains, have recently been identified from archaeobotanical materials recovered during the excavation of house sites (Lewis et al. 2016). Several pieces of cane stem and fragments of tuber suggest sugarcane and most probably sweet potato were consumed and plausibly grown at Kuk a few hundred years ago. Jon Hather (pers. comm. to Jack Golson) previously identified some tuber fragments as lesser yam *(Dioscorea esculenta)*, although recent analyses suggest these are sweet potato. The location of domestication for lesser yam is unknown, although on the basis of its geographical distribution it is considered a 'recent' introduction to New Guinea (Allen 2005).

The combined presence of bottle gourd, wax gourd, *kudzu* and sweet potato, as well as domesticated animals, are suggestive of periodic eastward movements and introductions to New Guinea over the last 5000 years or so. The evidence is significant, because it shows that the interior of New Guinea was not isolated from Island Southeast Asia during the mid-to-late Holocene. The same exchange pathways that enabled the introduction of plants to New Guinea also enabled the periodic westward transfer of cultivars from New Guinea into Island Southeast Asia over the same time period. These plants included edible banana diploids (AA), *Saccharum robustum* (precursor for sugarcane domestication), *Metroxylon sagu* and plausibly some yams (*Dioscorea* spp.).

Table 11.1 Selected list of exotic food plants likely introduced to the interior of New Guinea in the pre-ethnographic past

Plant	Antiquity (cal BP)	Details	Primary References
Gourd (Cucurbitaceae)	2700–2000 c. 5700–5000 3400–800 (uncertain)	Exocarp fragments originally identified as possible *Lagenaria siceraria* from Warrawau, Upper Wahgi Valley Site MSI on Ruti Flats, Lower Jimi Valley	Golson et al. 1967; Powell 1970a, 1982a; Gillieson et al. 1985; Golson 2002
Cultivated wax gourd *(Benincasa hispida)*	2950–2000	Exocarp fragments and seeds from Kana, Middle Wahgi Valley	Matthews 2003; Muke and Mandui 2003
Possible lesser yam (cf. *Dioscorea esculenta*)	60–280	Charred tuber fragments from domestic contexts at Kuk Swamp, Upper Wahgi Valley	Jon Hather personal communication to Jack Golson
Probable sweet potato (Prob. *Ipomoea batatas*)	60–280	Charred tuber fragments from domestic contexts at Kuk Swamp, Upper Wahgi Valley	Lewis et al. 2016
Manioc *(Manihot esculenta)*	< 150	Starch residues on stone tools from domestic contexts, Tagali River Valley	Author's unpublished research
Kudzu *(Pueraria montana* var. *lobata)*	?	Ethnographically documented tuber used as food source during famine	Watson 1968; Strathern 1969
Betelnut *(Areca catechu)*	?	Widely grown and chewed for psychoactive properties around the Indian Ocean and ISEA at the time of European exploration	Swadling et al. 1991; Fairbairn and Swadling 2005

Casuarina tree fallowing

Tree planting in fallows is a practice used by highlanders to 'fix' nitrogen and increase carbon in the soil, thereby enriching it for subsequent cultivation (Bourke and Allen 2009: 245–247; this section reproduced in amended form from Denham 2013b: 114). The most common practice is to transplant seedlings of *Casuarina oligodon* into plots near the end of the cropping cycle (see Figure 4.14). These trees grow for about 8–12 years before being cut down, ring-barked or pollarded prior to subsequent cultivation. Fast-growing *Casuarina* is also an important source of firewood and timber for construction.

Although *Casuarina* has been a feature of highland flora from at least the beginning of the Holocene (Haberle et al. 2012), Haberle (2007) has inferred that tree fallowing commenced around 1200 cal BP based on increased frequencies of *Casuarina* in pollen diagrams from the highlands. Subsequent palaeoecological research at Ambra Crater noted increases in woody disturbance taxa, *Casuarina*, fern spores and Cyperaceae from 1690–1420 cal BP; these were interpreted to represent increased intensity of human settlement, potentially including the beginning of tree fallowing (Sniderman et al. 2009: 456).

Archaeologically, *Casuarina* wood has been identified sporadically in different stratigraphic contexts at Kuk (Denham 2003b; cf. Powell 1982b) and more frequently in the fills of ditches dating to c. 700–c. 290 years ($n = 33$) and 250–50 years ago ($n = 185$) (Denham 2003a: Table G1.3). *Casuarina* was not present in the fills of securely identified ditches dating to c. 2000–1230/970 years ago or earlier. Thus at Kuk, *Casuarina* tree fallowing seems to have been initiated after the fall of Olgaboli tephra at c. 1190–970 cal BP.

Tree fallowing, as well as the planting of legume rotations, for example, winged bean *(Psophocarpus tetragonolobus)*, are techniques designed to restore soil fertility, thereby increasing productivity on valley slopes. Even though it was invented at least 1000 years ago, *Casuarina* tree fallowing is still practised in a relatively limited geographical area and has spread very slowly across the highlands (Figure 4.13; Bourke and Allen 2009).

Ethnography, ethnohistory and local-scale exchange: a Kalam case study

Ideas and things moved along local-scale, social networks that cumulatively resulted in their net transfer from the coast to the highlands. Although these social networks seem to have existed since the Pleistocene, they became more intensive and extensive during the late Holocene (Hughes 1977). For example, marine shells have been traded into the interior of the island for thousands of years (White 1972), and were a major source of wealth and status in highland societies by the mid-twentieth century (Glasse and Meggitt 1969). Numerous other introductions, or forms of indirect contact, occurred ahead of direct contact between highlanders and outsiders, including the net transfer of many ideas and things exotic to a region or the island as a whole. Introduced goods, such as steel axes (Salisbury 1962), were traded as valuables through the patchwork social mosaic of the New Guinea interior (Hughes 1977). Here, these processes are exemplified with reference to ethnographic and ethnohistorical records for the Kalam of the Upper Kaironk Valley, Bismarck Mountain Range.

Kalam oral traditions suggest that their ancestors were originally hunter-gatherers. They did not reside in a single place, but roamed the forest. At some point, they became swidden horticulturalists; they began to clear and plant gardens and build permanent homes. Kalam oral traditions can be differentiated into *sosm* and *kesm* (R. Bulmer 1967: 24): *sosm* are 'traditional tales referring to events outside the experience of living persons or specific remembered ancestors', whereas *kesm* consist of 'narratives relating the putatively true experiences of known people'. Things and practices that the Kalam always had correspond to the time depth of *sosm*, whereas numerous introductions are documented in *kesm*.

Kalam state that their ancestors planted lima beans *(Phaseolus lunatas)*, canola or rape *(Brassica napa)* and four types of sweet potato *(Ipomoea batatas)*, none of which are indigenous to the island (Table 11.2; R. Bulmer 1982: 283). Further they always had the bamboo 'Jew's harp', the dog *(Canis familiaris)*, the turtle-backed house and war magic. A number of cultivars are stated by Kalam to have been traditionally grown. The antiquity of taro *(Colocasia esculenta)* cultivation has been inferred for Kalam from its central role in ceremonies and because 'many sweet potato varieties are said to have been introduced within living memory, whereas this is true of only a small number of varieties of taro' (R. Bulmer 1967: 23).

The first domesticated pig was introduced to Kalam in the middle of the nineteenth century, shortly after the introduction of tobacco *(Nicotania tobacum)*. A man named Sabep-nd is accredited with bringing the first pig to the Kalam of the Upper Kaironk Valley (Majnep and Bulmer 1977: 19). Approximately three generations ago, the large-scale clearance of forest in the Upper Kaironk Valley and other areas occurred. The mid-altitude grasslands found throughout Kalam territory were formed only comparatively recently. These denuded landscapes are perceived as resulting from the invasion of old garden sites by grasses.

Other than tobacco and pig, nineteenth century introductions to the Kalam of the Upper Kaironk Valley include sword grass *(Miscanthus floridis)* and the planting of casuarinas *(Casuarina oligodon)*. Sword grass was brought in at the same time as the pig, and the neighbouring Maring are accredited with being the source (Majnep and Bulmer 1977: 23). Sword grass

Table 11.2 Ethnohistorical adoptions, innovations and introductions among the Kalam

Antiquity	Item/Practice
Ancestral[1]	Sweet potato *(Ipomoea batatas)*
Ancestral[1]	Lima bean *(Phaseolus lunatas)*
Ancestral[1]	Rape *(Brassica napa)*
Ancestral	Dog *(Canis familiaris)*
Ancestral	Jew's harp
Ancestral	Magic
Ancestral	Turtle-backed house
Traditional[1]	Banana *(Musa* cvs.)
Traditional[1]	Sugarcane *(Saccharum officinarum)*
Traditional[1]	Pitpit *(Saccharum* sp., *Setaria* sp.)
Traditional[1]	Greens (e.g., *Rungia* sp.)
Traditional[1]	Cucumbers *(Cucuminis stivus)*
Traditional[1]	Taro *(Colocasia esculenta)*
Traditional[2]	Gourds *(Lagenaria siceraria, Benincasa hispida)*
Mid-19th century[3]	Sword grass *(Miscanthus floridus)*
Mid-19th century[2]	Tobacco *(Nicotania tobacum)*
Mid-19th century[4]	Pig *(Sus scrofa)*
Mid-19th century[5]	Spears and shields
1860s[4]	Fighting
1880–1890[4]	Fencing of gardens
Late 19th century[5]	Witchcraft accusations
Late 19th century[3]	Casuarina planting *(Casuarina oligodon)*
1920s[6]	Steel axe heads
1940s[1]	Taro *konkon (Xanthosoma sagittifolium)*
1940s[1]	Maize *(Zea mays)*
1940s[1]	Manioc *(Manihot esculenta)*
1950s[1]	Pumpkin *(Cucurbita maxima)*
Late 1950s[1]	Sweet potato varieties
Late 1950s[1]	Irish potato *(Solanum tuberosum)*
Late 1950s[1]	Runner bean *(Phaseolus multiflorus)*
Late 1950s[1]	French bean *(Phaseolus vulgaris)*
Late 1950s[1]	Shallott *(Allium cepa)*
Late 1950s[1]	Cabbage *(Brassica oleracea)*
1960s[1]	Passion fruit *(Passiflora edulis)*
1960s[1]	Watercress *(Nasturtium officinale)*
1963[1]	Peanut *(Arachis hypogaea)*
1963/1964[1]	Choko *(Sechiumedule* sp.)
Mid-1960s[1]	Coffee *(Coffea arabica)*
Late 1960s[1]	Eucalyptus *(Eucalyptus calaphylla)*
Recent	Cardamon *(Elettaria cardamomum)*
Recent	Avocado *(Persea americana)*
Recent	Lime *(Tilia* sp.)
Recent	Coconut *(Cocos nucifera)*
Recent	Chilli pepper *(Capsicum* sp.)
Recent	Betelnut *(Areca catechu)*
Recent	Cattle *(Bos taurus)*
Recent	Duck *(Anatidae)*

Source: After Denham 1996: Table 2.3
Notes on sources:
1 For the Upper Kaironk Valley (R. Bulmer 1982)
2 For the Kalam in general (R. Bulmer 1967)
3 For the Upper Kaironk Valley (Majnep and Bulmer 1977)
4 For the Upper Kaironk Valley (Majnep and Bulmer 1977; Riebe 1974)
5 For the Upper Kaironk Valley (Riebe 1974)
6 For the Kalam in general (R. Bulmer 1968)

is now widespread and considered a pest. The grass invades old gardens preventing forest regrowth. The dense stands of grass are difficult to cultivate and Kalam state a preference to cut and clear primary and secondary forest.

Casuarina planting is stated as beginning at the turn of the century. Casuarinas are planted in old garden sites for they are rapidly growing, nutrient enriching trees that allow a shorter fallow period than would be necessary under mixed forest growth (Brookfield and Brown 1963: 50–51; Majnep and Bulmer 1977: 23–24). The relatively recent origins for the planting of casuarinas is curious, because the Kalam living in the Upper Kaironk Valley readily identify with this practice and refer to themselves as 'the casuarina planters' (Riebe 1974).

Between 1880–1890, the first fence was constructed around a garden in the Kaironk Valley in order to keep wild pigs out (Majnep and Bulmer 1977: 24). Prior to the introduction of the domestic pig, ancestors did not hunt wild pigs in the Upper Kaironk Valley, for there were none. Before the adoption of pig husbandry, wild game was an important source of protein, as well as being required for offerings at ceremonies. Only recently have pigs become the main animal consumed at important ceremonies; formerly game mammals, particularly wallabies, were the preferred meat.

During the first half of the twentieth century, a variety of artefacts and crops were introduced ahead of direct European contact. The introduction of *Xanthosoma* taro, maize *(Zea mays)* and manioc *(Manihot esculenta)* represented significant alterations to the traditional Kalam diet. Shortly thereafter during the 1950s, the first Europeans entered parts of Kalam territory. As with many groups in the highlands, the first Europeans were initially believed to be dead ancestors.

The time depth of oral knowledge reported thus far for Kalam is significant because it likely pertains to the last 200 years or so, in terms of named historical individuals and events. As a result, many plants that have been introduced to New Guinea in the last several hundred years, such as sweet potato, are considered to be ancestral plants that have always been cultivated. Although the translation of these oral accounts to chronological time is problematic, the various Kalam narratives convey a relative sense of time depth for different practices, animals and plants.

The time depth for the incorporation of animals and plants among other communities in the interior of New Guinea can be directly inferred from oral traditions and from their incorporation in ceremonies and rituals. For example, Febi in the Upper Strickland Valley recount how the first people arose from sago *(Metroxylon sagu*; personal fieldwork 2008) and Myanmin refer to themselves as the taro people (Morren 1986; personal fieldwork 2010). Animal and plant domesticates that have only recently been introduced do not usually feature in oral accounts, especially those related to the origins of a named group, nor hold social significance; namely, they are not incorporated into significant ceremonies and rituals. Through time, different ideas and things become incorporated into oral traditions and cultural practices, as witnessed for the Kalam and documented throughout the highlands in recent decades.

Sweet potato, pigs and big men

> No ethnographer has the right to assume that what [s]he observes … after … contact is a true image of the traditional economic and social systems. … Many of the present confusions in our understanding of Melanesian social systems rest on a failure to apprehend the consequences of these … changes, and to set ethnographic data in its historical context.
>
> (R. Bulmer 1965: 236)

Of all the introductions, arguably the most significant for the development of ethnographically recorded 'big men' societies in the highlands are sweet potato *(Ipomoea batatas)* and pig *(Sus* sp.) husbandry. Pigs *(Sus scrofa)* are intimately associated with ethnographic portrayals of highlander societies (this section reproduced in amended form from Denham 2013b: 114–115). Traditionally, pigs were a means through which highlanders could store surplus and are implicated in the emergence of big men societies (Modjeska 1982; Golson and Gardner 1990). In part, big men were able to accumulate prestige through control of the circulation of pigs and other valuables in traditional exchange ceremonies, such as the Hagen *moka* (Strathern 1971).

Pigs may have only come to dominate highlander social and economic life following the introduction of another exotic, the South American domesticate, sweet potato *(Ipomoea batatas)* (Weissner 2005). Even though it is frost intolerant, sweet potato is well adapted to grow in the highlands because it outperforms taro in poorer soils at altitude (Brookfield 1964; Ballard et al. 2005). Today, sweet potato is a near ubiquitous staple in the highlands, where it replaced banana, taro and yams as the principal crops of highland cultivation within the last 400 years (Golson 1982a: 120; Bayliss-Smith and Golson 1992a: 21). Sweet potato was a Polynesian introduction to the Pacific, sometime after 1000 cal BP, although it is generally considered to be a post-Magellan introduction to New Guinea (post-AD 1500s; Roullier et al. 2013).

Even more intriguing than seeking to date the introduction of these two exotics is the speed at which they were incorporated into highlander socio-economic life. The integration of sweet potato into existing systems of cultivation seems relatively straightforward and probably occurred rapidly; it is a vegetatively propagated tuberous plant like taro and yams. Conversely, highlanders, like the rest of New Guinea's inhabitants, had not previously domesticated any indigenous animals. The only other experiences with domesticated animals were the dog *(Canis familiaris)* and chicken *(Gallus gallus)*, both of Asian derivation and both introduced during the late Holocene (Sutton et al. 2009).

The adoption of pigs would have required a major shift in thinking and land use. Rather than hunting animals or possibly capturing them for later trade or consumption, people would have kept, bred and raised pigs. Pigs need to be fed, fenced and protected. The adoption of pig husbandry was probably slow initially following introduction to highlanders; their number and social significance expanded following the widespread adoption of sweet potato over the last 500 years.

Why adopt?

This chapter, like the majority of this book, is focussed on presenting how agriculture emerged and changed in the highlands of New Guinea. Here, the major chronology for the emergence of agriculture in the highlands is augmented by a supplementary history of later innovations, introductions and adoptions. Thus far, there has been little discussion of why people embraced change through the adoption of practices, animal domesticates and cultivars. A starting point is to consider the ways in which the agricultural history of New Guinea has been interpreted previously.

Initial portrayals of the agricultural sequence in the highlands of New Guinea reflected dualistic thinking about forms of plant exploitation and the respective roles of people and the environment. Earlier practices (namely, those pre-dating 2000 years ago) were considered horticultural in that 'each is structurally diversified in ways suggesting that provision was being made for crops of different edaphic and hydrologic requirements growing side-by-side in the same plantings' (Golson 1997b: 45). Later practices (post-dating 2000 years ago) were agricultural in the sense that they 'exhibit standardised patterns based on long, straight ditches parallel and at right angles to each other that enclosed square to rectangular plots, thought to be planting areas for a single crop' (Golson 1997b: 45).

Although the systematisation of drainage networks within the last 2000 years is clear at Kuk, and likely occurred elsewhere, such a chronology is not relevant to understanding the history of agricultural practices across the island of New Guinea, or even in the highlands. Agricultural practices across New Guinea are highly variable in the recent past and present (Bourke and Harwood 2009), and were certainly variable in the distant past (Denham 2006, 2011). The primary differences in Golson's schema between horticulture and agriculture are polyculture versus monoculture, respectively, and the nature of the field system. Yet any distinction between horticulture and agriculture is not clear-cut in the New Guinea context (Golson 1997b; Leach 1999; Denham 2005c). Agriculture refers to the cultivation of plants in prepared ground, and horticulture refers to the cultivation of plants in gardens. For traditional New Guinean practices, the terms are effectively synonymous. Gardens are the plots in which people cultivate plants for food, whether single or multi-cropped and whether single-planted/long fallow (swidden cultivation) or successively planted/short fallow (intensive cultivation) (see Bourke 2001). Some gardens are located within settlements (i.e., house gardens), whereas others are more dispersed (e.g., swidden plots). In the New Guinean context, gardens are equivalent to fields; they are tracts of land or plots used for cultivation.

In terms of people's engagement with their environment, forms of plant exploitation pre-dating 2000 years ago have been characterised as formative adaptation to natural climatic and edaphic factors, whereas later cultivation practices were considered intensifications triggered by the social environment (Golson 1982b: 301; 1990: 145; implicit in Golson and Gardner 1990: 405–406). Wetland agricultural practices were conceived as existing in a dialectical relationship to an increasingly altered environment (Golson 1982b: 301). Nearly all interpretations viewed wetland practices as part of much broader land use strategies, which included both dryland agriculture as well as hunting and gathering practices in uncleared areas (e.g., Golson 1990: 140; cf. Denham and Haberle 2008). Drainage and cultivation of wetlands were triggered by stresses in the dryland productive sphere, whether soil nutrient depletion, drought, overpopulation and so on (Golson 1977c, 1982b; Gorecki 1979, 1986; Bayliss-Smith 2007). Wetland use was abandoned following a series of putative innovations in dryland cultivation practices – such as tillage, *Casuarina* tree fallowing and sweet potato – that rendered ongoing and labour-intensive exploitation of wetlands redundant (Golson 1977c). Abandonment of drainage networks over the last 2500 years was hastened by aeolian tephra events that clogged drainage networks.

A shift in interpretative emphasis away from neo-Boserupian thinking followed more socially oriented readings of the archaeological findings at Kuk by Modjeska and Gorecki. Modjeska (1977, 1982) viewed swamp cultivation as localised centres of high productivity that yielded a greater abundance of resources than required for subsistence. The surplus was used for cultural exchange and facilitated the development of limited political structures, which intensified with the adoption of intensive sweet potato cultivation together with pig husbandry. Gorecki viewed the construction of ditch networks as a social phenomenon reflecting communal organisation and management in contrast to the more individual practices of earlier phases (1986: 163). In the light of Modjeska's and Gorecki's reappraisals, Golson and co-workers increasingly considered the social contexts of wetland subsistence (Golson 1982a: 127–136; Golson and Gardner 1990: 398–408 and Bayliss-Smith and Golson 1999: 206–212, respectively).

In contrast, recent interpretations have deliberately sidelined questions of why, or causation (Denham 2005c, 2006, 2007a, 2009, 2011):

> In attempting to address questions of 'why?' there is a continual interplay between what happened in the past and how that past is made meaningful in the present. Given this hermeneutic trap, we restrict our focus to *what* people were doing and *how* they were using resources across the landscape.
>
> (Denham and Haberle 2008: 485, emphasis in original)

Effectively, understanding why people adopt can be interpreted with respect to a gamut of off-the-peg theories that exist to understand people's actions in the past. These extend from the idiosyncratic, which reify individual decision marking, to the structural, which seemingly deny the historical efficacy of human agency.

Ultimate causation is extremely difficult to elicit in historical research; there are always alternative interpretations. It seems unlikely that a single cause, or *causa efficiens*, was important at all times in the past. Rather different combinations of purpose and circumstance may apply for the adoption of a particular practice in given social, namely geographical and historical, contexts (Denham 2013b: 116).

To illustrate, the increased drainage of wetlands over the last 4000 years could be correlated with and attributed to climatic stresses accompanying intensification of ENSO during the late Holocene. El Niño years give rise to droughts in parts of the highlands, which have caused major food shortages (Allen 2000). At these times, wetlands became centres of agricultural productivity due to the collapse in dryland production; so people maintain wetland plots, in part, as a resource diversification and risk reduction strategy (Ballard 1995). Even though the inception of ditch digging around 4400–4000 years ago and the increasingly widespread adoption of wetland drainage technology within the last 2750–2150 years could be linked to climate, this is not necessarily the sole or primary explanation. There were other proximate and distant influences upon the innovation of wetland drainage using ditches, as well as upon its adoption by some groups and not be others. Although synchronicities and patterns may be identified in the timing of particular events, correlations should not be uncritically read as causation (Gould 1970).

Any attempt to elicit rationales for innovations, introductions and adoptions in agricultural practices in the highlands requires a consideration of multiple, intersecting factors (Denham 2013b: 116). Environmental phenomena include climatic fluctuations through time (Haberle and David 2004), differences in climates across the highlands (McAlpine et al. 1983), the extent of deforestation and forest degradation (Haberle 2007) and at least six volcanic ash fall events (Coulter et al. 2009). Social phenomena include demography, transformations in political structures, encounters with other groups, the exchange of new ideas and things and so on. Most factors lie at the nexus of human-environmental relationships, such as landscape degradation, which includes forest clearance and disturbance, together with soil erosion, resource depletion and alteration of the environment for cultivation; these environmental transformations may have necessitated an increasing dependence on cultivation. Just as the rationale for the inception of agriculture in different parts of the world should be evaluated contextually, albeit within comparative frames of reference, so too the reasons for the adoption of an agricultural practice within any given locale in the highlands requires consideration of the detailed multidisciplinary evidence for that place.

Part IV

Taking a broader view

12 Historical resilience of agriculture in the highlands

The best prophet of the future is the past.

(traditional quote attributed to Lord Byron)

The agricultural history documented for the highlands is not only of intrinsic value for archaeology, agronomy and long-term history; it has broader *relevance* for several contemporary issues in New Guinea and beyond. These issues are only touched upon here, and each would warrant separate investigation. Yet work on agricultural history in the highlands has, in one way or another, contributed to an understanding of these more applied issues.

Learning from the past

The long-term history of agriculture in the highlands of New Guinea, which spans the Holocene, is a dynamic history. In no sense can this history be seen as static or be conceived in terms of a unilinear trajectory; rather it is ever-changing and characterised by high degrees of regional variability, expanding and contracting repertoires of plant exploitation, and the occasional abandonment or adoption of practices and plants. The types of agricultural practice, the crops grown and the extent of environmental transformations have changed and transformed through time. This long-term history of continuity and discontinuity provides the historical context for interpreting agricultural practices across the highlands today.

Contemporary agricultural practices in the highlands – as well as for New Guinea as a whole – clearly exhibit this dynamism. South American domesticates – including sweet potato and maize – were introduced to the island and had dispersed to the highlands before direct European contact there in the early twentieth century (see Table 11.1). These plants had dispersed westwards across the Pacific with European explorers and colonists, as well as potentially with earlier Polynesian and Chinese voyagers. Indeed, sweet potato had become the dominant staple across the majority of the highlands by the time of European exploration. Within the last few centuries, numerous other plants have been adopted and integrated into cultivation practices, including tobacco, manioc, peanut, coffee and rice, as well as new varieties of traditional crops, such as taro (see Table 11.2).

Although much literature has focussed on the Ipomoean Revolution, or Sweet Potato Revolution, as a decisive period in highlands proto-history (from Watson 1965 and Brookfield and White 1968 to Modjeska 1982 and Ballard et al. 2005), New Guinea cultivators have been experimenting with local plants and have exhibited a willingness to adopt newly introduced species into their crop plant repertoire throughout the Holocene. Although some of these plants became major staples, other indigenous plants have become less important through time (Barrau 1965). For instance, it has long been presumed that taro (Watson 1968; Bayliss-Smith 1985), or a combination of taro-greater yam-banana (Bourke 2009), were more important staples prior to the widespread adoption and cultivation of sweet potato. Several other plants were probably also more important in the distant past, but these are poorly known ecologically, ethnographically and archaeologically today. They include *kudzu*, highland pitpit, some yam species and castanopsis nuts.

Cultivators in the highlands did not exist in isolation from broader social practices in the lowlands, the circum-New Guinea region or developments in Southeast Asia. Plants, practices and ideas diffused through the socially constructed maritime landscapes of Island Southeast Asia and New Guinea over the last 3000–5000 years (Denham 2004b, 2010; Donohue and Denham 2010) and potentially earlier (Bulbeck 2008). As things were introduced to the coast, they dispersed along local and longer-distance riverine exchange pathways (following Hughes 1977). Such interactions and exchanges are exhibited in the material cultural record of shell ornaments, obsidian (White 1967), pottery (Gaffney, Summerhayes et al. 2015) and pigs (Sutton et al. 2009), yet they are only glimpsed in the archaeobotanical records of mid-Holocene bottle gourd, wax gourd and other poorly dated introductions.

Today, some agronomic practices are relatively ubiquitous, whereas others are more localised. For example, *Casuarina* tree fallowing originated over a thousand years ago, yet it is still practised in only a restricted region of the highlands. Similarly, mounds and raised bed cultivation have long antiquities in the highlands and contemporary distributions exhibit some regional concentrations and differentiation. These distributions do not appear to solely reflect environmental constraints; they also reflect cultural preferences and dispositions. Understanding how these distributions came to be requires a long-term historical perspective, rather than just a synchronic understanding of social practices and environmental processes.

If we seek to understand agriculture in the highlands in the recent past and project into the PresentFuture (Denham and Mooney 2008), a term coined to connote how the future is always cointended in the present, we need to have a long-term understanding of the historical processes involved. Such an understanding requires more than just a focus on agronomy in the past, which has necessarily been the focus of this book. Rather, a broader vision is needed, in which any transformations in agricultural practice in the present or the past are seen to make 'social sense'.

Agricultural sustainability

Today, New Guinea highlanders are renowned innovators. People will make use of the available resources to solve a particular problem in the present or with a view to potential benefits in the future. This perspective applies to using an old tin can to fix a broken exhaust pipe (fieldwork, Minj, 1990) or to planting a new variety of taro (fieldwork, Karimui, 2008). Neither adoption can be seen in separation from the broader social milieu pervading a given historical context.

The reasons why people do certain things cannot be reduced to a formula of people as ecological maximisers *(Homo ecologicus)* or economic rationalists *(Homo economicus)*. People in contemporary societies do not usually behave in such one-dimensional ways either as individuals or communities. Just think of the people you know and the societies you live in. Why would we anticipate that people in the past would behave in a simpler or more predictable way just because their technology was different?

People's behaviour is directed and regulated by numerous social factors, of which ecology and economy are only two. For example, history and the contemporary world are littered with examples of societies engaging in practices that proved, or are proving, to be ultimately unsustainable, whether ecologically or economically (Diamond 2005; United Nations 2016). Although people's behaviour can best be understood with respect to individual or group actions within a given historical situation, which is not usually of their own choosing (following Marx 1852), the reasons why people did things are rarely transparent. Indeed, the last 150 years are replete with numerous social theories that have sought to clarify the character of human action in terms of agency-structure dualisms or dualities (Giddens 1984), or subsume such dichotomies within new concepts, such as *habitus* (Bourdieu 1990).

Globally, the adoption and expansion of agriculture to the present is considered one of the most environmentally damaging innovations of humans (Ruddiman et al. 2015). Agriculture has resulted in vast and permanent environmental transformations on all the inhabited continents (Boivin et al. 2016), although the timing of its inception and spread vary greatly (Larson et al. 2014). Agriculture has enabled unparalleled demographic growth, which in turn has necessitated further agricultural expansion or intensification and further environmental degradation.

The long-term resilience of agricultural practices in the highlands has been a product of human practices spanning millennia. These practices have seen extensive regions of montane rainforest cleared for cultivation, whereas other forests have been degraded through hunting, burning, planting of favoured species and the collection of plant materials that have all reduced biodiversity and forest structure (Faith et al. 2001; Haberle 2007). In some regions, soils have also been degraded, as fallow cycles have been foreshortened and cultivation prolonged, albeit with diminishing returns. The cumulative effects of agricultural practices in the highlands are not environmentally neutral, either in terms of environmental degradation or in terms of the continued sustainability and viability of certain types of cultivation.

Having said this, some regions exhibit prolonged chronologies of cultivation – such as the Upper Wahgi Valley – that clearly demonstrate how people have been able to adapt cultivation practices to accommodate new social pressures and environmental degradation in the past (Golson 1982a, Modjeska 1982; Golson and Gardner 1990). People have been cultivating in the Upper Waghi Valley for at least 6400–7000 years, which is testament to the long-term sustainability of cultivation practices there. As undoubtedly happened numerous times in the past, long-term sustainability is today being jeopardised by new social demands resulting from unprecedented population increase, a desire for cash cropping and participation in the market economy and the periodic adoption of 'inappropriate' agricultural technologies.

In the New Guinea highlands, the environmental effects of agriculture and occupation vary greatly – from long-term cultivation and settlement on some of the major valley floors, to occasional visitation and temporary encampments in other valleys. The long-term effects of human occupation can only be inferred from archaeological and palaeoecological records that provide a long-term foundation for understanding human impacts on the environment. These records provide a much longer-term, although more coarse-grained, basis for inferring environmental transformations in the highlands than, as well as providing a supplement to, ethnohistorical and longitudinal ecological studies (Willis et al. 2004). In theory, these archaeo- and palaeo-records enable inferences to be made regarding stepped changes in environmental indices from less intensive practices in the Pleistocene, to the emergence and spread of agriculture in the Holocene, to the rapid transformations of the last 100 years. As yet, there has been no serious attempt to interlink and compare these ancient records with more recent ethnohistorical and ecological records to understand the long-term effects of human practices on the environment for any location in the highlands.

Crop diversity and food security

Arising from a consideration of the long-term reduction in biodiversity is a cognate recent concern with a reduction in crop diversity with a view to crop improvement and food security (Gepts et al. 2012). Most of the crop plants grown in the highlands have not been extensively studied ecologically or genetically. Over the last few decades, attempts have been made to improve agriculture in the highlands through the importation of exotic varieties of traditional cultivars from Southeast Asia and further afield. As a result, in some communities only a few people – usually old – grow the named, traditional varieties of a crop that emerged as a result of millennia of selection on New Guinea for different phenotypic traits, including colour, taste, acridity, shape, texture and so on.

Although anecdotal, based on 20 years of fieldwork across mainland Papua New Guinea, there is a general impression that vast numbers of cultivars for some traditional crops are falling out of cultivation or have already been abandoned (see Figure 5.9). Once these varieties stop being grown, they are likely to go extinct unless they survive as feral garden escapees. This represents a dramatic reduction in crop diversity, with losses of ecological (namely, the range of ecological adaptations and tolerances of cultivars), phenotypic (namely, behavioural and morphological attributes resulting from prolonged selection) and genotypic (namely, genetic loci reflecting cumulative human selection) diversity, as well as attendant ethnographic information on cultivation, processing and culinary practices.

Conversely to the increasing losses of cultivar diversity at the local level, there is simultaneously an increasing global awareness that the reduction in genetic diversity of cultivars for staple crops is creating major problems for the long-term sustainability of local subsistence, as well as global, cash-cropping practices. To exemplify, the major commercial banana cultivar globally is the Cavendish, an AAA triploid (*Musa* cv). The Cavendish is derived from three *Musa acuminata* 'parents', including *Musa acuminata* ssp. *banksii* from the New Guinea region. Like its predecessor Gros Michel (also AAA), which was the main globally cultivated plantation banana prior to 1950s when it was decimated by Panama disease *(Fusarium oxysporum)*, the Cavendish is on borrowed time. The Cavendish is clonally propagated, has a very narrow genetic base globally and will likely succumb to new strains of Panama disease that are spreading through plantations across the world. In the face of this threat, there has been an international effort to investigate the genetic history of banana domestication, including wild plants and locally important cultivars under traditional forms of cultivation across Southeast Asia and New Guinea, in order to develop a comparable AAA cultivar that is similar to the Cavendish in terms of fruit quality for consumers yet resistant to new Panama strains (Perrier et al. 2011). If local and traditional cultivar diversity has declined too far, it would not be possible to reconstruct the domestication history for the banana in order to secure food production, and the same is true for other traditionally cultivated plants that are now either important subsistence or commercial crops around the world.

In addition to conserving current cultivar diversity, there is an urgent need to reappraise famine foods, potentially abandoned foods and wild food plants to ensure food security into the future. The reasons for the marginal status of these food plants today are not known, but this does not mean they were not important in the past or no longer have relevance for food security into the future. Renewed agronomic, nutritional and genetic investigation may yet yield evidence for the importance of these plants for future food security.

Adapting to climate change

People living in the highlands of New Guinea, like those elsewhere in the world, have always adapted to changing climates. From first arrival on the island over 45,000 years ago, people had to accommodate glacial cooling, the Last Glacial Maximum, post-glacial climatic amelioration, the Younger Dryas cold reversal, more stable climates of the Holocene, intensification of ENSO during the late Holocene, and recent global warming, to name but a few periods and events. Significantly, people living in different places of the world – and even in different parts of the highlands – responded to these challenges in different ways.

Current literature on agricultural origins makes much of stable climates from the beginning of the Holocene as a major causal factor for the inception of agriculture (Richerson et al. 2001). Recent investigations of plant and animal domestication show that more caution is needed. Only a few places on earth engaged in nascent agricultural and domesticatory practices at different times during the early Holocene (before c. 8000 cal BP; Larson et al. 2014; Fuller et al. 2014); the highlands of New Guinea is one of them.

The invocation of stable climates as a primary motor of agricultural innovation does not greatly assist in understanding why agriculture emerged in some places, whereas it did not in others. Any climatic causation for early agriculture soon shifts sideways from climate as primary motor to the biophysical realm in order to understand the suitability of resources and environments in different places for agriculture. In effect, argumentation shifts to specific socio-environmental factors to explain why certain people with certain technologies in certain places with certain resources and certain climates developed agriculture, whereas others did not. Following this line of argument, although climate was initially posited as a fundamental cause, it becomes a relatively widespread precondition, that in turn becomes a relatively benign context that has no 'causal' or 'explanatory' power for understanding why agriculture emerged in some places and not in others.

Although climate is not necessarily a causal panacea to understand agricultural innovation; people in the past, just like the present, undoubtedly have needed to respond to climatic fluctuations mediated through animal and plant resource availability, environmental transformations, susceptibility to fire and so on. An example of this may be the initiation and expansion of ditch networks to drain wetland margins for cultivation between 4000–2000 years ago. On the basis of the experience of El Niño events over the last few decades, wetlands in the highlands are less impacted during these drought years and provide stable loci of food production. This would have been true in the past. People may have initially extended production onto wetland margins as part of diverse food strategies, which extended back into the early and mid-Holocene. From around 4000 years ago, wetland cultivation became an increasingly important buffer to maintain food security as ENSO events increased in magnitude and frequency. Ditched drainage of wetlands enabled people to expand the area under cultivation in wetlands and provide more secure inter-annual food production.

In this scenario, climate cannot be presumed to be the fundamental 'cause' of ditched drainage. Such an interpretation is too simplistic. Rather, people in certain places responded to periodic droughts and resultant famines by innovating and digging ditches into wetlands to drain them for cultivation. Ditches probably already existed as a technology, yet it still took almost 2000 years for the practice of wetland drainage for cultivation to spread more widely across the highlands.

Just as climates have changed in the past – most notably with climatic amelioration at the end of the Pleistocene and intensification of ENSO in the late Holocene – they continue to change today. Global warming has led to the upward altitudinal migration of many crop plants, as well as some mosquito-borne diseases in the highlands. The upper altitudinal limit of several crop plants has increased by a couple of hundred metres over the last 40 years (Bourke pers. comm. 2014). This has enabled a corresponding altitudinal extension of the main zone of highland settlement. At the same time, though, malaria has extended to higher altitudes, thereby making the extensive floors of some highland valleys, even up to c. 1200–1400 m, less favourable for settlement. The optimal zone for highland agriculture and settlement (following Brookfield 1964) is shifting upwards. These altitudinal shifts in production and malaria will create major social problems for dense and highly territorial populations living on the floors of the major inter-montane valleys; there is less room at the top (after Golson 1977c).

Making social sense

The persistence of agricultural practices in the highlands of New Guinea is noteworthy. Some practices have persisted for thousands of years – including mounding, ditch digging, tillage and so on – and many crops were originally exploited in the early Holocene – including bananas, taro and yam. Despite these long chronologies of similarity and continuity, highly regional and local traditions of agricultural practice and cultivar preference have emerged. The reasons for the emergence and persistence of regional and local diversity are unclear; most arguments vacillate between environmental and cultural causation, yet all seem unsatisfactory.

Rather than seeking to present a causal explanation for regional diversity in agricultural practices and crop assemblages, it is sufficient here to allude to the ways people, plants and place are intertwined within highland landscapes. Here, we seek to delve below the macro-scale of agricultural history – beyond the chronologies, maps, tables and dates – to infer the nuances of social history. Such an exercise necessitates a creative leap.

One way in which we can enrich agricultural history is to begin to go beyond one-dimensional portrayals of people in our understanding of human-environment interactions in the past. People as historical agents are more than summed radiocarbon dates or charcoal frequency distributions (Figure 12.1). The human role can be reduced to this one-dimensional characterisation, which obviates any need to really get to grips with human agency, practices and the importance of place in the past. It is more fulfilling to 'populate pollen diagrams' and begin to creatively envisage how people were engaged in the landscape based on the evidence at hand. This may require a creative 'spatialising' of human

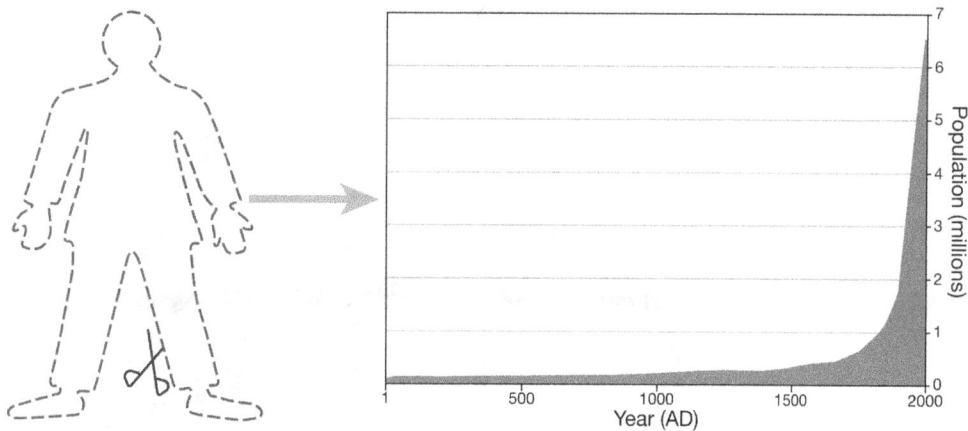

Figure 12.1 Caricature of one-dimensional portrayals of human agency in the past

practices across the landscape, as done for the Upper Wahgi Valley for different time periods (Figures 4.20 and 4.21). Such diagrams are not to be read literally; rather, they create an impression of the range of practices occurring in different places in the landscape at different times.

Another way to enrich agricultural history is to conceptualise how aspects of vegetative cultivation and social relationships are intertwined (Figure 12.2). Although past kinship relationships are currently invisible in highlands archaeology, vegetative propagation seems to have always been the dominant form of cultivation there. Perhaps, in the absence of contrary evidence, we can suppose that there were also some similarities in the nature of social relationships in the past. The intertwining of social relationships with crop plant life cycles – as well as the self-identification of groups with plants – provides a window on why vegetative propagation has persisted as the dominant mode of crop reproduction in the highlands and why some groups have retained certain crops and practices.

Many societies across New Guinea have mapped their social relationships and worldviews onto the landscape in diverse ways. Across these landscapes, places are named after people who have lived there, after quasi-historical apical ancestors or after 'mythical' beings who are often attributed with bringing the world and its people into existence. These categories are not discrete, they often blur further back in time, yet all represent the spatial expression of an historical identity that is fundamental to the creation of place. One of the key materialities that informs the ways communities construct their social relationships, as well as project their group identity across space, is their relationship to plants. Not just individual plants, but also the stands and groves that people have planted, tended or marked. Indeed, these physical acts of intervention in plant phenology are essential characteristics of how people have traditionally formed social relationships, inscribed their identities in the landscape and created their territories in New Guinea. Arguably, people's orientation to plants, or vegetative disposition, has helped frame people's relationships to each other and to their landscapes.

If people's social identities are so closely entwined with vegetative planting cycles or specific plants, any transformations to agricultural practice in the past would only have been adopted if they made 'social sense'. Namely, a transformation would need to be accommodated within the existing social milieu, as well as being perceived as agronomically, nutritionally or socially beneficial. It has been argued that the resistance of some vegeculturalists to cereal adoption on Borneo has been driven by these types of social resilience or inertia (Barton and Denham 2011; Barton 2012). Similarly, it would not make social sense for the *Casuarina* planters (Kalam) or taro people (Min) to abandon tree fallowing or their staple crop, respectively. Such an agronomic transformation would fundamentally undermine each group's identity and not make 'social sense'.

Today, many highlanders are keen to abandon traditional practices in order to become more integrated into the market economy. The cultural shock experienced with First Contact and subsequent relative deprivation undermined traditional identities in numerous ways. Since First Contact, traditional beliefs and worldviews have been systematically and sometimes deliberately eroded through colonial and post-colonial administrations, missionisation, education, mass media, cash economy and so on. Yet despite all this, traditional cultivation practices and associated social sense exhibit considerable resilience and persistence; they are still primarily based on vegetative principles.

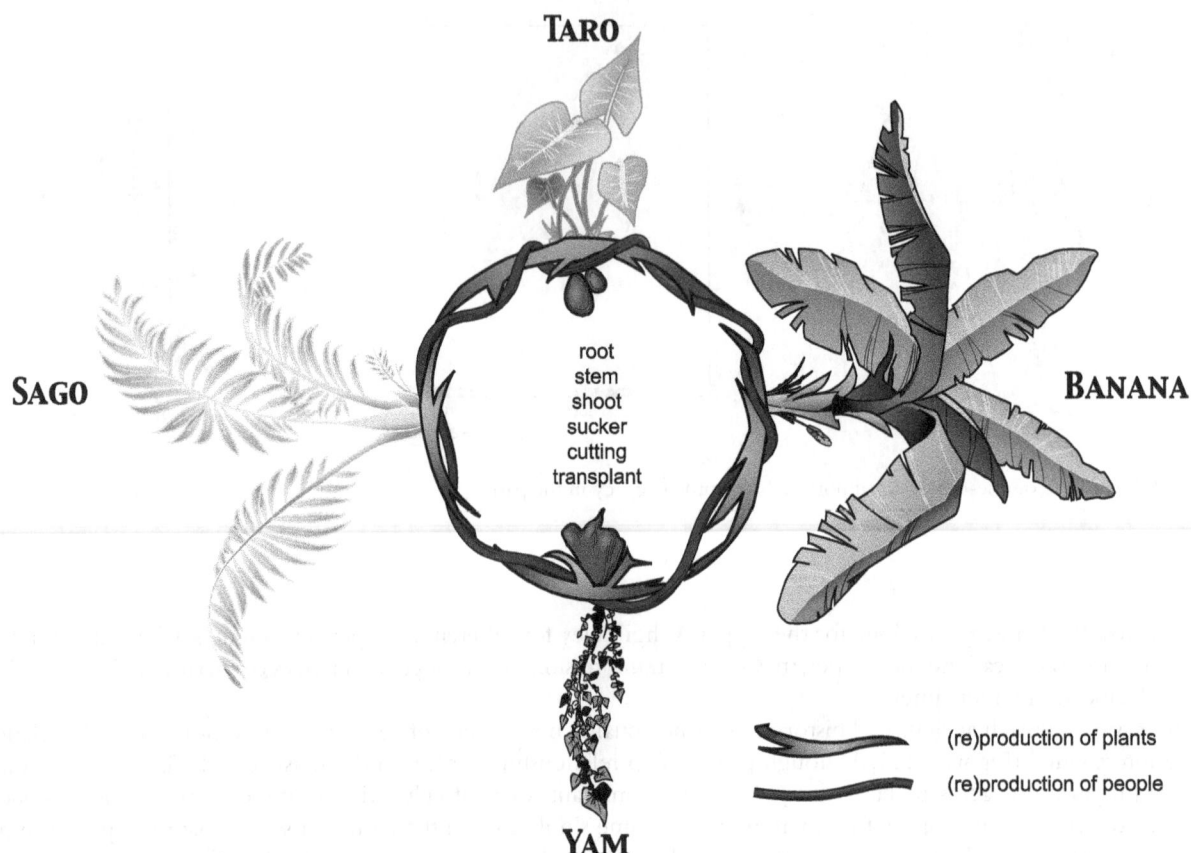

TARO

SAGO

root
stem
shoot
sucker
cutting
transplant

BANANA

(re)production of plants

(re)production of people

YAM

Figure 12.2 Mutual entanglement of the (re)production of plants and the (re)production of social relationships under vegeculture in New Guinea

Note: Designed by Tim Denham following conversations with John Muke

Source: Barton and Denham 2017: Figure 1, original drafted by Phil Scamp

A look to the future

Agriculture in the highlands of New Guinea is facing a number of stark challenges. Highland populations are growing rapidly and people's expectations are higher. There are more mouths to feed and people are growing more cash crops for sale at local or regional markets in order to participate in the market economy through the purchase of goods and food-stuffs, payment of school fees, health services, travel costs, domestic services (electricity, mobile phones), rents and so on. Similar increases in population and social demands may have occurred in the past; namely, they plausibly accompanied or resulted from the widespread adoption of the pig and sweet potato and the initiation of large-scale ritual exchanges in the highlands (following Modjeska 1982).

These new challenges are on a different scale to those encountered previously. For the first time, there is an increasing interdependence between highland production and the world economy. Fortuitously, the vast majority of land in the highlands and across Papua New Guinea is still held under traditional forms of ownership, namely it has not been alien-ated and there is no land market. As a consequence, most communities are able to fall back on local subsistence cultivation as a buffer against the vagaries of the global economy. Eventually, though, a new ceiling on production will be reached.

For groups living on the densely populated valley floors today, there is limited scope to expand their territories. They are hemmed in on all sides. Their only means to increase production is to get smarter, namely to innovate and intensify production. Agricultural intensification brings its own risks, such as increased environmental and soil degradation. Tradi-tional practices may persist, but they will be increasingly supplemented by artificial fertilisers and pesticides. People will need to continue to innovate to make agriculture sustainable into the future.

A reliance by increasing populations on increasingly intensive agriculture within defined territorial groups makes highland communities much more vulnerable to major climatic events. There is less margin for error. The 1997 El Niño year witnessed widespread food shortages in the highlands. A climatic event of similar magnitude today would have even greater short-to-medium-term consequences, both in terms of scale and in terms of social recovery.

In retrospect, the long-term history of agriculture in the highlands of New Guinea is highly relevant to understand the many challenges faced by people living there today. These agronomic challenges are demographic, socio-economic, environmental and climatic. Without an understanding of how people have responded to similar challenges in the past, there is little hope of planning for a successful future.

13 The global significance of early agriculture on New Guinea

The emergence of agriculture in the highlands during the early Holocene is significant to the people who live there today. Agricultural history is taught in schools across Papua New Guinea. Not only do highlanders take pride in the fact that they developed their own form of agriculture, they also view it as significant that they practised agriculture before Europeans and most other places in the world. The significance of early agricultural history within Papua New Guinea is demonstrated by the successful nomination of the Kuk Early Agricultural Site to the World Heritage List in 2008 (DEC 2007; Muke et al. 2007; Denham 2012, 2013c). It is Papua New Guinea's first and only World Heritage Site. The agricultural history of the highlands is globally significant in other ways.

Telling a story from the margins

The New Guinea chronology is amongst the oldest evidence for agriculture in the world. The record of early agriculture in the highlands is comparable in antiquity to that reconstructed for Southwest Asia, China and Central America (Bellwood 2005; Barker 2006). In contrast to most regions, agriculture in New Guinea was based on the vegetative propagation of a whole range of food plants rather than on the seed-based domestication of cereals, legumes and tree crops. Few other places on earth can demonstrate a semi-continuous agricultural history going back to the inception of cultivation in the early Holocene and extending to the present day, such as that documented for the Upper Wahgi Valley. Only a handful of landscapes globally can provide such specific and interwoven lines of evidence. Comparable records include Abu Hureyra in Syria (Moore et al. 2000), the Lower Jordan Valley in the Occupied Territories of Palestine (Kislev et al. 2006; Weiss et al. 2006), the Lower Yangtze Valley (Fuller et al. 2009; Zhao 2011) and the Oaxaca Valley in Mexico (Flannery 1986; Perry and Flannery 2007).

Despite the demonstrated antiquity and significance of the record from the highlands, some remain uncertain about the veracity of the agricultural chronology there and others view it peripherally in their accounts of early agriculture globally (e.g., Smith 1998b; Bellwood 2005; Barker 2006). Certainly, the New Guinea record is unexpected and strange. However, the multidisciplinary evidence clearly demonstrates that the island was not a 'backwater' at the margins of agricultural developments on Eurasia. So why are so many reticent to accept the evidence? Essentially, the New Guinea record challenges many of the preconceptions people hold about early agriculture.

First, early agriculture is often associated with Neolithic cultural traits – such as polished axe-adzes, processing technologies, pottery, sedentism and so on. Although the more widespread adoption of ground axe-adzes is associated with the clearest evidence for early agriculture in the Upper Wahgi Valley, the majority of agricultural tools are relatively 'simple' and include wooden digging sticks, wooden spades and potentially stone hoes (Golson 1977b, 2005). There is no clear 'neolithic' signature in the highlands (Denham 2006); namely, there is not a dramatic shift in material culture preserved in the archaeological record of the highlands during the Holocene that is coincident with the advent of agriculture.

Second, early agriculture in the highlands is not associated with the development of socio-political hierarchies, so-called 'complex' societies and 'the rise of civilisation'. Rather, highland societies are well-known globally for their big men social institutions, in which leaders are communally acknowledged due to their skill at oration, accumulating wealth and status through exchange, fighting or retention of traditional knowledge (Sahlins 1967; also see Godelier 1986). Big men lead through consensus and persuasion rather than by direct command. The traditional societies documented ethnographically in the highlands are relatively small-scale, egalitarian, focussed on reciprocity and mutual balance and broadly anarchic in socio-political structure. At the same time, highland communities are able to engage in large-scale communal projects, whether the construction of major drainage networks in wetlands (Ballard 1995), major exchange ceremonies (Strathern 1971) or warfare (see Clastres 1977 for comparable discussion in South America). Thus, the New Guinea record forces a re-evaluation of lingering unilinear and neo-Darwinian conceptions that often wrap early agriculture into rungs on the evolutionary ladder (critiqued in Golson 1977a), pseudo-scientific concepts of human nature (Diamond

1997), and of the global applicability of models of early agricultural development originally designed for Eurasia (critiqued in Denham et al. 2007); namely, it requires us to fully engage with *difference*.

Third, early agriculture in New Guinea, and vegetative systems generally, are often considered to lack expansive capacity (Harris 2002). Such an argument runs counter to all evidence. Crops domesticated in the New Guinea region spread under cultivation throughout Island Southeast Asia and potentially beyond before c. 3000 years ago. Importantly, several of these vegetatively propagated crops had spread right across the subtropical and tropical Old World – from West Africa to the eastern Pacific – before the age of European exploration. At that time, they had a greater longitudinal range than any other crop plants dispersed by people under cultivation (although see bottle gourd as the possible exception; Kistler et al. 2014). Bananas, taro, sugarcane and yam were fundamental to Polynesian colonisation of the Pacific, perhaps one of the most rapid and widespread geographical colonisation by people ever seen. These same crops had spread westward around the Indian Ocean to Africa before European voyagers entered the Indian Ocean in the late fifteenth century AD. Although the timings of crop introduction to the African continent are poorly known (Murdock 1959), bananas may have reached West Africa over 2300 years ago (Mbida Mindzie et al. 2001; Perrier et al. 2011).

In short, the New Guinea evidence challenges us to confront many of our preconceptions, or prejudices (after Gadamer 1976), about the world and the way that world came to be. These preconceptions say more about ourselves and our cultural background than they do about early agriculture in the past. The New Guinea evidence does not seem to 'fit' into many pre-established frames of reference about early agriculture, societal development and the current world order. But, then, why should it? We need to be careful that we do not eternalise our present when we look back into the past.

We cannot completely avoid our prejudices in the way we construct the past. We can, though, be open to difference in the past instead of collapsing difference using our own frames of reference, whether these focus on 'social evolution', 'human nature' or 'Eurasian models of early agriculture'. As such, New Guinea provides a decentred history that challenges the dominant or 'master' narratives of early agriculture. At the same time, it forces a decentring of global debates about early agriculture to encompass former margins. Early agriculture need not be entwined with discourses on pottery, animal domesticates, sedentism, socio-political hierarchies, civilisation, cereals and sexual reproduction of crops. Instead, early agriculture can emerge with relative continuities in material culture, with dispersed and temporary settlement patterns, with relatively egalitarian societies that persist through time and with vegetative propagation of a range of crop plants. Thus, the long-term history of agriculture in the highlands of New Guinea is an enabling discourse; it opens up new horizons for the interpretation of the past.

Is New Guinea unique within the Asia-Pacific region?

As yet, the New Guinea highlands provide the only record of early agricultural development in the Southeast Asian and Australasian regions. This does not preclude the possibility that early agricultural developments occurred elsewhere in these regions, whether on Borneo (Barker 2014) or northern Australia (Jones and Meehan 1989). At this time, there is no other regional record of plant exploitation that provides a similar level of multidisciplinary detail to that documented in the Upper Wahgi Valley of Papua New Guinea.

In pursuing this theme, there were possibly mosaics of plant exploitation across Southeast Asia and Australasia from the Pleistocene to the recent past. Initially, the first modern human inhabitants of these regions exhibited similar behaviours that reflected how people variably adapted to the places they encountered. Through time, and especially during the Holocene, initial differences in emphasis in plant exploitation practices accumulated to become differences in kind across Island Southeast Asia, lowland Melanesia, the highlands of New Guinea and northern Australia (Denham, Fullagar et al. 2009). These differences in emphasis resulted from differences in practical knowledge, plant resources and cultural preferences in different places. The resultant plant exploitation mosaics likely included forms of planting and cultivation that are archaeologically invisible. These mosaics enabled the westward dispersal of vegetatively propagated plants – such as banana – from New Guinea into Island Southeast Asia and beyond (Denham 2010, 2013d).

Similar mosaics may have extended into northern Australia – where banana, taro and greater yam populations are well established today, although largely assumed to be native or recent introductions (although see Denham, Donohue et al. 2009 for alternative hypotheses). However, it is plausible that communities in northern Australia experimented with some forms of cultivation either through indigenous innovation or through contacts with voyagers from Island Southeast Asia, New Guinea, the Torres Strait and possibly Polynesia before more recent Macassan and European influences. At present, there is no solid archaeobotanical evidence to indicate cultivation in the distant past on the Australian mainland, although few studies have systematically applied the suite of techniques required – primarily parenchyma, phytolith and starch grain analysis. In the absence of data, we should keep an open mind.

Importantly, rather than searching for 'agricultural origins' in other places in Southeast Asia or Australasia, the focus should be upon establishing robust regional records of plant exploitation and developing a common framework for interpretation. The common ground advocated here is practice based, because it focuses upon the constituent practices people engaged in, which may often be common to multiple forms of plant exploitation. In doing so, the monolithic categories of 'foraging' or 'farming' can be avoided; these top-down categories tend to skew interpretation and constrain discourse within inherited frameworks. Instead, a focus on practices enables regional chronologies of plant exploitation to be built from the bottom-up, as done for the Upper Wahgi Valley, and provides a more porous or 'open' framework for interpreting plant exploitation in the past.

Plant exploitation in the tropics is different

Modern humans have been able to adapt to and live in almost every type of environment, from deserts to ice caps (this section reproduced in amended form from Denham 2016b: 412–413). Yet there is a lingering suspicion that people could not have inhabited tropical rainforests on a permanent basis prior to agriculture without altering them to any large extent, or without access to coastal, lacustrine or riverine resources (Bailey et al. 1989; Bailey and Headland 1991). Echoes of this 'green desert' hypothesis persist, but are without conceptual or evidential foundation.

Foremost, on a conceptual level, it makes little sense to set up a scenario that asserts people could not have lived in rainforests unless they significantly modified their environment. This line of argument is artificial and unworldly. Wherever people have gone, they deliberately or inadvertently transformed their environment, and this is also true of non-agricultural societies. People have altered species compositions by fishing, hunting and collecting. People burned vegetation to create niches for food procurement and to intensify the density of favoured resources in the landscape, whether encouraging plant food production or to aid hunting, as well as to create certain kinds of landscapes and for pleasure. People have also been responsible for redistributing species around the landscape, perhaps often inadvertently through food discard and rubbish heaps, as well as through the creation of favourable ecological niches around settlements and through the deliberate translocation of plants and animals.

Wherever people have gone, they have created cultural landscapes. Rainforests are no longer recognised as 'virgin' or pristine environments (Willis et al. 2004); rather, they are cultured, some might say 'domesticated' landscapes (Yen 1989; Terrell et al. 2003). Nowadays, the biodiversity of rainforests is acknowledged to be partially anthropic (Haberle 2007); they are palimpsests that bear the cumulative imprint of human activities from first settlement to the present.

On New Guinea, palaeoecological records indicate that people have disturbed montane and lowland tropical rainforests for millennia, potentially since first colonisation (Groube 1989). Although the stone tool kit of early colonisers, including the waisted blade, was not suitable for chopping trees down, it would have enabled people to kill trees by ring-barking, clear and smash undergrowth and prevent regrowth. Together with fire, this tool kit was enough for people to start transforming their rainforest environment. The more widespread adoption of fully ground adzes during the mid-Holocene would have enabled people to chop down trees (Christensen 1975a). Accelerated rates of cumulative degradation and clearance of some montane valleys can be traced to this time (Hope and Haberle 2005; Haberle et al. 2012). Certainly, large-scale transformations to tropical rainforests in the interior of New Guinea, including for agriculture, had occurred before the advent of metal tools within the last few hundred years.

Having recognised that people have transformed rainforests for as long as they have been living in them, it should be noted that tropical rainforests are not necessarily abundant in 'ready-at-hand' edible resources. People have learned to live in rainforests – a 'landscape learning' process that extends over tens of millennia. Food availability in tropical rainforests is certainly in 'the eye of the beholder'. Whereas a stranger to a rainforest in the highlands of New Guinea may be first drawn to edible fruits, nuts and tubers – and much archaeological literature focuses on these plant groups – this misses a whole range of other edible plants, including: palms, ferns and grasses rich in sago, or edible pith; grasses that yield sugar-rich juices or are cooked as vegetables; the various ferns and leaves that can be eaten cooked as green vegetables or eaten raw as salad; and the numerous flowers, inflorescences and seeds (Powell 1976; Kocher Schmid 1991). Broad spectrum plant exploitation practices provided the range of nutrients to sustain human diets in the New Guinea rainforest prior to a dependence on cultivation.

Starch-, protein- and oil-rich plants tend to be more visible in the archaeological record, either through macrobotanical remains of nut shells, fruit stones and the occasional charred tuber, or through microfossil analysis of preserved phytoliths and starch grains. Although these food groups are fundamental to human diet, a carbohydrate fixation has emerged in some archaeological literature – most closely associated with Optimal Foraging Theory – that fails to adequately encompass broad spectrum plant exploitation (see consideration with respect to New Guinea by Denham and Barton 2006). People in the past would have exploited a wider range of plants than most present-day horticulturalists. Most contemporary agricultural societies have a relatively narrow diet breadth, especially in terms of staples and vegetables; consequently, they are not good analogues to understand plant exploitation in the distant past.

New possibilities for understanding how people adapted to tropical rainforests in the past emerge once we move beyond several common misconceptions, namely, that rainforests were green deserts, people needed metal tools to clear tropical rainforests and people focussed on carbohydrate-rich plants. The rainforests of New Guinea are a cultural landscape; they bear the imprint of a human presence that has lasted tens of millennia. Today most groups living in these rainforests are reliant on cultivation and the rearing of pigs, yet it can be envisaged that people exploited a broader array of faunal and floral resources in the past, even though many of these practices would be hard to trace archaeologically. These broad spectrum practices enabled people to live permanently within the tropical rainforests of the highlands, even though population densities were probably low and mobility high (Denham and Barton 2006).

People in several wet tropical regions – whether the lowland neotropics (Piperno and Pearsall 1998), West Africa (Mbida Mindzie et al. 2001), southern India (Fuller et al. 2004), Southeast Asia (Higham 2002; Barker 2014) and New Guinea – developed or adopted agriculture. Cultivation in these wet tropical regions exhibit several commonalities with each other and some differences with cereal-based cultivation (Harris 1969, 1972, 2002; Hildebrand 2007). Foremost, they tend to focus on vegetative propagation of a wide range of plant types, including herbs, grasses, root-crop, shrubs and trees-palms-pandanus. Additionally, early cultivation occurred within small clearings in the rainforest, in disturbed gaps or along ecotones. Plots were usually cultivated for a few years before abandonment due to rapid nutrient cycling and depletion of soils within tropical rainforests. Slash-and-burn or slash-and-mulch practices were commonly adopted to release nutrients from cleared vegetation into the soil and prolong fertility for crops. Although the palaeoecological visibility of these practices is limited and hard to differentiate from previous disturbance regimes, they often had long-term influences on vegetation composition, including forest degradation and replacement with secondary and grass taxa (Haberle 1994, 2007; Haberle et al. 2012). These commonalities reflect a shared orientation of anatomically modern humans to their world, even though this orientation was differentially expressed in specific historical and geographic settings.

Agricultural history for each region of the world needs to be considered in its own terms. The investigation of early agriculture in the highlands of New Guinea has required the adoption of an innovative suite of multidisciplinary techniques, the development of a new practice-based methodological approach and the formulation of a new conception of early agriculture. The range of techniques applied to the investigation of early agriculture includes mixed-method archaeobotany, palaeoecology and stratigraphic analyses. A practice-based approach focuses on establishing a chronology of constituent practices that are contextually bundled into forms of plant exploitation in the past. A new conceptual framework encapsulates the multidimensional character of early agriculture, as manifest in each region of the world.

As discussed with respect to highland New Guinea, the character of early agriculture in tropical rainforests requires a different approach than the investigation of cereal-based agro-ecosystems. Rather than focussing upon morphological changes in plant species or a reliance on domesticates, multiple lines of evidence are needed to try and infer the dependence of people upon cultivated food in the past. Sometimes these traces are ambiguous or faint, such as the 10,000-year-old activities on the wetland margin at Kuk, or are likely to be destroyed, such as archaeological evidence for shifting cultivation. In other cases, the consilience of multiple lines of evidence can be used to clearly infer what people were doing in the past and whether this represents a form of cultivation. The application of an innovative suite of archaeobotanical and palaeoecological techniques – principally phytolith, pollen and microcharcoal analyses in concert with residue analysis – as well as renewed archaeological and mixed-method stratigraphic investigations, clearly identified the mixed cultivation of bananas and other crops using mounds along the wetland margin at Kuk approximately 7000–6400 years ago.

The search for origins: a regulative idea

> We can only find and posit the origin in relation to that which has already become. For if we speculate, we can certainly construct an unmitigated and absolute beginning in which we could devoutly believe; but we could not prove it historically. . . . Furthermore, if the results are presented in the form of narration, they will have been assigned an origin ad hoc which, therefore, can only be a relative one. This is important to note because, in explaining the origin of that which has become, the genetic mode of narration consistently leads one to the misconception that it is possible to verify historically why things had to come into being and why they had to become what they became.
>
> (Johann Gustav Droysen, *The Investigation of Origins,* in Mueller-Vollmer 1985: 124–125)

In this quote, Droysen indicates that the search for origins is relative and somewhat arbitrary. This is clearly seen in debates concerning early agriculture. Depending upon the definition proffered – whether based on domesticates, dependence upon domesticates, environmental transformation and social dependence on cultivation and so on – different origins can be brought forth from the past. Ultimately, narratives about origins are narratives about the present. We seek to find the origins of something in the past because that particular trait is significant to us in the present. We rarely ask the question: why is this trait so important?

From this perspective, debates about agriculture are not fundamentally solely to do with plants, domestication and cultivation. Agriculture is a cipher for something else. It is a marker that differentiates – farmer from forager, agriculturalist from hunter-gatherer, and 'us' from 'them'. A concern with agriculture is important because it feeds into debates about how we conceive relationships between people today and in the recent, historical past.

Agriculturalists are often portrayed as dynamic, more 'advanced', more civilised and more dominant, whereas non-agriculturalists are unchanging, less advanced, less civilised and more marginal. These are not new themes; they recur through centuries, if not millennia, of history and are translated into contemporary political and social arenas. To some degree, debates about early agriculture are normalising discourses that legitimate inequality and dominance in the present.

Importantly, early agriculture in New Guinea subverts this type of reasoning. The New Guinea highlands provide an example of early agriculture that does not fit the traditional paradigm. It is a voice from the margins that challenges orthodox views and hopefully makes people reflect upon the presuppositions embedded in their reconstructions of the past. As such, it has reflexive, as well as substantive, value.

The investigation of early agriculture in the highlands of New Guinea is a challenging and unusual story that enriches our understanding of the human journey. Although the search for origins may be arbitrary, it is not futile. Rather than being viewed as a precise point in space or time that lies waiting to be uncovered, 'origins' are regulative ideas that drive debate forward. A regulative idea is something that is needed to start a line of enquiry and is continually being worked towards, even though it may never be reached. It forever remains beyond our grasp. Thus, the true value of investigations into early agriculture in any region of the world is not in finding the origin, it is in the searching.

Bibliography

Abraham, K. and P. Gopinathan Nair 1991. Polyploidy and sterility in relation to sex in *Dioscorea alata* L. (Dioscoreaceae). *Genetica* 83: 93–97.

Alexander, J. and D.G. Coursey 1969. The origins of yam cultivation. In P.J. Ucko and G.W. Dimbledy (eds.), *The Domestication and Exploitation of Plants and Animals*, pp. 405–425. London: Gerald Duckworth.

Allaby, R.G. 2014. Domestication syndrome in plants. In C. Smith (ed.), *Encyclopedia of Global Archaeology*, pp. 2182–2184. New York: Springer.

Allen, B. 2000. The 1997–98 Papua New Guinea drought: perceptions of disaster. In R.H. Grove and J. Chappell (eds.), *El Niño – History and Crisis*, pp. 109–122. Cambridge: White Horse Press.

Allen, B. 2005. The place of agricultural intensification in Sepik foothills prehistory. In A. Pawley, R. Attenborough, J. Golson and R. Hide (eds.), *Papuan Pasts: Cultural, Linguistic and Biological Histories of Papuan-Speaking Peoples*, pp. 585–623. Pacific Linguistics 572. Canberra: Pacific Linguistics, Research School of Pacific and Asian Studies, The Australian National University.

Allen, B. and C. Ballard 2001. Beyond intensification: reconsidering agricultural transformations. *Asia Pacific Viewpoint* 42(2–3): 157–162.

Allen, B. and R.M. Bourke 2009. People, land and environment. In R.M. Bourke and T. Harwood (eds.), *Food and Agriculture in Papua New Guinea*, pp. 27–127. Canberra: ANU E Press.

Allen, J. 1970. Prehistoric agricultural systems in the Wahgi Valley – a further note. *Mankind* 7: 177–183.

Allen, J. 1972. The first decade in New Guinea archaeology. *Antiquity* XLVI: 180–190.

Ambrose, W. 1975. Stabilizing degraded swamp wood by freeze drying. Reprint 75/8/4, *ICOM Committee for Conservation, 4th Triennial Meeting*, Venice.

Ambrose, W. 1990. Application of freeze-drying to archaeological wood. In R.M. Rowell and R.J. Barbour (eds.), *Archaeological Wood: Properties, Chemistry and Conservation*, pp. 235–261. Advances in Chemistry Series No. 225. Washington, DC: American Chemical Society.

Ambrose, W. 1991. Manus, mortars and the kava concoction. In A. Pawley (ed.), *Man and a Half: Essays in Pacific Anthropology and Ethnobiology in Honour of Ralph Bulmer*, pp. 461–469. Auckland: The Polynesian Society.

Anderson, A. 2009. The rat and the octopus: Initial human colonization and the prehistoric introduction of domestic animals to Remote Oceania. *Biological Invasions* 11: 1503–1519.

Anderson, A. and G. Summerhayes 2008. Edge-ground and waisted axes in the western Pacific Islands: implications for an example from the Yaeyama Islands, southernmost Japan. *Asian Perspectives* 47: 45–58.

Aplin, K. 1981. *Kamapuk fauna: a late Holocene vertebrate faunal sequence from the Western Highlands District, Papua New Guinea with implications for palaeoecology and archaeology*. Unpublished BA (Hons.) thesis, Australian National University.

Argent, G.C.G. 1976. The wild bananas of Papua New-Guinea. *Notes of the Royal Botanical Gardens of Edinburgh* 35: 77–114.

Arnau, G., K. Abraham, M.N. Sheela, H. Chair, A. Sartie and R. Asiedu 2010. Yams. In J.E. Bradshaw (ed.), *Root and Tuber Crops*, pp. 127–148. London: Springer.

Arnaud, E. and J.P. Horry (eds.) 1997. *Musalogue: A Catalogue of Musa Germplasm: Papua New Guinea Collecting Missions, 1988–1990*. Montpellier: International Network for the Improvement of Banana and Plantain.

Asch, D.L. and J.P. Hart 2004. Crop domestication in prehistoric eastern North America. In *Encyclopedia of Plant and Crop Science*, pp. 314–319. New York: Marcel Dekker.

Asouti, E. and D.Q. Fuller 2013. A contextual approach to the emergence of agriculture in Southwest Asia. *Current Anthropology* 54: 299–345.

Bailey, R.C., G. Head, M. Jenike, B. Owen, R. Rechtman and E. Zechenter 1989. Hunting and gathering in tropical rain forest: is it possible? *American Anthropologist* 91: 59–82.

Bailey, R.C. and T.N. Headland 1991. The tropical rainforest: is it a productive environment for human foragers? *Human Ecology* 19: 261–285.

Balée, W. 1994. *Footprints of the Forest: Ka'apor Ethnobotany*. New York: Columbia University Press.

Ballard, C. 1995. *The death of a great land: ritual, history and subsistence revolution in the Southern Highlands of Papua New Guinea*. Unpublished PhD thesis, Australian National University.

Ballard, C. 2000. Condemned to repeat history? ENSO-related drought and famine in Irian Jaya, Indonesia, 1997–98. In R.H. Grove and J. Chappell (eds.), *El Nino, History and Crisis: Studies From the Asia-Pacific Region*, pp. 132–148. Cambridge: The White Horse Press.

Ballard, C. 2001. Wetland drainage and agricultural transformations in the Southern Highlands of Papua New Guinea. *Asia Pacific Viewpoint* 42: 287–304.

Ballard, C. 2017. The wetland field systems of the New Guinea highlands. In J. Golson, T.P. Denham, P.J. Hughes, P. Swadling and J. Muke (eds.), *Ten Thousand Years of Cultivation at Kuk Swamp in the Highlands of Papua New Guinea*, pp. 65–83. Terra Australis 46. Canberra: ANU E Press.

Ballard, C., P. Brown, R.M. Bourke and T. Harwood (eds.) 2005. *The Sweet Potato in Oceania: A Reappraisal.* Ethnology Monographs 19 and Oceania Monograph 56. Pittsburgh and Camperdown: University of Pittsburgh and University of Sydney.

Ballard, C., T.P. Denham and S.G. Haberle 2013. Wetland archaeology in the highlands of New Guinea. In F. Menotti and A. O'Sullivan (eds.), *The Oxford Handbook of Wetland Archaeology,* pp. 231–248. Oxford: Oxford University Press.

Balme, J. 2000. Excavations revealing 40,000 years of occupation at Mimbi Caves, south central Kimberley, Western Australia. *Australian Archaeology* 5: 1–5.

Barker, G. 2006. *The Agricultural Revolution in History.* Oxford: Oxford University Press.

Barker, G. (ed.) 2014. *Rainforest Foraging and Farming in Island Southeast Asia.* Cambridge: McDonald Institute Monographs.

Barrau, J. 1955. *Subsistence Agriculture in Melanesia.* Noumea: New Caledonia.

Barrau, J. 1963. Introduction. In J. Barrau (ed.), *Plants and the Migrations of Pacific Peoples: A Symposium,* pp. 1–6. Honolulu: Bishop Museum Press.

Barrau, J. 1965. Witnesses of the past: notes on some food plants of Oceania. *Ethnology* 4(3): 282–294.

Barrett, J. 1994. *Fragments From Antiquity: An Archaeology of Social Life 2900–1200 BC.* Oxford: Blackwell.

Barton, H. 2012. The reversed fortunes of sago and rice *Oryza sativa* in the rainforests of Sarawak Borneo. *Quaternary International* 249: 96–104.

Barton, H. and T.P. Denham 2011. Prehistoric vegeculture and social life in Island Southeast Asia and Melanesia. In G. Barker and M. Janowski (eds.), *Why Cultivate? Anthropological and Archaeological Approaches to Foraging-Farming Transitions in Southeast Asia,* pp. 17–25. McDonald Institute Monograph. Cambridge: McDonald Institute for Archaeological Research.

Barton, H. and T.P. Denham 2017. Vegecultures and the social-biological transformations of plants and people. *Quaternary International.*

Barton, H., T.P. Denham, K. Neumann and M. Arroyo-Kalin 2012. Long-term perspectives on human occupation of tropical rainforests: An introductory overview. *Quaternary International* 249: 1–3.

Barton, H. and V. Paz 2007. Subterranean diets in the tropical rain forests of Sarawak, Malaysia. In T.P. Denham, J. Iriarte and L. Vrydaghs (eds.), *Rethinking Agriculture: Archaeological and Ethnoarchaeological Perspectives,* pp. 50–77. Walnut Creek: Left Coast Press.

Barton, H., P. Piper and R. Rabett 2009. Composite hunting technologies from the Terminal Pleistocene and Early Holocene, Niah Cave, Borneo. *Journal of Archaeological Science* 36: 1708–1714.

Bates, D.M. and R.W. Robinson 1995. Cucumbers, melons and water-melons: Cucumis and Citrillus (Cucurbitaceae). In J. Smartt and N.W. Simmonds (eds.), *Evolution of Crop Plants,* pp. 89–96. Harlow: Longman.

Bayliss-Smith, T. 1985. Pre-Ipomoean agriculture in the New Guinea Highlands above 2000 metres: some experimental data on taro cultivation. In I. Farrington (ed.), *Prehistoric Intensive Agriculture in the Tropics,* pp. 285–320. Oxford: British Archaeological Reports, International Series 232, Part I.

Bayliss-Smith, T. 1988. Prehistoric agriculture in the New Guinea Highlands: problems of defining the altitudinal limits to growth. In J.L. Bintliff, D.A. Davidson and E.G. Grant (eds.), *Conceptual Issues in Environmental Archaeology,* pp. 153–160. Edinburgh: Edinburgh University Press.

Bayliss-Smith, T. 1996. People-plant interactions in the New Guinea highlands: agricultural hearthland or horticultural backwater? In D.R. Harris (ed.), *The Origins and Spread of Agriculture and Pastoralism in Eurasia,* pp. 499–523. London: University College London Press.

Bayliss-Smith, T. 2007. The meaning of ditches: interpreting the archaeological record using insights from ethnography. In T.P. Denham, J. Iriarte and L. Vrydaghs (eds.), *Rethinking Agriculture: Archaeological and Ethnoarchaeological Perspectives,* pp. 126–148. Walnut Creek: Left Coast Press.

Bayliss-Smith, T. and J. Golson 1992a. Wetland agriculture in New Guinea Highlands prehistory. In B. Coles (ed.), *The Wetland Revolution in Prehistory,* pp. 15–27. Exeter: The Prehistoric Society and Wetland Archaeological Research Project.

Bayliss-Smith, T. and J. Golson 1992b. A Colocasian revolution in the New Guinea Highlands? Insights from Phase 4 at Kuk. *Archaeology in Oceania* 27: 1–21.

Bayliss-Smith, T. and J. Golson 1999. The meaning of ditches: deconstructing the social landscapes of drainage in New Guinea, Kuk, Phase 4. In C. Gosden and J. Hather (eds.), *The Prehistory of Food: Appetites for Change,* pp. 199–231. London: Routledge.

Bayliss-Smith, T., J. Golson and P.J. Hughes 2017a. Phase 4: Major disposal channels, slot-like ditches and grid-patterned fields. In J. Golson, T.P. Denham, P.J. Hughes, P. Swadling and J. Muke (eds.), *Ten Thousand Years of Cultivation at Kuk Swamp in the Highlands of Papua New Guinea,* pp. 239–268. Terra Australis 46. Canberra: ANU E Press.

Bayliss-Smith, T., J. Golson and P.J. Hughes 2017b. Phase 5: Retreating forests, flat-bottomed ditches and raised fields. In J. Golson, T.P. Denham, P.J. Hughes, P. Swadling and J. Muke (eds.), *Ten Thousand Years of Cultivation at Kuk Swamp in the Highlands of Papua New Guinea,* pp. 269–296. Terra Australis 46. Canberra: ANU E Press.

Bayliss-Smith, T., J. Golson and P.J. Hughes 2017c. Phase 6: Impact of the sweet potato on swamp landuse, pig rearing and exchange relations. In J. Golson, T.P. Denham, P.J. Hughes, P. Swadling and J. Muke (eds.), *Ten Thousand Years of Cultivation at Kuk Swamp in the Highlands of Papua New Guinea,* pp. 297–324. Terra Australis 46. Canberra: ANU E Press.

Bellwood, P. 2005. *First Farmers.* Oxford: Blackwell.

Bellwood, P. and C. Renfrew (eds.) 2002. *Examining the Farming/Language Dispersal Hypothesis.* Cambridge: McDonald Institute for Archaeological Research.

Bender, B. 1978. Gatherer-hunter to farmer: a social perspective. *World Archaeology* 10: 204–222.

Berding, N. and H. Koike 1980. Germplasm conservation of the *Saccharum* complex: A collection from the Indonesian Archipelago. *Hawaiian Planters Record* 59: 87–176.

Berkelhammer, M., A. Sinha, L. Stott, H. Cheng, F.S.R. Pausata and K. Yoshimura 2013. An abrupt shift in the Indian monsoon 4000 years ago. In L. Giosan, D.Q. Fuller, K. Nicoll, R.K. Flad and P.D. Clift (eds.), *Climates, Landscapes, and Civilizations,* pp. 75–87. Washington, DC: American Geophysical Union.

Bird, M.I., C. Hunt and D. Taylor 2005. Palaeoenvironments of insular Southeast Asia during the Last Glacial Period: a savanna corridor in Sundaland? *Quaternary Science Reviews* 24: 2228–2242.

Blackwood, B. 1950. *The Technology of a Modern Stone Age People in New Guinea*. Oxford: Oxford University Press.

Bleeker, P. 1983. *Soils of Papua New Guinea*. Canberra: CSIRO and ANU Presses.

Blong, R.J. 1986. Pleistocene volcanic debris avalanche from Mount Hagen, Papua New Guinea. *Australian Journal of Earth Sciences* 33: 287–294.

Blong, R.J. 2017. Tibito tephra, *taim tudak* and the impact of thin tephra falls. In J. Golson, T.P. Denham, P.J. Hughes, P. Swadling and J. Muke (eds.), *Ten Thousand Years of Cultivation at Kuk Swamp in the Highlands of Papua New Guinea*, pp. 133–144. Terra Australis 46. Canberra: ANU E Press.

Blong, R.J., T. Wagner and J. Golson 2017. Volcanic ash at Kuk. In J. Golson, T.P. Denham, P.J. Hughes, P. Swadling and J. Muke (eds.), *Ten Thousand Years of Cultivation at Kuk Swamp in the Highlands of Papua New Guinea*, pp. 117–132. Terra Australis 46. Canberra: ANU E Press.

Boivin, N.L., M.A. Zeder, D.Q. Fuller, A. Crowther, G. Larson, J.M. Erlandson, T.P. Denham and M.D. Petraglia 2016. Ecological consequences of human niche construction: Examining long-term anthropogenic shaping of global species distributions. *Proceedings of the National Academy of Sciences (USA)* 113: 6388–6396.

Booth, R.K., S.T. Jackson, S.L. Forman, J.E. Kutzbach, I. Bettis, J. Kreig and D.K. Wright 2005. A severe centennial-scale drought in midcontinental North America 4200 years ago and apparent global linkages. *Holocene* 15: 321–328.

Bourdieu, P. 1990. *The Logic of Practice*. Cambridge: Polity Press.

Bourke, R.M. 1989. *Altitudinal limits of 220 economic crop species in Papua New Guinea*. Unpublished manuscript. Canberra: Department of Human Geography, Research School of Pacific and Asian Studies, The Australian National University.

Bourke, R.M. 1996. Edible indigenous nuts in Papua New Guinea. In M.L. Stevens, R.M. Bourke and B.R. Evans (eds.), *South Pacific Indigenous Nuts*, pp. 45–55. Canberra: Australian Centre for International Agricultural Research.

Bourke, R.M. 2001. Intensification of agricultural systems in Papua New Guinea. *Asia Pacific Viewpoint* 42: 219–236.

Bourke, R.M. 2009. History of agriculture in Papua New Guinea. In R.M. Bourke and T. Harwood (eds.), *Food and Agriculture in Papua New Guinea*, pp. 10–26. Canberra: ANU E Press.

Bourke, R.M. 2017. Environment and food production in Papua New Guinea. In J. Golson, T.P. Denham, P.J. Hughes, P. Swadling and J. Muke (eds.), *Ten Thousand Years of Cultivation at Kuk Swamp in the Highlands of Papua New Guinea*, pp. 51–64. Terra Australis 46. Canberra: ANU E Press.

Bourke, R.M. and B. Allen 2009. Village food production systems. In R.M. Bourke and T. Harwood (eds.), *Food and Agriculture in Papua New Guinea*, pp. 193–269. Canberra: ANU E Press.

Bourke, R.M., C. Camarotto, E.J. D'Souza, K. Nema, T.N. Tarepe and S. Woodhouse 2004. *Production Patterns of 180 Economic Crops in Papua New Guinea*. Canberra: Coombs Academic Publishing.

Bourke, R.M. and T. Harwood (eds.) 2009. *Food and Agriculture in Papua New Guinea*. Canberra: ANU E Press.

Bowdery, D. 1999. Phytoliths from tropical sediments: reports from Southeast Asia and Papua New Guinea. *Bulletin of the Indo-Pacific Prehistory Association* 18: 159–168.

Bowers, N. 1964. A further note on a recently reported root crop from the New Guinea highlands. *Journal of the Polynesian Society* 73: 333–335.

Bowers, N. 1968. *The ascending grasslands: an anthropological study of ecological succession in a high mountain valley of New Guinea*. Unpublished PhD thesis, Columbia University. Ann Arbor, MI: University Microfilms International.

Bowler, J.M., H. Johnston, J.M. Olley, J.R. Prescott, R.G. Roberts, W. Shawcross and N.A. Spooner 2003. New ages for human occupation and climatic change at Lake Mungo, Australia. *Nature* 421: 837–840.

Bowman, D.M.J.S., M. Garde and A. Saulwick 2001. *Kunj-ken makka man-wurrk* (fire is for kangaroos) landscape burning in central Arnhem land seen through an ethnographic lens. In A. Anderson, I. Lilley and S. O'Connor (eds.), *Histories of Old Ages: Essays in Honour of Rhys Jones*, pp. 61–78. Canberra: Pandanus Books.

Bradley, R. 1978. Prehistoric field systems in Britain and northwest Europe – a review of some recent work. *World Archaeology* 9: 265–280.

Bronk Ramsey, C. 1995. Radiocarbon calibration and analysis of stratigraphy: The OxCal program. *Radiocarbon* 37: 425–430.

Brookfield, H.C. 1964. The ecology of Highland settlement: some suggestions. *American Anthyropologist* 66(4, Part 2): 20–38.

Brookfield, H.C. 1989. Frost and drought through time and space, Part III: what were conditions like when the high valleys were first settled? *Mountain Research and Development* 9: 306–321.

Brookfield, H.C. and P. Brown 1963. *Struggle for Land*. Melbourne: Oxford University Press.

Brookfield, H.C. and D. Hart 1971. *Melanesia*. London: Methuen.

Brookfield, H.C. and J.P. White 1968. Revolution or evolution in the prehistory of the New Guinea highlands. *Ethnology* 7: 43–52.

Bruno, M. 2009. Practice and history in the transition to food production. *Current Anthropology* 50: 703–706.

Bulbeck, D. 2007. Where the river meets the sea: a parsimonious model for *Homo sapiens* colonisation of the Indian Ocean rim and Sahul. *Current Anthropology* 48: 315–321.

Bulbeck, D. 2008. An integrated perspective on the Austronesian diaspora: the switch from cereal agriculture to maritime foraging in the colonisation of Island Southeast Asia. *Australian Archaeology* 67: 31–51.

Bulmer, R.N.H. 1964. Edible seeds and prehistoric stone mortars in the Highlands of east New Guinea. *Man* 64: 147–150.

Bulmer, R.N.H. 1965. Beliefs concerning the propagation of new varieties of sweet potato in two New Guinea highland societies. *Journal of the Polynesian Society* 74: 237–239.

Bulmer, R.N.H. 1966. Birds as possible agents in the propagation and dispersal of sweet potato. *Emu* 65: 165–182.

Bulmer, R.N.H. 1967. Why is the cassowary not a bird? A problem of zoological taxonomy among the Karam of the New Guinea highlands. *Man* (NS) 2: 5–25.

Bulmer, R.N.H. 1968. Worms that croak and other mysteries of Karam natural history. *Mankind* 6: 621–639.

Bulmer, R.N.H. 1982. Crop introductions and their consequences in the Upper Kaironk Valley, Simbai area, Madang Province. In R.M. Bourke and V. Kesavan (eds.), *Proceedings of the Second Papua New Guinea Food Crops Conference*, volume 2, pp. 282–288. Port Moresby: DPI.

Bulmer, R.N.H. and J.I. Menzies 1972. Karam classification of marsupials and rodents. *Journal of the Polynesian Society* 81: 472–499.

Bulmer, R.N.H. and J.I. Menzies 1973. Karam classification of marsupials and rodents: Part 2. *Journal of the Polynesian Society* 82: 86–107.

Bulmer, R.N.H., J.I. Menzies and F. Parker 1975. Kalam classification of reptiles and fishes. *Journal of the Polynesian Society* 84: 267–308.

Bulmer, S. 1964. Prehistoric stone implements from the New Guinea Highlands. *Oceania* 34: 246–268.

Bulmer, S. 1966. *The prehistory of the Australian New Guinea Highlands: a discussion of archaeological field survey and excavations, 1959–60*. Unpublished MA thesis, University of Auckland.

Bulmer, S. 1973. *Notes on 1972 excavations at Wañelek, an open settlement site in the Kaironk Valley, Papua New Guinea*. Department of Anthropology Working Paper 29. Auckland: University of Auckland.

Bulmer, S. 1975. Settlement and economy in prehistoric Papua New Guinea: a review of the archaeological evidence. *Journal de la Société des Océanistes* 31: 7–75.

Bulmer, S. 1977a. Between the mountain and the plain: prehistoric settlement and environment in the Kaironk Valley. In J.H. Winslow (ed.), *The Melanesian Environment*, pp. 61–73. Canberra: ANU Press.

Bulmer, S. 1977b. Waisted blades and axes. In R.V.S. Wright (ed.), *Stone Tools as Cultural Markers: Change, Evolution and Complexity*, pp. 40–59. Canberra: Australian Institute of Aboriginal Studies.

Bulmer, S. 1991. Variation and change in stone tools in the highlands of Papua New Guinea: the witness of Wanelek. In A. Pawley (ed.), *Man and a Half: Essays in Pacific Anthropology and Ethnobiology in Honour of Ralph Bulmer*, pp. 470–478. Auckland: The Polynesian Society.

Bulmer, S. 2005. Reflections in stone: axes and the beginnings of agriculture in the central highlands of New Guinea. In A. Pawley, R. Attenborough, J. Golson and R. Hide (eds.), *Papuan Pasts: Cultural, Linguistic and Biological Histories of Papuan-Speaking Peoples*, pp. 387–450. Pacific Linguistics 572. Canberra: Pacific Linguistics, Research School of Pacific and Asian Studies, The Australian National University.

Bulmer, S. and R.N.H. Bulmer 1964. The prehistory of the Australian New Guinea Highlands. *American Anthropologist* 66(4, Pt.2): 39–76.

Burkill, I.H. 1935. *A Dictionary of the Economic Products of the Malay Peninsula. 2 Volumes*. London: Crown Agents.

Burton, J. 1984. *Axe makers of the Wahgi: Pre-colonial industrialists of the Papua New Guinea highlands*. PhD thesis, The Australian National University. Ann Arbor, MI: University Microfilms International.

Caballero, J. 2004. Patterns in human – plant interaction: an evolutionary perspective. Paper presented at the *International Society of Ethnobiology, Ninth International Congress*, Canterbury, England. Canterbury, 13–17 June 2004.

Carreel, F., D. Gonzalez de Leon, P. Lagoda, C. Lanaud, C. Jenny, J-P. Horry and H. Tezenas du Montcel 2002. Ascertaining maternal and paternal lineage within *Musa* chloroplast and mitochondrial DNA RFLP analyses. *Genome* 45: 679–692.

Castle, D.A., J. McCunnall and I.M. Tring 1984. *Field Drainage: Principles and Practices*. London: Batsford.

Cauvin, J. 2000. *The Birth of the Gods and the Origins of Agriculture*. Cambridge: Cambridge University Press.

Chaïr, H., R.E. Traore, M.F. Duval, R. Rivallan, A. Mukherjee, L.M. Aboagye, W.J. Van Rensburg, V. Andrianavalona, M.A.A. Pinheiro de Carvalho, F. Saborio, M. Sri Prana, B. Komolong, F. Lawac and V. Lebot 2016. Genetic diversification and dispersal of taro (*Colocasia esculenta* (L.) Schott). *PLOS ONE* 11(6): e0157712. doi: 10.1371/journal.pone.0157712

Chappell, J. 2005. Geographic changes of coastal lowlands in the Papuan past. In A. Pawley, R. Attenborough, J. Golson and R. Hide (eds.), *Papuan Pasts: Cultural, Linguistic and Biological Histories of Papuan-Speaking Peoples*, pp. 525–540. Pacific Linguistics 572. Canberra: Pacific Linguistics, Research School of Pacific and Asian Studies, The Australian National University.

Chartres, C.J. and C.F. Pain 1984. A climosequence of soils in Late Quaternary volcanic ash in Highland New Guinea. *Geoderma* 32: 131–155.

Chikwendu, V.E. and C.E.A. Okezie 1989. Factors responsible for the enoblement of African yams: inferences from experiments in yam domestication. In D.R. Harris and G.C. Hillman (eds.), *Foraging and Farming: The Evolution of Plant Domestication*, pp. 344–357. London: Unwin Hyman.

Childe, V.G. 1936. *Man Makes Himself*. London: Watts.

Christensen, O.A. 1975a. Hunters and horticulturalists: a preliminary report of the 1972–4 excavations in the Manim Valley, Papua New Guinea. *Mankind* 10: 24–36.

Christensen, O.A. 1975b. A tanged blade from the New Guinea Highlands. *Mankind* 10: 37–39.

Clarke, A.C., M.K. Burtenshaw, P.A. McLenachan, D.L. Erickson and D. Penny 2006. Reconstructing the origins and dispersal of the Polynesian bottle gourd (*Lagenaria siceraria*). *Molecular Biology and Evolution* 23: 893–900.

Clarke, W.C. 1971. *Place and People: An Ecology of a New Guinea Community*. Berkeley: University of California Press.

Clarkson, C., Z. Jacobs, B. Marwick, R. Fullagar, L. Wallis, M. Smith, R.G. Roberts, E. Hayes, K. Lowe, X. Carah, S.A. Florin, J. McNeill, D. Cox, L.J. Arnold, Q. Hua, J. Huntley, H.E.A. Brand, T. Manne, A. Fairbairn, J. Shulmeister, L. Lyle, M. Salinas, M. Page, K. Connell, G. Park, K. Norman, T. Murphy and C. Pardoe 2017. Human occupation of northern Australia by 65,000 years ago. *Nature* 547: 306–310.

Clastres, P. 1977. *Society Against the State*. R. Hurley, trans. New York: Zone Books.

Clement, C.R., M. de Cristo-Araújo, G.C. d'Eeckenbrugge, A.A. Pereira and D. Picanço-Rodridgues 2010. Origins and domestication of Native American crops. *Diversity* 2: 72–106.

Condit, I.J. 1947. *The Fig*. Waltham: Chronica Botanica.

Connolly, B. and R. Anderson 1987. *First Contact: New Guinea's Highlanders Encounter the Outside World*. New York: Viking Penguin.

Cook, C.D. 1999. Pandanus agroforestry of the Amungme in Irian Jaya, Indonesia. *Forest, Farm, and Community Tree Research Reports* 4: 95–103.

Cook, C.D. and J. Webster 2016. *Amua-Gaig-e. The Ethnobotany of the Amungme of Papua, Indonesia*. Ottawa: Canadian Science Publishing.

Corlett, R.T. 1984. Human impact on the subalpine vegetation of Mount Wilhelm, Papua New Guinea. *Journal of Ecology* 72: 841–854.

Cosgrove, R. 1995. *The Illusion of Riches: Scale, Resolution and Explanation in Tasmanian Pleistocene Human Behaviour*. British Archaeological Reports International Series, vol. 608. Oxford: BAR.

Coulter, S., T.P. Denham, C. Turney and V. Hall 2009. The geochemical characterisation and correlation of locally distributed Late Holocene tephras layers at Ambra Crater and Kuk Swamp, Papua New Guinea. *Geological Journal* 44: 568–592.

Coupaye, L. 2013. *Growing Artefacts, Displaying Relationships: Yams, Art and Technology Among the Nyamikum Abelam of Papua New Guinea*. New York and Oxford: Berghahn Books.

Coursey, D.G. 1972. The civilizations of the yam: interrelationships of man and yams in Africa and the Indo-Pacific region. *Archaeology and Physical Anthropology in Oceania* VII(1): 215–233.

Coursey, D.G. 1976. Yams: Dioscorea spp. (Dioscoreaceae). In N.W. Simmonds (ed.), *Evolution of Crop Plants*, pp. 70–74. London: Longman.

Croft, J.R. 1982. *Ferns and Man in New Guinea*. Lae/Canberra: Papua New Guinea National Herbarium/Australian National Herbarium, Centre for Plant Biodiversity Research.

Daniels, J. and C. Daniels 1993. Sugarcane in prehistory. *Archaeology in Oceania* 28: 1–7.

David, B. and T.P. Denham 2006. Unpacking Australian prehistory. In B. David, B. Barker and I. McNiven (eds.), *The Social Archaeology of Indigenous Societies*, pp. 52–71. Canberra: Aboriginal Studies Press.

David, B., R. Roberts, C. Tuniz, R. Jones and J. Head 1997. New optical and radiocarbon dates from Ngarrabullgan Cave, a Pleistocene archaeological site in Australia: implications for the comparability of time clocks and for the human colonization of Australia. *Antiquity* 71: 183–188.

De Candolle, A. 1884. *Origin of Cultivated Plants*. London: Kegan Paul.

De Langhe, E. and P. de Maret 1999. Tracking the banana: its significance in early agriculture. In C. Gosden and J. Hather (eds.), *The Prehistory of Food: Appetites for Change*, pp. 377–396. London: Routledge.

De Langhe, E., X. Perrier, M. Donohue and T.P. Denham 2015. The original banana split: Multidisciplinary implications of the generation of African and Pacific Plantains in Island Southeast Asia. *Ethnobotany Research and Applications* 14: 299–312.

De Langhe, E., L. Vrydaghs, P. de Maret, X. Perrier and T.P. Denham 2009. Why bananas matter: An introduction to the history of banana domestication. *Ethnobotany Research and Applications* 7: 165–177.

Deitrich, O., M. Huen, J. Notroff and K. Schmidt 2012. The role of cult and feasting in the emergence of Neolithic communities: New evidence from Gobekli Tepe, south-eastern Turkey. *Antiquity* 86: 674–695.

Denevan, W.M. 2001. *Cultivated Landscapes of Native Amazonia and the Andes*. Oxford: Oxford University Press.

Denham, T.P. 1996. *Understanding the Kalam, understanding ourselves: reflections on the re-presentation of the Kalam, Bismarck Mountain Range, Papua New Guinea*. Unpublished MS thesis, Department of Geography, Pennsylvania State University.

Denham, T.P. 2003a. *The Kuk Morass: multi-disciplinary investigations of early to Mid Holocene Plant exploitation at Kuk Swamp, Wahgi Valley, Papua New Guinea*. 2 volumes. Unpublished PhD thesis. Canberra: School of Archaeology and Anthropology, The Australian National University.

Denham, T.P. 2003b. Archaeological evidence for mid-Holocene agriculture in the interior of Papua New Guinea: a critical review. *Archaeology in Oceania* 38: 159–176.

Denham, T.P. 2004a. Early agriculture in the Highlands of New Guinea: an assessment of Phase 1 at Kuk Swamp. In V. Attenbrow and R. Fullagar (eds.), *A Pacific Odyssey: Archaeology and Anthropology in the Western Pacific. Papers in Honour of Jim Specht*. Records of the Australian Museum, Supplement 29: 47–57.

Denham, T.P. 2004b. The roots of agriculture and arboriculture in New Guinea: looking beyond Austronesian expansion, Neolithic packages and Indigenous origins. *World Archaeology* 36: 610–620.

Denham, T.P. 2005a. Agricultural origins and the emergence of rectilinear ditch networks in the highlands of New Guinea. In A. Pawley, R. Attenborough, J. Golson and R. Hide (eds.), *Papuan Pasts: Cultural, Linguistic and Biological Histories of Papuan-Speaking Peoples*, pp. 329–361. Pacific Linguistics 572. Canberra: RSPAS, ANU.

Denham, T.P. 2005b. Food for thought. *NatureAustralia* 28: 50–55.

Denham, T.P. 2005c. Envisaging early agriculture in the Highlands of New Guinea: landscapes, plants and practices. *World Archaeology* 37: 290–306.

Denham, T.P. 2006. The origins of agriculture in New Guinea: evidence, interpretation and reflection. In I. Lilley (ed.), *Blackwell Guide to Archaeology in Oceania: Australia and the Pacific Islands*, pp. 160–188. Oxford: Blackwell.

Denham, T.P. 2007a. Early to mid-Holocene plant exploitation in New Guinea: towards a contingent interpretation of agriculture. In T.P. Denham, J. Iriarte and L. Vrydaghs (eds.), *Rethinking Agriculture: Archaeological and Ethnoarchaeological Perspectives*, pp. 78–108. Walnut Creek: Left Coast Press.

Denham, T.P. 2007b. Early agriculture: Recent conceptual and methodological developments. In T.P. Denham and J.P. White (eds.), *The Emergence of Agriculture: A Global View*, pp. 1–25. London: Routledge.

Denham, T.P. 2007c. Exploiting diversity: Plant exploitation and occupation in the interior of New Guinea during the Pleistocene. *Archaeology in Oceania* 42: 41–48.

Denham, T.P. 2007d. Early fig domestication, or gathering of wild parthenocarpic figs? *Antiquity* 81: 457–461.

Denham, T.P. 2008a. Environmental archaeology: Interpreting practices-in-the-landscape through geoarchaeology. In B. David and J. Thomas (eds.), *Handbook of Landscape Archaeology*, pp. 468–481. Walnut Creek: Left Coast Press.

Denham, T.P. 2008b. Traditional forms of plant exploitation in Australia and New Guinea: the search for common ground. *Vegetation History and Archaeobotany* 17: 245–248.

Denham, T.P. 2009. A practice-centred method for charting the emergence and transformation of agriculture. *Current Anthropology* 50: 661–667.

Denham, T.P. 2010. From domestication histories to regional prehistory: Using plants to re-evaluate early and mid-Holocene interaction between New Guinea and Southeast Asia. *Food and History* 8: 3–22.

Denham, T.P. 2011. Early agriculture and plant domestication in New Guinea and Island Southeast Asia. *Current Anthropology* 52(S4): S379–S395.

Denham, T.P. 2012. Building institutional and community capacity for World Heritage in Papua New Guinea: the Kuk Early Agricultural Site and beyond. In A. Smith (ed.), *World Heritage in a Sea of Islands*, pp. 98–103. Paris: UNESCO.

Denham, T.P. 2013a. Ancient and historic dispersals of sweet potato in Oceania. *Proceedings of the National Academy of Sciences (USA)* 110: 1982–1983.

Denham, T.P. 2013b. A long-term history of horticultural innovation and introduction in the highlands of Papua New Guinea. In D. Frankel, J. Webb and S. Lawrence (eds.), *Archaeology in Environment and Technology: Intersections and Transformation*, pp. 101–122. Oxford: Routledge.

Denham, T.P. 2013c. *Traim tasol* . . . Cultural heritage management in Papua New Guinea. In S. Brockwell, S. O'Connor and D. Byrne (eds.), *Transcending the Culture-Nature Divide in Cultural Heritage: Views From the Asia-Pacific Region*, pp. 117–127. Terra Australis 36. Canberra: ANU E Press.

Denham, T.P. 2013d. Early farming in Island Southeast Asia: an alternative hypothesis. *Antiquity* 87: 250–257.

Denham, T.P. 2014. New Guinea during the Holocene. In P. Bahn and C. Renfrew (eds.), *The Cambridge World Prehistory. Volume I*, pp. 578–597. Cambridge: Cambridge University Press.

Denham, T.P. 2015. The swamp cultivators at Kuk: Early agriculture in the highlands of New Guinea. In G. Barker and C. Goucher (eds.), *The Cambridge World History. Volume II. A World With Agriculture*, pp. 445–471. Cambridge: Cambridge University Press.

Denham, T.P. 2016a. Revisiting the past: Sue Bulmer's contribution to the archaeology of Papua New Guinea. *Archaeology in Oceania* 51(S1): 5–10.

Denham, T.P. 2016b. Socio-environmental adaptation to the montane rainforests of New Guinea. In M. Oxenham and H. Buckley (eds.), *The Routledge Handbook of Bioarchaeology in Southeast Asia and the Pacific*, pp. 409–426. Abingdon, Oxford: Routledge.

Denham, T.P. 2017. Domesticatory relationships in the New Guinea Highlands. In J. Golson, T.P. Denham, P.J. Hughes, P. Swadling and J. Muke (eds.), *Ten Thousand Years of Cultivation at Kuk Swamp in the Highlands of Papua New Guinea*, pp. 39–50. Terra Australis 46. Canberra: ANU E Press.

Denham, T.P., J. Atchison, J. Austin, S. Bestel, D. Bowdery, A. Crowther, N. Dolby, A. Fairbairn, J. Field, A. Kennedy, C. Lentfer, C. Matheson, S. Nugent, J. Parr, M. Prebble, G. Robertson, J. Specht, R. Torrence, H. Barton, R. Fullagar, S. Haberle, M. Horrocks, T. Lewis and P. Matthews 2009. Archaeobotany in Australia and New Guinea: practice, potential and prospects. *Australian Archaeology* 68: 1–10.

Denham, T.P. and C. Ballard 2003. Jack Golson and the investigation of prehistoric agriculture in Highland New Guinea: recent work and future prospects. In T.P. Denham and C. Ballard (eds.), *Perspectives on Prehistoric Agriculture in New Guinea. Archaeology in Oceania*, Special Issue 38: 129–134.

Denham, T.P. and H. Barton 2006. The emergence of agriculture in New Guinea: continuity from pre-existing foraging practices. In D.J. Kennett and B. Winterhalder (eds.), *Behavioral Ecology and the Transition to Agriculture*, pp. 237–264. Berkeley: University of California Press.

Denham, T.P. and H. Barton 2014. Vegeculture: General principles. In C. Smith et al. (eds.), *Encyclopedia of Global Archaeology*, pp. 7608–7611. New York: Springer.

Denham, T.P., C. Bronk Ramsey and J. Specht 2012. Dating the appearance of Lapita pottery in the Bismarck Archipelago and its dispersal to Remote Oceania. *Archaeology in Oceania* 47: 39–46.

Denham, T.P. and M. Donohue 2009. Pre-Austronesian dispersal of banana cultivars west from New Guinea: linguistic relics from eastern Indonesia. *Archaeology in Oceania* 44: 18–28.

Denham, T.P., M. Donohue and S. Booth 2009. Revisiting an old hypothesis: Horticultural experimentation in northern Australia. *Antiquity* 83: 634–648.

Denham, T.P., R. Fullagar and L. Head 2009. Plant exploitation on Sahul: From colonisation to the emergence of regional specialisation during the Holocene. *Quaternary International* 202: 29–40.

Denham, T.P., J. Golson and P.J. Hughes 2004. Reading early agriculture at Kuk (Phases 1–3), Wahgi Valley, Papua New Guinea: the wetland archaeological features. *Proceedings of the Prehistoric Society* 70: 259–298.

Denham, T.P., J. Golson and P.J. Hughes 2017a. Phase 1: The case for 10,000-year-old agriculture at Kuk. In J. Golson, T.P. Denham, P.J. Hughes, P. Swadling and J. Muke (eds.), *Ten Thousand Years of Cultivation at Kuk Swamp in the Highlands of Papua New Guinea*, pp. 187–200. Terra Australis 46. Canberra: ANU E Press.

Denham, T.P., J. Golson and P.J. Hughes 2017b. Phase 2: Mounded cultivation during the mid Holocene. In J. Golson, T.P. Denham, P.J. Hughes, P. Swadling and J. Muke (eds.), *Ten Thousand Years of Cultivation at Kuk Swamp in the Highlands of Papua New Guinea*, pp. 201–220. Terra Australis 46. Canberra: ANU E Press.

Denham, T.P., J. Golson and P.J. Hughes 2017c. Phase 3: The emergence of ditches. In J. Golson, T.P. Denham, P.J. Hughes, P. Swadling and J. Muke (eds.), *Ten Thousand Years of Cultivation at Kuk Swamp in the Highlands of Papua New Guinea*, pp. 221–238. Terra Australis 46. Canberra: ANU E Press.

Denham, T.P. and E. Grono 2017. Sediments or soils? Multi-scale geoarchaeological investigations of stratigraphy and early cultivation practices at Kuk Swamp, highlands of Papua New Guinea. *Journal of Archaeological Science* 77: 160–171.

Denham, T.P. and S.G. Haberle 2008. Agricultural emergence and transformation in the Upper Wahgi valley during the Holocene: Theory, method and practice. *The Holocene* 18: 481–496.

Denham, T.P., S.G. Haberle and C. Lentfer 2004. New evidence and interpretations for early agriculture in Highland New Guinea. *Antiquity* 78: 839–857.

Denham, T.P., S.G. Haberle, C. Lentfer, R. Fullagar, J. Field, M. Therin, N. Porch and B. Winsborough 2003. Origins of agriculture at Kuk Swamp in the Highlands of New Guinea. *Science* 301: 189–193.

Denham, T.P., S.G. Haberle and A. Pierret 2009. A multi-disciplinary method for the investigation of early agriculture: Learning lessons from Kuk. In A. Fairbairn, S. O'Connor and B. Marwick (eds.), *New Directions in Archaeological Science*, pp. 139–154. Terra Australis 28. Canberra, ANU E Press.

Denham, T.P., J. Iriarte and L. Vrydaghs (eds.) 2007. *Rethinking Agriculture: Archaeological and Ethnoarchaeological Perspectives*. Walnut Creek: Left Coast Press (hardback edition in 2007; paperback edition in 2009).

Denham, T.P. and S. Mooney 2008. Human-environment interactions in Australia and New Guinea during the Holocene. *The Holocene* 18: 373–379.

Denham, T.P. and M-J. Mountain 2016. Resolving some chronological problems at Nombe rock shelter in the highlands of Papua New Guinea. *Archaeology in Oceania* 51(S1): 73–83.

Denham, T.P., K. Sniderman, K. Saunders, B. Winsborough and A. Pierret 2009. Contiguous multi-proxy analyses (X-radiography, diatom, pollen and microcharcoal) of Holocene archaeological features at Kuk Swamp, Upper Wahgi valley, Papua New Guinea. *Geoarchaeology* 24: 715–742.

Department of Environment and Conservation (DEC) (prepared by T.P. Denham, J. Muke, L. Salas, V. Genorupa and others) 2007. *The Kuk early agricultural site: a cultural landscape*. World Heritage Nomination (successful). Port Moresby: Department of Environment and Conservation, Government of Papua New Guinea.

Derrida, J. 1978. Structure, sign, and play in the discourse of the human sciences. In J. Derrida, A. Bass (trans.), *Writing and Difference*. London: Routledge.

Dewar, R.E. 2003. Rainfall variability and subsistence systems in Southeast Asia and the western Pacific. *Current Anthropology* 44: 369–388.

Diamond, J. 1997. *Guns, Germs and Steel*. London: Norton and Company.

Diamond, J. 2005. *Collapse: How Societies Choose to Fail or Survive*. Melbourne: Penguin.

Dobney, K., T. Cucchi and G. Larson 2008. The pigs of Island Southeast Asia and the Pacific: new evidence for taxonomic status and human-mediated dispersal. *Asian Perspectives* 47: 59–74.

Donoghue, D. 1988. *Pandanus and changing site use: a study from Manim Valley, Papua New Guinea*. Unpublished BA (Hons.) thesis, University of Queensland.

Donoghue, D. 1989. Carbonised plant fossils. In W. Beck, A. Clarke and L. Head (eds.), *Plants in Australian Archaeology*, pp. 90–100. St Lucia: University of Queensland. Tempus 1.

Donohue, M. and T.P. Denham 2010. Farming and language in Island Southeast Asia: reframing Austronesian history. *Current Anthropology* 51: 223–256.

Droysen, J.G. 1880 (1985). The investigation of origins. In K. Mueller-Vollmer (ed.), *The Hermeneutic Reader*, pp. 124–126. New York: Continuum.

Dumont, R. and P. Vernier 2000. Domestication of yams (*Dioscorea cayenensis-rotundata*) within the Bariba ethnic group in Benin. *Outlook on Agriculture* 29: 137–142.

Dwyer, P.D. and M. Minnegal 1991. Hunting in lowland, tropical rain forest: towards a model of non-agricultural subsistence. *Human Ecology* 19: 187–212.

Earle, T.K. 1978. *Economic and Social Organisation of a Complex Chiefdom: The Halelea District, Kauai, Hawai'i*. Volume 63, Museum of Anthropology. Ann Arbor: University of Michigan Press.

Eckert, C.G. 2002. The loss of sex in clonal plants. *Evolutionary Ecology* 15: 501–520.

Edelman, C.H. and R. Brinkman 1962. Physiography of gilgai soils. *Journal of Soil Science* 94: 366–370.

Edmonds, J.M. and J.A. Chweya 1997. *Black Nightshades: Solanum nigrum L. and Related Species*. Promoting the Conservation and Use of Underutilized and Neglected Crops 15. Rome: IGPRI.

Egloff, B. 2008. *Bones of the Ancestors: The Ambun Stone*. Lanham: Alta Mira Press.

Erikson, D.L., B.D. Smith, A.C. Clarke, D.H. Sandweiss and N. Tuross 2005. An Asian origin for a 10,000 year-old domesticated plant in the Americas. *Proceedings of the National Academy of Sciences (USA)* 102: 18315–18320.

Evans, B. and M-J. Mountain 2005. *Pasin bilong tumbuna*: archaeological evidence for early human activity in the highlands of Papua New Guinea. In A. Pawley, R. Attenborough, J. Golson and R. Hide (eds.), *Papuan Pasts: Cultural, Linguistic and Biological Histories of Papuan-Speaking Peoples*, pp. 363–386. Pacific Linguistics 572. Canberra: RSPAS, ANU.

Fairbairn, A.S. 2005. An archaeobotanical perspective on Holocene plant use practices in lowland northern New Guinea. *World Archaeology* 37: 487–502.

Fairbairn, A.S., G.S. Hope and G.R. Summerhayes 2006. Pleistocene occupation of New Guinea's highland and subalpine environments. *World Archaeology* 38: 371–386.

Fairbairn, A.S. and P. Matthews 2017. Plant microfossils: seeds, fruits and wood. Textbox 10.1. In J. Golson, T.P. Denham, P.J. Hughes, P. Swadling and J. Muke (eds.), *Ten Thousand Years of Cultivation at Kuk Swamp in the Highlands of Papua New Guinea*, pp. 164–167. Terra Australis 46. Canberra: ANU E Press.

Fairbairn, A.S. and P. Swadling 2005. Re-dating mid-Holocene betelnut (*Areca catechu* L.) and other plant use at Dongan, Papua New Guinea. *Radiocarbon* 47: 377–382.

Faith, D.P., C.R. Margules and P.A. Walker 2001. A biodiversity conservation plan for Papua New Guinea based on biodiversity trade-offs analysis. *Pacific Conservation Biology* 6: 304–324.

FAO (Food and Agriculture Organization of the United Nations) 2012. *FAO Statistical Yearbook 2012: World Food and Agriculture*. Rome: FAO.

Feil, D. 1987. *The Evolution of Highland Papua New Guinea Societies*. Cambridge: Cambridge University Press.

Field, J., M. Fillios and S. Wroe 2008. Chronological overlap between humans and megafauna in Sahul (Pleistocene Australia – New Guinea): A review of the evidence. *Earth Science Reviews* 89: 97–115.

Fifield, L.K., M.I. Bird, C.S.M. Turney, P.A. Hausladen, G.M. Santos and M.L. di Tada 2001. Radiocarbon dating of the human occupation of Australia prior to 40 ka BP – success and pitfalls. *Radiocarbon* 43: 1139–1145.

Flannery, K.V. 1969. Origins and ecological effects of early domestication in Iran and the Near East. In P.J. Ucko and G.W. Dimbleby (eds.), *The Domestication and Exploitation of Plants and Animals*, pp. 73–100. London: Duckworth.

Flannery, K.V. 1986. *Guilá Naquitz*. Orlando: Academic Press.

Flannery, T.F. 1992. New Pleistocene marsupials (Macropodidae, Diprotodontidae) from subalpine habitats in Irian Jaya, Indonesia. *Alcheringa* 16: 321–331.

Flannery, T.F. 1995. *Mammals of New Guinea*. Sydney: The Australian Museum, Robert Brown & Associates.

Flannery, T.F., M-J. Mountain and K. Aplin 1983. Quaternary kangaroos (Macropodidae: Marsupialia) from Nombe rockshelter, Papua New Guinea: with comments on the nature of the megafaunal extinction in the New Guinea Highlands. *Proceedings of the Linnean Society of New South Wales* 107: 77–97.

Flannery, T.F. and M. Plane 1986. A new late Pleistocene diprotodontid (Marsupialia) from Pureni, Southern Highlands Province, Papua New Guinea. *Journal of Australian Geology and Geophysics* 10: 65–76.

Fleming, A. 1978. The prehistoric landscape of Dartmoor: Part 1: South Dartmoor. *Proceedings of the Prehistoric Society* 44: 97–123.

Fleming, A. 1987. Coaxial field systems: some questions of time and space. *Antiquity* 61: 188–202.

Fleming, A. 1989. The genesis of coaxial field systems. In R. Torrence and S. van der Leeuw (eds.), *What's New? A Closer Look at the Process of Innovation*, pp. 63–81. London: Unwin Hyman.

Flenley, J. 1979. *The Equatorial rainforest: A Geological History*. London: Butterworths.

Ford, R.I. 1985. The processes of plant production in prehistoric North America. In R.I. Ford (ed.), *Prehistoric Plant Production in North America*, pp. 1–18. Anthropological Paper 75, Museum of Anthropology. Ann Arbor: University of Michigan.

Fredericksen, C.F.K., M. Spriggs and W. Ambrose 1993. Pamwak rockshelter: a Pleistocene site on Manus Island, Papua New Guinea. In M.A. Smith, M. Spriggs and B. Fankhauser (eds.), *Sahul in Review: Pleistocene Archaeology in Australia, New Guinea and Island Melanesia*, pp. 144–152. Occasional Papers in Prehistory 24. Canberra: Department of Prehistory, Research School of Pacific Studies, Australian National University.

French, B.R. 1986. *Food Plants of Papua New Guinea: A Compendium*. Privately published book.

French, B.R. 2006. *Food Plants of Papua New Guinea: A Compendium*. Revised edition. Burnie, Tasmania: Food Plants International, Inc.

Fullagar, R., J. Field, T.P. Denham and C. Lentfer 2006. Early and mid Holocene processing of taro (*Colocasia esculenta*) and yam (*Dioscorea* sp.) at Kuk Swamp in the Highlands of Papua New Guinea. *Journal of Archaeological Science* 33: 595–614.

Fullagar, R., J. Field and L. Kealhofer 2008. Grinding stones and seeds of change: starch and phytoliths as evidence of plant food processing. In Y.M. Rowan and J.R. Ebling (eds.), *New Approaches to Old Stones*, pp. 159–172. London: Equinox Press.

Fullagar, R. and J. Golson 2017. Kuk stone artefacts: technology, usewear and residues. In J. Golson, T.P. Denham, P.J. Hughes, P. Swadling and J. Muke (eds.), *Ten Thousand Years of Cultivation at Kuk Swamp in the Highlands of Papua New Guinea*, pp. 373–402. Terra Australis 46. Canberra: ANU E Press.

Fuller, D.Q., T.P. Denham, M. Arroyo-Kalin, L. Lucas, C. Stevens, L. Qin, R.G. Allaby and M.D. Purugganan 2014. Convergent evolution and parallelism in plant domestication revealed by an expanding archaeological record. *Proceedings of the National Academy of Sciences (USA)* 111: 6147–6152.

Fuller, D.Q., R. Korisettar, P.C. Venkatasubbaiah and M.K. Jones 2004. Early plant domestication in southern India: some preliminary archaeo-botanical results. *Vegetation History and Archaeobotany* 13: 115–129.

Fuller, D.Q., L. Qin, Y. Zheng, Z. Zhao, X. Chen, L.A. Hosoya and G-P. Sun 2009. The domestication process and domestication rate in rice: Spikelet bases from the Lower Yangtze. *Science* 323: 1607–1610.

Gadamer, H-G. 1976. *Philosophical Hermeneutics*. D. Linge, ed. and trans. Berkeley: University of California Press.

Gaffney, D., A. Ford and G.R. Summerhayes 2015. Crossing the Pleistocene-Holocene transition in the New Guinea highlands: evidence from the lithic assemblage of Kiowa rockshelter. *Journal of Anthropological Archaeology* 39: 223–246.

Gaffney, D., A. Ford and G.R. Summerhayes 2016. Sue Bulmer's legacy in highland New Guinea: a re-examination of the Bulmer collection and future directions. *Archaeology in Oceania* 51(S1): 23–32.

Gaffney, D., G.R. Summerhayes, A. Ford, J. Scott, T.P. Denham, J. Field and W.R. Dickinson 2015. Earliest pottery on New Guinea mainland reveals Austronesian influences in highland environments 3000 years ago. *PLOS ONE* 10(9): e0134497. doi: 10.1371/journal.pone.0134497.

Gagné, W.C. 1982. Staple crops in subsistence agriculture. In J.L. Gressitt (ed.), *Biogeography and Ecology of New Guinea. Volume 1*, pp. 229–259. The Hague: Junk.

Gardner, R. and K.G. Heider 1969. *Gardens of War: Life and Death in the New Guinea Stone Age*. New York: Random House.

Geneste, J-M., B. David, H. Plisson, J-J. Delannoy and F. Petchey 2012. The origins of ground-edge axes: New findings from Nawarla Gabarn-mang, Arnhem Land (Australia) and global implications for the evolution of fully modern humans. *Cambridge Archaeological Journal* 22: 1–17.

Gepts, P. 2002. A comparison between crop domestication, classical plant breeding, and genetic engineering. *Crop Science* 42: 1780–1790.

Gepts, P., R. Bettinger, S. Brush, T. Famula, P. McGuire, C. Qualset and A. Damania. (eds.) 2012. *Biodiversity in Agriculture: Domestication, Evolution, and Sustainability.* Cambridge: Cambridge University Press.

Giddens, A. 1984. *The Constitution of Society.* Cambridge: Polity.

Gillieson, D., P. Gorecki and G. Hope 1985. Prehistoric agricultural systems in a lowland swamp, Papua New Guinea. *Archaeology in Oceania* 20: 32–37.

Gillieson, D., P. Gorecki, J. Head and G. Hope 1987. Soil erosion and agricultural history in the central Highlands of Papua New Guinea. In V. Gardiner (ed.), *International Geomorphology Part II*, pp. 507–522. London: John Wiley and Sons.

Gillieson, D., G. Hope and J. Luly 1989. Environmental change in the Jimi Valley. In P. Gorecki and D.S. Gillieson (eds.), *A Crack in the Spine: Prehistory and Ecology of the Jimi-Yuat River, Papua New Guinea*, pp. 103–122. Queensland: JCU.

Gillieson, D., F. Oldfield and A. Krawiecki 1986. Records of prehistoric soil erosion from rock-shelter sites in Papua New Guinea. *Mountain Research and Development* 6: 315–324.

Gillison, A.N. 1972. The tractable grasslands of Papua New Guinea. In M.W. Ward (ed.), *Change and Development in Rural Melanesia*, pp. 161–172. Port Moresby and Canberra: University of Papua New Guinea and Australian National University.

Glasse, R.M. and M.J. Meggitt (eds.) 1969. *Pigs, Pearlshells, and Women: Marriage in the New Guinea Highlands.* Englewood Cliffs: Prentice-Hall.

Godelier, M. 1986. *The Making of Great Men: Male Domination and Power among the New Guinea Baruya.* Trans. R. Swyer. Cambridge: Cambridge University Press.

Golson, J. 1976a. Archaeology and agricultural history in the New Guinea Highlands. In G. de G. Sieveking, I.A. Longworth and K.E. Wilson (eds.), *Problems in Economic and Social Archaeology*, pp. 201–220. London: Duckworth.

Golson, J. 1976b. *Archaeological investigations at Kuk Tea Research Station, Mount Hagen.* Unpublished manuscript.

Golson, J. 1977a. *The Ladder of Social Evolution: Archaeology and the Bottom Rungs.* Sydney: Sydney University Press.

Golson, J. 1977b. Simple tools and complex technology: Agriculture and agricultural implements in the New Guinea Highlands. In R.V.S. Wright (ed.), *Stone Tools as Cultural Markers: Change, Evolution and Complexity*, pp. 154–161. Atlantic Highlands: Humanities Press Inc.

Golson, J. 1977c. No room at the top: agricultural intensification in the New Guinea Highlands. In J. Allen, J. Golson and R. Jones (eds.), *Sunda and Sahul: Prehistoric Studies in Southeast Asia, Melanesia and Australia*, pp. 601–638. London: Academic Press.

Golson, J. 1977d. The making of the New Guinea Highlands. In J.H. Winslow (ed.), *The Melanesian Environment*, pp. 45–56. Canberra: Australian National University Press.

Golson, J. 1981. New Guinea agricultural history: a case study. In D. Denoon and C. Snowden (eds.), *A Time to Plant and a Time to Uproot*, pp. 55–64. Port Moresby: Institute of Papua New Guinea Studies.

Golson, J. 1982a. The Ipomoean revolution revisited: society and sweet potato in the upper Wahgi Valley. In A. Strathern (ed.), *Inequality in New Guinea Highland Societies*, pp. 109–136. Cambridge: Cambridge University Press.

Golson, J. 1982b. Kuk and the history of agriculture in the New Guinea Highlands. In R.J. May and H. Nelson (eds.), *Melanesia: Beyond Diversity*, pp. 297–307. Canberra: Research School of Pacific and Asian Studies, Australian National University.

Golson, J. 1985. Agricultural origins in Southeast Asia: a view from the east. In V.N. Misra and P. Bellwood (eds.), *Recent Advances in Indo-Pacific Prehistory*, pp. 307–314. Leiden: E. J. Bill.

Golson, J. 1989. The origins and development of New Guinea agriculture. In D.R. Harris and G.C. Hillman (eds.), *Foraging and Farming: The Evolution of Plant Exploitation*, pp. 678–687. London: Unwin Hyman.

Golson, J. 1990. Kuk and the development of agriculture in New Guinea: retrospection and introspection. In D.E. Yen and J.M.J. Mummery (eds.), *Pacific Production Systems: Approaches to Economic Prehistory*, pp. 139–147. Canberra: Australian National University.

Golson, J. 1991a. The New Guinea Highlands on the eve of agriculture. *Bulletin of the Indo-Pacific Prehistory Association* 11: 82–91.

Golson, J. 1991b. Bulmer Phase II: early agriculture in the New Guinea Highlands. In A. Pawley (ed.), *Man and a Half: Essays in Pacific Anthropology and Ethnobiology in Honour of Ralph Bulmer*, pp. 484–491. Auckland: The Polynesian Society.

Golson, J. 1991c. Introduction: transitions to agriculture in the Pacific agriculture. *Bulletin of the Indo-Pacific Prehistory Association* 11: 48–53.

Golson, J. 1996. New Guinea: the making of a prehistory. In M. Julien, M. Orliac and C. Orliac (eds.), *Memoire de Pierre, Memoire d'Homme: Tradition et Archaeologie en Oceanie*, pp. 153–179. Paris: Publications de la Sorbonne.

Golson, J. 1997a. The Tambul spade. In H. Levine and A. Ploeg (eds.), *Work in Progress: Essays in New Guinea Highlands Ethnography in Honour of Paula Brown Glick*, pp. 142–171. Frankfurt: Peter Lang.

Golson, J. 1997b. From horticulture to agriculture in the New Guinea Highlands. In P.V. Kirch and T.L. Hunt (eds.), *Historical Ecology in the Pacific Islands: Prehistoric Environmental and Landscape Change*, pp. 39–50. New Haven and London: Yale University Press.

Golson, J. 2000. A stone bowl fragment from the Early Middle Holocene of the Upper Wahgi Valley, Western Highlands Province, Papua New Guinea. In A. Anderson and T. Murray (eds.), *Australian Archaeologist: Collected Papers in Honour of Jim Allen*, pp. 231–248. Canberra: Coombs Academic Publishing, ANU.

Golson, J. 2001. New Guinea, Australia and the Sahul connection. In A. Anderson, I. Lilley and S. O'Connor (eds.), *Histories of Old Ages: Essays in Honour of Rhys Jones*, pp. 185–210. Canberra: Pandanus Books, Australian National University.

Golson, J. 2002. Gourds in New Guinea, Asia and the Pacific. In S. Bedford, C. Sand and D. Burley (eds.), *Fifty Years in the Field. Essays in Honour and Celebration of Richard Shutler Jr.'s Archaeological Career*, pp. 69–78. New Zealand Archaeological Journal Monograph 25, Auckland: Auckland Museum.

Golson, J. 2005. The middle reaches of New Guinea history. In A. Pawley, R. Attenborough, J. Golson and R. Hide (eds.), *Papuan Pasts: Cultural, Linguistic and Biological Histories of Papuan-Speaking Peoples*, pp. 451–492. Pacific Linguistics 572. Canberra: RSPAS, ANU.

Golson, J. 2007. Unravelling the story of early plant exploitation in highland Papua New Guinea. In T.P. Denham, J. Iriarte and L. Vrydaghs (eds.), *Rethinking Agriculture: Archaeological and Ethnoarchaeological Perspective*, pp. 109–125. Walnut Creek: Left Coast Press.

Golson, J. 2017a. Artefacts of wood. In J. Golson, T.P. Denham, P.J. Hughes, P. Swadling and J. Muke (eds.), *Ten Thousand Years of Cultivation at Kuk Swamp in the Highlands of Papua New Guinea*, pp. 359–372. Terra Australis 46. Canberra: ANU E Press.

Golson, J. 2017b. Houses in and out of the swamp. In J. Golson, T.P. Denham, P.J. Hughes, P. Swadling and J. Muke (eds.), *Ten Thousand Years of Cultivation at Kuk Swamp in the Highlands of Papua New Guinea*, pp. 325–350. Terra Australis 46. Canberra: ANU E Press.

Golson, J., T.P. Denham, P.J. Hughes, P. Swadling and J. Muke (eds.) 2017. *Ten Thousand Years of Cultivation at Kuk Swamp in the Highlands of Papua New Guinea*. Terra Australis 46. Canberra: ANU E Press.

Golson, J. and D. Gardner 1990. Agriculture and sociopolitical organisation in New Guinea Highlands prehistory. *Annual Review of Anthropology* 19: 395–417.

Golson, J. and P.J. Hughes 1980. The appearance of plant and animal domestication in New Guinea. *Journal de la Société des Océanistes* 36: 294–303.

Golson, J., R.J. Lampert, J.M. Wheeler and W.R. Ambrose 1967. A note on carbon dates for horticulture in the New Guinea Highlands. *Journal of the Polynesian Society* 76: 369–371.

Golson, J. and A. Steensberg 1985. The tools of agricultural intensification in the New Guinea Highlands. In I. Farrington (ed.), *Prehistoric Intensive Agriculture in the Tropics*, pp. 347–384. Oxford: British Archaeological Reports, International Series 232, Part I.

Golson, J. and P. Ucko 1994. Foreword. In J.G. Hather (ed.), *Tropical Archaeobotany: Applications and New Developments*, pp. xiv–xix. London: Routledge.

Gorecki, P. 1978. Further notes on prehistoric wooden spades from the New Guinea Highlands. *Tools and Tillage* III: 185–190.

Gorecki, P. 1979. Population growth and abandonment of swamplands: A New Guinea example. *Journal de la Société des Océanistes* 35: 97–107.

Gorecki, P. 1982. *Ethnoarchaeology at Kuk: problems in site formation process*. Unpublished PhD dissertation, Department of Anthropology, University of Sydney.

Gorecki, P. 1986. Human occupation and agricultural development in the Papua New Guinea Highlands. *Mountain Research and Development* 6: 159–166.

Gorecki, P., M. Mabin and J. Campbell 1991. Archaeology and geomorphology of the Vanimo coast, Papua New Guinea: preliminary results. *Archaeology in Oceania* 26: 119–122.

Gott, B. 2005. Aboriginal fire management in south-eastern Australia: aims and frequency. *Journal of Biogeography* 32: 1203–1208.

Gould, P. 1970. Is *statistix inferens* the geographical name for a wild goose? *Economic Geography* 46: 439–448.

Gregory, D.J. 2000. Time geography. In R.J. Johnston, D.J. Gregory, G. Pratt and M.J. Watts (eds.), *The Dictionary of Human Geography*, pp. 830–833. Oxford: Blackwell.

Gregory, K.J. and D.E. Walling 1973. *Drainage Basin Form and Process: A Geomorphological Approach*. London: Edward Arnold.

Gremillion, K.J., L. Barton and D.R. Piperno 2014. Particularism and the retreat from theory in the archaeology of agricultural origins. *Proceedings of the National Academy of Sciences (USA)* 111: 6171–6177.

Gremillion, K.J. and D.R. Piperno 2009. Human behavioural ecology, phenotypic (developmental) plasticity and agricultural origins. *Current Anthropology* 50: 615–619.

Grivet, L., C. Daniels, J.C. Glaszman and A. D'Hont 2004. A review of recent molecular genetics evidence for sugarcane evolution and domestication. *Ethnobotany Research and Applications* 2: 9–17.

Groube, L. 1989. The taming of the rainforests: a model for Late Pleistocene forest exploitation in New Guinea. In D.R. Harris and G.C. Hillman (eds.), *Foraging and Farming: The Evolution of Plant Exploitation*, pp. 292–304. London: Unwin Hyman.

Groube, L., J. Chappell, J. Muke and D. Price 1986. A 40,000 year-old human occupation site at Huon Peninsula, Papua New Guinea. *Nature* 324: 453–455.

Guddemi, P. 1992. When horticulturalists are like hunter-gatherers: the Sawiyano of Papua New Guinea. *Ethnology* 31: 303–314.

Haantjens, H.A., J.R. McAlpine, E. Reiner, R.G. Robbins and J.C. Saunders 1970. *Lands of Goroka – Mount Hagen Area, Territory of Papua and New Guinea*. Land Research Series No. 27. Melbourne: Commonwealth Scientific and Industrial Research Organisation.

Haberle, S.G. 1993. *Late quaternary environmental history of the Tari Basin, Papua New Guinea*. Unpublished PhD thesis, Australian National University.

Haberle, S.G. 1994. Anthropogenic indicators in pollen diagrams: problems and prospects for late Quaternary palynology in New Guinea. In J.G. Hather (ed.), *Tropical Archaeobotany: Applications and New Developments*, pp. 172–201. London: Routledge.

Haberle, S.G. 1995. Identification of cultivated *Pandanus* and *Colocasia* in pollen records and the implications for the study of early agriculture in New Guinea. *Vegetation History and Archaeobotany* 4: 195–210.

Haberle, S.G. 1998. Late Quaternary change in the Tari Basin, Papua New Guinea. *Palaeogeography, Palaeoclimatology, Palaeoecology* 137: 1–24.

Haberle, S.G. 2003. The emergence of an agricultural landscape in the highlands of New Guinea. *Archaeology in Oceania* 38: 149–158.

Haberle, S.G. 2007. Prehistoric human impact on rainforest biodiversity in highland New Guinea. *Philosophical Transactions of the Royal Society. Series B, Biological Sciences* 362: 219–228.

Haberle, S.G. and Chepstow-Lusty, A. 2000. Can climate influence cultural development? A view through time. *Environment and History* 6: 349–369.

Haberle, S.G. and B. David 2004. Climates of change: human dimensions of Holocene environmental change in low latitudes of the PEPII transect. *Quaternary International* 118–119: 165–179.

Haberle, S.G., G.S. Hope and Y. de Fretes 1991. Environmental change in the Baliem Valley, Montane Irian Jaya, Republic of Indonesia. *Journal of Biogeography* 18: 25–40.

Haberle, S.G., G.S. Hope and S. van der Kaars 2001. Biomass burning in Indonesia and Papua New Guinea: natural and human induced fire events in the fossil record. *Palaeogeography, Palaeoclimatology, Palaeoecology* 171: 259–268.

Haberle, S.G., C. Lentfer, S. O'Donnell and T.P. Denham 2012. The palaeoenvironments of Kuk Swamp from the beginnings of agriculture in the highlands of Papua New Guinea. *Quaternary International* 249: 129–139.

Hägerstrand, T. 1970. What about people in regional science? *Papers of the Regional Science Association* 24: 7–21.

Hallsworth, E.G., G.K. Robertson and F.R. Gibbons 1955. Studies in pedogenesis in New South Wales: VII. The 'gilgai' soils. *Journal of Soil Science* 6: 1–31.

Hamm, G., P. Mitchell, L.F. Arnold, G.J. Prideaux, D. Questiaux, N.A. Spooner, V.A. Levchenko, E.C. Foley, T.H. Worthy, B. Stephenson, V. Coulthard, C. Coulthard, S. Wilton and D. Johnston 2016. Cultural innovation and megafauna interaction in the early settlement of arid Australia. *Nature* 539: 280–283.

Harris, D.R. 1969. Agricultural systems, ecosystems and the origins of agriculture. In P.J. Ucko and G.W. Dimbleby (eds.), *The Domestication and Exploitation of Plants and Animals*, pp. 3–15. London: Duckworth.

Harris, D.R. 1972. The origins of agriculture in the tropics. *American Scientist* 60: 180–193.

Harris, D.R. 1973. The prehistory of tropical agriculture: an ethnoecological model. In C. Renfrew (ed.), *The Explanation of Culture Change: Models in Prehistory*, pp. 391–417. London: Duckworth.

Harris, D.R. 1989. An evolutionary continuum of people-plant interaction. In D.R. Harris and G.C. Hillman (eds.), *Foraging and Farming: The Evolution of Plant Exploitation*, pp. 11–26. London: Unwin Hyman.

Harris, D.R. 1990. Vavilov's concept of centres of origin of cultivated plants: its genesis and its influence on the study of agricultural origins. *Biological Journal of the Linnean Society* 39: 7–16.

Harris, D.R. 1995. Early agriculture in New Guinea and the Torres Strait divide. *Antiquity* 69(Special Number 265): 848–854.

Harris, D.R. 1996a. Introduction: themes and concepts in the study of early agriculture. In D.R. Harris (ed.), *The Origins and Spread of Agriculture and Pastoralism in Eurasia*, pp. 1–9. London: University College London Press.

Harris, D.R. 1996b. The origins and spread of agriculture and pastoralism in Eurasia: an overview. In D.R. Harris (ed.), *The Origins and Spread of Agriculture and Pastoralism in Eurasia*, pp. 552–573. London: University College London Press.

Harris, D.R. 1998. Introduction: the multi-disciplinary study of cross-cultural plant exchange. In H.D.V. Pendergast, N.L. Etkin, D.R. Harris and P.J. Houghton (eds.), *Plants for Food and Medicine*, pp. 85–91. Kew: Royal Botanic Gardens.

Harris, D.R. 2002. The expansion capacity of early agricultural systems: a comparative perspective on the spread of agriculture. In P. Bellwood and C. Renfrew (eds.), *Examining the Farming/Language Dispersal Hypothesis*, pp. 31–40. Cambridge: McDonald Institute for Archaeological Research.

Harris, D.R. 2007. Agriculture, cultivation and domestication: exploring the conceptual framework of early food production. In T.P. Denham, J. Iriarte and L. Vrydaghs (eds.), *Rethinking Agriculture: Archaeological and Ethnoarchaeological Perspectives*, pp. 16–35. Walnut Creek: Left Coast Press.

Harris, D.R. and D.Q. Fuller 2014. Agriculture: definition and overview. In C. Smith (ed.), *Encyclopedia of Global Archaeology*, pp. 104–113. New York: Springer.

Harris, E.C. 1977. *Hed mound: a New Guinea house site.* Unpublished manuscript.

Harris, E.C. and P.J. Hughes 1978. An early agricultural system at Mugumamp Ridge, Western Highlands Province, Papua New Guinea. *Mankind* 11: 437–445.

Hastorf, C. 1998. The cultural life of early domestic plant use. *Antiquity* 72: 773–782.

Hastorf, C. 2016. *The Social Archaeology of Food: Thinking About Eating From Prehistory to the Present.* Cambridge: Cambridge University Press.

Hather, J.G. 1992. The archaeobotany of subsistence in the Pacific. *World Archaeology* 24(1): 70–81.

Hather, J.G. 1996. The origins of tropical vegeculture: Zingiberaceae, Araceae and Dioscoreaceae in Southeast Asia. In D.R. Harris (ed.), *The Origins and Spread of Agriculture and Pastoralism in Eurasia*, pp. 538–550. London: University College London Press.

Hather, J.G. 2000. *Archaeological Parenchyma.* London: Archaeotype Publications.

Hawkes, J.G. 1983. *The Diversity of Crop Plants.* Cambridge, MA: Harvard University Press.

Hays, T.E. 2005. Vernacular names for staple foods in Irian Jaya. In A. Pawley, R. Attenborough, J. Golson and R. Hide (eds.), *Papuan Pasts: Cultural, Linguistic and Biological Histories of Papuan-Speaking Peoples*, pp. 625–670. Pacific Linguistics 572. Canberra: RSPAS, ANU.

Hedges, R.E.M., R.A. Housley, C.R. Bronk and G.J. Van Klinken 1995. Radiocarbon dates from the Oxford AMS system: Archaeometry datelist 20. *Archaeometry* 37: 417–430.

Heidegger, M. 1962. *Being and Time.* J. Macquarrie and E. Robinson, trans. Oxford: Blackwell.

Heinsohn, T.E. 2010. Marsupials as introduced species: long-term anthropogenic expansion of the marsupial frontier and its implications for zoogeographic interpretation. In S.G. Haberle, J. Stevenson and M. Prebble (eds.), *Altered Ecologies: Fire, Climate and Human Influence on Terrestrial Landscapes*, pp. 133–176. Terra Australis 32. Canberra: ANU EPress.

Higham, C. 2002. *Early Cultures of Mainland Southeast Asia.* Bangkok: Art Media Resources.

Hildebrand, E.A. 2007. A tale of two tuber crops: how attributes of enset and yams may have shaped prehistoric human – plant interactions in southwest Ethiopia. In T.P. Denham, J. Iriarte and L. Vrydaghs (eds.), *Rethinking Agriculture: Archaeological and Ethnoarchaeological Perspectives*, pp. 273–298. Walnut Creek: Left Coast Press.

Hillman, G.C. 1981. Crop husbandry: evidence from macroscopic remains. In I. Simmons and M. Tooley (eds.), *The Environment in British Prehistory*, pp. 183–191. London: Duckworth.

Hillman, G.C. 1996. Late Pleistocene changes in wild plant-foods available to hunter-gatherers of the northern Fertile Crescent: possible preludes to cereal cultivation. In D.R. Harris (ed.), *The Origins and Spread of Agriculture and Pastoralism in Eurasia*, pp. 159–203. London: University College London Press.

Hillman, G.C. and M.S. Davies 1990. Measured domestication rates in wild wheats and barley under primitive cultivation, and their archaeological implications. *Journal of World Prehistory* 4: 157–222.

Hillman, G.C. and M.S. Davies 1999. Domestication rate in wild wheats and barley under primitive cultivation. In P.C. Anderson (ed.), *Prehistory of Agriculture: New Experimental and Ethnographic Approaches*, pp. 70–102. Monograph 40. Los Angeles: Institute of Archaeology, UCLA.

Hiscock, P., S. O'Connor, J. Balme and T. Maloney 2016, World's earliest ground-edge axe production coincides with human colonisation of Australia. *Australian Archaeology* 82: 2–11.

Hitchcock, G. 2010. Mound-and-ditch taro gardens of the Bensbach or Torassi River area, southwest Papua *New Guinea. The Artefact* 33: 70–90.

Hodder, I. 1999. *The Archaeological Process: An Introduction*. Oxford: Blackwell.

Hodder, I. 2010. *Religion in the Emergence of Civilisation: Çatalhöyük as Case Study*. Cambridge: Cambridge University Press.

Hope, G.S. 1983. The vegetational changes of the last 20,000 years at Telefomin, Papua New Guinea. *Singapore Journal of Tropical Geography* 4: 25–33.

Hope, G.S. 1998. Early fire and forest change in the Baliem Valley, Irian Jaya, Indonesia. *Journal of Biogeography* 25: 453–461.

Hope, G.S. 2009. Environmental change and fire in the Owen Stanley Ranges, Papua New Guinea. *Quaternary Science Reviews* 28: 2261–2276.

Hope, G.S., T.G. Flannery and Boeardi 1993. A preliminary report of changing Quaternary mammal faunas in subalpine New Guinea. *Quaternary Research* 40: 117–126.

Hope, G.S., D. Gillieson and J. Head 1988. A comparison of sedimentation and environmental change in New Guinea shallow lakes. *Journal of Biogeography* 15: 603–618.

Hope, G.S. and J. Golson 1995. Late Quaternary change in the mountains of New Guinea. *Antiquity* 69(Special Number 265): 818–830.

Hope, G.S., J. Golson and J. Allen 1983. Palaeoecology and prehistory in New Guinea. *Journal of Human Evolution* 12: 37–60.

Hope, G.S. and S.G. Haberle 2005. The history of the human landscapes of New Guinea. In A. Pawley, R. Attenborough, J. Golson and R. Hide (eds.), *Papuan Pasts: Cultural, Linguistic and Biological Histories of Papuan-Speaking Peoples*, pp. 541–554. Pacific Linguistics 572. Canberra: RSPAS, ANU.

Hope, G.S. and J.H. Hope 1976a. Man on Mount Jaya. In G.S. Hope, J.A. Peterson, U. Radak and I. Allison (eds.), *The Equatorial Glaciers of New Guinea*, pp. 225–238. Rotterdam: AA Balkena.

Hope, J.H. and G.S. Hope 1976b. Palaeoenvironments for man in New Guinea. In R.L. Kirk and A. Thorne (eds.), *The Origins of the Australasians*, pp. 29–53. Canberra: Australian Institute of Aboriginal Studies.

Horrocks, M., S. Bulmer and R.O. Gardner 2008. Plant microfossils in prehistoric archaeological deposits from Yuku rock shelter, Western Highlands, Papua New Guinea. *Journal of Archaeological Science* 35: 290–301.

Huff, J. 2016. Revisiting NFB: ceramic technology in the eastern highlands of Papua New Guinea at 3200 cal BP. *Archaeology in Oceania* 51(S1): 84–90.

Hughes, I.M. 1977. *New Guinea Stone Age Trade: The Geography and Ecology of Traffic in the Interior*. Terra Australis 3. Canberra: Department of Prehistory, Research School of Pacific Studies, The Australian National University.

Hughes, P.J. 1985. Prehistoric man-induced soil erosion: examples from Melanesia. In I. Farrington (ed.), *Prehistoric Intensive Agriculture in the Tropics*, pp. 393–408. Oxford: British Archaeological Reports, International Series 232, Part I.

Hughes, P.J., T.P. Denham and J. Golson 2017. Kuk Swamp. In J. Golson, T.P. Denham, P.J. Hughes, P. Swadling and J. Muke (eds.), *Ten Thousand Years of Cultivation at Kuk Swamp in the Highlands of Papua New Guinea*, pp. 87–116. Terra Australis 46. Canberra: ANU E Press.

Hughes, P.J., M.E. Sullivan and D. Yok 1991. Human induced erosion in a Highlands catchment in Papua New Guinea: the prehistoric and contemporary records. *Zeitschrift für Geomorphologie Suppl.* 83: 227–239.

Hyndman, D.C. 1984. Ethnobotany of Wopkaimin *Pandanus*: significant Papua New Guinea plant resource. *Economic Botany* 38: 287–303.

Hynes, R. and A.K. Chase 1982. Plants, sites and domiculture: Aboriginal influence on plant communities. *Archaeology in Oceania* 17: 138–150.

Ingold, T. 1996. Growing plants and raising animals: an anthropological perspective on domestication. In D.R. Harris (ed.), *The Origins and Spread of Agriculture and Pastoralism in Eurasia*, pp. 12–24. London: University College London Press.

Ingold, T. 2000. *The Perception of the Environment*. London: Routledge.

Ishige, N. 1977. The world of the highlanders. In Kyoto University Botanical Research Association (ed.), *Collected Reports of the Kyoto University West Irian Scholars Research Group 1963–1964*, pp. 114–135. Tokyo: Kado ta Hideo, Asahi Shimbunsha. [In Japanese.]

Jones, M.J. and T. Brown 2007. Selection, cultivation and reproductive isolation: a reconsideration of the morphological and molecular signals of domestication. In T.P. Denham, J. Iriarte and L. Vrydaghs (eds.), *Rethinking Agriculture: Archaeological and Ethnoarchaeological Perspectives*, pp. 36–49. Walnut Creek: Left Coast Press.

Jones, M.K. and S. Colledge 2001. Archaeobotany and the transition to agriculture. In D.R. Brothwell and A.M. Pollard (eds.), *Handbook of Archaeological Sciences*, pp. 393–401. Chichester: John Wiley and Sons.

Jones, R. 1969. Fire-stick farming. *Australian Natural History* 16: 224–228.

Jones, R. and B. Meehan 1989. Plant foods of the Gidjingal: ethnographic and archaeological perspectives from northern Australia on tuber and seed exploitation. In D.R. Harris and G.C. Hillman (eds.), *Foraging and Farming: The Evolution of Plant Exploitation*, pp. 120–135. London: Unwin Hyman.

Jussuret, S. 2010. Socializing geoarchaeology: Insights from Bourdieu's theory of practice applied to Neolithic and Bronze Age Crete. *Geoarchaeology* 25: 675–708.

Kahlheber, S. and K. Neumann 2007. The development of cultivation in semi-arid West Africa. In T.P. Denham, J. Iriarte and L. Vrydaghs (eds.), *Rethinking Agriculture: Archaeological and Ethnoarchaeological Perspectives*, pp. 320–346. Walnut Creek: Left Coast Press.

Kasprus, A. 1973. *The Tribes of the Middle Ramu and the Upper Keram Rivers, North-East New Guinea*. Bonn: Verlag Des. Anthropos-Institut.

Keim, A.P. 2012. The pandan flora of Foja-Mamberamo Game Reserve and Baliem Valley, Papua-Indonesia. *Reinwardtia* 13: 271–297.

Kennedy, J. 2008. Pacific bananas: complex origins, multiple dispersals? *Asian Perspectives* 47: 75–94.

Kennedy, J. 2009. Bananas and people in the homeland of genus *Musa*: not just pretty fruit. *Ethnobotany Research and Applications* 7: 179–197.

Kennedy, J. and W.C. Clarke 2004. *Cultivated landscapes of the Southwest Pacific*. Resource Management in the Asia-Pacific (RMAP) Working Paper 50. Canberra: RMAP, Australian National University.

Kershaw, A.P., M.B. Bush, G.S. Hope, K-F. Weiss, J.G. Goldammer and R. Sanford 1997. The contribution of humans to past biomass burning in the tropics. In J. Clark, H. Cachier, J.G. Goldammer and B. Stocks (eds.), *Sediment Records of Biomass Burning and Global Change*, pp. 413–442. Berlin: Springer Verlag.

Kershaw, A.P., S. van der Kaars, P. Moss, B. Opdyke, F. Guichard, S. Rule and C. Turney 2006. Environmental change and the arrival of people in the Australian region. *Before Farming* 2006/1, article 2.

Ketan, J. 1998. *An Ethnohistory of Kuk: A Retrospective View of the World Heritage Project, Land Shortage, and Population Pressure Among the Kawelka People of Mount Hagen, Papua New Guinea*. A special NRI report. Boroko: The National Research Institute of Papua New Guinea.

Kirch, P.V. 1997. *The Lapita Peoples*. Oxford: Blackwell.

Kirch, P.V. and D.E. Yen 1982. *Tikopia: The Prehistory and Ecology of a Polynesian Outlier*. B.P. Bishop Museum Bulletin 238. Honolulu: Bishop Museum Press.

Kislev, M.E., A. Hartmann and O. Bar-Yosef 2006. Early domesticated fig in the Jordan Valley. *Science* 312: 1372–1374.

Kistler, L., A. Montenegro, B.D. Smith, J.A. Gifford, R.E. Green, L.A. Newsom and B. Shapiro 2014. Transoceanic drift and the domestication of African bottle gourds in the Americas. *Proceedings of the National Academy of Sciences (USA)* 111: 2937–2941.

Kjaer, A., A. Barford, C. Asmussen and O. Seberg 2004. Investigation of genetic and morphological variation in the sago palm (*Metroxylon sagu*; Arecaceae) in Papua New Guinea. *Annals of Botany* 94: 109–117.

Knighton, D. 1998. *Fluvial Forms and Processes*. London: Edward Arnold.

Kocher Schmid, C. 1991. *Of People and Plants: A Botanical Ethnography of Nokopo Village, Madang and Morobe Provinces, Papua New Guinea*. Basel: Verlag.

Kropotkin, P.A. 1902. *Mutual Aid: A Factor of Evolution*. London: Heinemann.

Ladizinsky, G. 1998. *Plant Evolution Under Domestication*. Dordrecht: Kluwer.

Lambeck, K. and J.C. Chappell 2001. Sea level change through the last glacial cycle. *Science* 292: 679–686.

Lampert, R.J. 1967. Horticulture in the New Guinea Highlands – C14 dating. *Antiquity* XLI: 307–309.

Lampert, R.J. 1970. *Archaeological report of the Minjigina site. Appendix 5 in J.M. Powell, The impact of man on the vegetation of the Mount Hagen region, New Guinea*. Unpublished PhD thesis, Australian National University.

Larson, G., T. Cucchi, M. Fujita, E. Matisoo-Smith, J. Robins, A. Anderson, B. Rolett, M. Spriggs, G. Dolman, T.-H. Kim, N.T.D. Thuy, E. Randi, M. Doherty, R.A. Due, R. Bollt, T. Djubiantono, B. Griffin, M. Intoh, E. Keane, P. Kirch, K.-T. Li, M. Morwood, L.M. Pedriña, P.J. Piper, R.J. Rabett, P. Shooter, G. Van den Bergh, E. West, S. Wickler, J. Yuan, A. Cooper and K. Dobney 2007. Phylogeny and ancient DNA of *Sus* provides insights into neolithic expansion in Island Southeast Asia and Oceania. *Proceedings of the National Academy of Sciences (USA)* 104: 4834–4839.

Larson, G., E. Karlsson, A. Perri, M. Webster, S. Ho, J. Peters, P. Stahl, P. Piper, F. Lingaas, F. Fredholm, K. Comstock, J. Modiano, C. Schelling, A. Agoulnik, P. Leegwater, K. Dobney, J. Vigne, C. Vila, L. Andersson and K. Lindblad-Toh 2012. Rethinking dog domestication by integrating genetics, archaeology, and biogeography. *Proceedings of the National Academy of Sciences (USA)* 109: 8878–8883.

Larson, G., D.R. Piperno, R.G. Allaby, M.D. Purugganan, L. Andersson, M. Arroyo-Kalin, L. Barton, C. Climer Vigueira, T.P. Denham, K. Dobney, A.N. Doust, P. Gepts, M.T. Gilbert, K.J. Gremillion, L. Lucas, L. Lukens, F.B. Marshall, K.M. Olsen, J.C. Pires, P. Richerson, R. Rubio de Casas, O.I. Sanjur, M.G. Thomas and D.Q. Fuller 2014. Current perspectives and the future of domestication studies. *Proceedings of the National Academy of Sciences (USA)* 111: 6139–6146.

Latham, M. 1980. *Données sur les sols et palesols de la Kuk Tea Research Station et sur leur utilisation pour la culture de la patate douce*. Unpublished manuscript, Kuk archive, ANH, RSPAS, ANU.

Latinis, K. 2000. The development of subsistence system models for island Southeast Asia and Near Oceania: the nature and role of arboriculture and arboreal-based economies. *World Archaeology* 32: 41–67.

Lea, D.A.M. 1966. Yam growing in the Maprik area. *Papua New Guinea Agricultural Journal* 18: 5–16.

Leach, H. 1999. Intensification in the Pacific: a critique of the archaeological criteria and their application. *Current Anthropology* 40: 311–321.

Leahy, M.J. 1936. The Central Highlands of New Guinea. *Geographical Journal* 87: 229–262.

Leahy, M.J. and M. Crain 1937. *The Land That Time Forgot: Adventures and Discoveries in New Guinea*. London: Hurst and Blackett.

Leavesley, M.G. 2004. *Trees to the sky: prehistoric hunting in New Ireland, Papua New Guinea*. Unpublished Ph.D. dissertation, Department of Archaeology and Anthropology, Australian National University, Canberra.

Leavesley, M.G., M.I. Bird, L.K. Fifield, P.A. Hausladen, G.M. Santos and M.L. di Tada 2002. Buang Merabak: early evidence for human occupation in the Bismarck Archipelago, Papua New Guinea. *Australian Archaeology* 54: 55–57.

Leavesley, M.G. and U. Troitzsch 2007. A preliminary study into the Lavongai rectilinear earth mounds: An XRD and phytolith analysis. In J. Specht and V. Attenbrow (eds.), *Archaeological Studies of the Middle and Late Holocene, Papua New Guinea*, Part VIII, Technical Reports of the Australian Museum Online 20: 243–254.

Lebot, V. 1999. Biomolecular evidence for plant domestication in Sahul. *Genetic Resources and Crop Evolution* 46: 619–628.

Lebot, V. 2009. *Root and Tuber Crops: Cassava, Sweet Potato, Yams and Aroids*. Crop Production Science in Horticulture, Volume 17. Wallingford: CABI Publishing.

Lebot, V., R. Malapa and K Abraham 2017. The Pacific yam (*Dioscorea nummularia* Lam.), an under-exploited tuber crop from Melanesia. *Genetic Resources and Crop Evolution* 64: 217–235.

Lebot, V., M.S. Prana, N. Kreike, H. van Heck, J. Pardales, T. Okpul, T. Gendua, M. Thongjiem, H. Hue, N. Viet and T.C. Yap 2004. Characterisation of taro (*Colocasia esculenta* (L.) Schott) genetic resources in Southeast Asia and Oceania. *Genetic Resources and Crop Evolution* 51: 381–392.

Lebot, V., B. Trilles, J.L. Noyer and J. Modesto 1998. Genetic relationships between *Dioscorea alata* L. cultivars. *Genetic Resources and Crop Evolution* 45: 499–509.

Lehmann J., D. Kern, B. Glaser and W. Woods 2003. *Amazonian Dark Earths: Origins, Properties and Management*. Dordrecht: Kluwer Press.

Lemonnier, P. 2002, Des vergers de *Pandanus* spp. comme poste avancé de la culture. Notule d'ethnobotanique Baruya. *Journal de la Société des Océanistes* 114–115: 159–164.

Lentfer, C.J. 2003. *Plants, people and landscapes in prehistoric Papua New Guinea: a compendium of Phytolith (and starch) analyses*. Unpublished PhD thesis. Lismore, NSW: Southern Cross University.

Lentfer, C.J. 2009. Tracing domestication and cultivation of bananas from phytoliths: An update from Papua New Guinea. *Ethnobotany Research and Applications* 7: 247–270.

Lentfer, C.J. and T.P. Denham 2017. The archaeobotany of Kuk. In J. Golson, T.P. Denham, P.J. Hughes, P. Swadling and J. Muke (eds.), *Ten Thousand Years of Cultivation at Kuk Swamp in the Highlands of Papua New Guinea*, pp. 163–184. Terra Australis 46. Canberra: ANU E Press.

Lentfer, C.J., C. Pavlides and J. Specht 2010. Natural and human impacts in a 35 000-year vegetation history in central New Britain, Papua New Guinea. *Quaternary Science Reviews* 29: 3750–3767.

Lentfer, C.J. and R. Torrence 2007. Holocene volcanic activity, vegetation succession, and ancient human land use: Unraveling the interactions on Garua Island, Papua New Guinea. *Review of Palaeobotany and Palynology* 143: 83–105.

Lévi-Strauss, C. 1969. *The Elementary Structures of Kinship*, J.H. Bell and J.R. von Sturmer, trans. and R. Needham, ed. Boston: Beacon Press.

Lewis, T., T.P. Denham and J. Golson 2016. A renewed archaeological and archaeobotanical assessment of house sites at Kuk Swamp in the highlands of Papua New Guinea. *Archaeology in Oceania* 51(S1): 91–103.

Lin, E. 2016. *The emergence of rectilinear ditch networks at Kuk Swamp: a soil micromorphological perspective*. Unpublished MArchSci (Advanced) sub-thesis, School of Archaeology and Anthropology, Australian National University.

Loy, T., M. Spriggs and S. Wickler 1992. Direct evidence for human use of plants 28,000 years ago: starch residues on stone artefacts from northern Solomon Islands. *Antiquity* 66: 898–912.

Majnep, I.S. and R.N.H. Bulmer 1977. *Birds of my Kalam Country*. Auckland: Auckland University Press.

Majnep, I.S. and R.N.H. Bulmer 2007. *Animals the Ancestors Hunted*. Belair, South Australia: Crawford House Publishing.

Malapa, R., G. Arnau, J.L. Noyer and V. Lebot 2005. Genetic diversity of the greater yam (*Dioscorea alata* L.) and relatedness to *D. nummularia* Lam. and *D. transversa* Br. as revealed with AFLP markers. *Genetic Resources and Crop Evolution* 52: 919–929.

Mangi, J. 1984. *Manim 2: 10 Years BP: A prehistory of Manim rockshelter, Western Highlands Province, Papua New Guinea*. Unpublished B.Litt. Thesis, Department of Prehistory and Anthropology, Australian National University.

Marshall, F., K. Dobney, T.P. Denham and J. Capriles 2014. Evaluating the roles of directed breeding and gene flow in animal domestication. *Proceedings of the National Academy of Sciences (USA)* 111: 6153–6158.

Martin, F.W. and A.M. Rhodes 1977. Intra-specific classification of *Dioscorea alata*. *Tropical Agriculture (Trinidad)* 54: 1–13.

Marx, K. 1852 (1987). *The Eighteenth Brumaire of Louis Bonaparte*. New York: International Publishers.

Matisoo-Smith, L. 2009. The commensal model for human settlement of the Pacific 10 years – what can we say and where to now? *The Journal of Island and Coastal Archaeology* 4: 151–163.

Matthews, P.J. 1991. A possible wild type taro: *Colocasia esculenta* var. *Aquatalis*. *Indo-Pacific Prehistory Association Bulletin* 11: 69–81.

Matthews, P.J. 1995. Aroids and Austronesians. *Tropics* 4: 105–126.

Matthews, P.J. 2003. Identification of *Benincasa hispida* (wax gourd) from the Kana archaeological site, Western Highlands Province, Papua New Guinea. *Archaeology in Oceania* 38: 186–191.

Matthews, P.J. 2004. Genetic diversity in taro, and the preservation of culinary knowledge. *Ethnobotany Research and Applications* 2: 55–71.

Matthews, P.J. 2014. *On the Trail of Taro: An Exploration of Natural and Cultural History*. Senri Ethnological Studies 88. Osaka: National Museum of Ethnography.

Mbida Mindzie, C., H. Doutrelepont, L. Vrydaghs, R.L. Swennen, R.J. Swennen, H. Beeckman, E. de Langhe and P. de Maret 2001. First archaeological evidence of banana cultivation in central Africa during the third millennium before present. *Vegetation History and Archaeobotany* 10: 1–6.

McApline, J.R., G. Keig and R. Falls 1983. *Climate of Papua New Guinea*. Canberra: CSIRO and ANU Press.

McConnell, K. 1998. The prehistoric use of Chenopodiaceae in Australia: evidence from Carpenter's Gap shelter 1 in the Kimberley, Australia. *Vegetation History and Archaeobotany* 7: 179–188.

McConnell, K. and S. O'Connor 1997. 40,000 years of food plants in the Southern Kimberley Ranges, Western Australia. *Australian Archaeology* 45: 20–31.

Mignouna, H.D. and A. Dansi 2003. Yam (*Dioscorea* spp.) domestication by the Nago and Fon ethnic groups in Benin. *Genetic Resources and Crop Evolution* 50: 519–528.

Modjeska, C.M. 1977. *Production among the Duna: aspects of horticultural intensification in central New Guinea*. Unpublished PhD thesis, Australian National University.

Modjeska, C.M. 1982. Production and inequality: perspectives from central New Guinea. In A. Strathern (ed.), *Inequality in New Guinea Highland Societies*, pp. 50–108. Cambridge: Cambridge University Press.

Moore, A.M.T., G.C. Hillman and A.J. Legge (eds.) 2000. *Village on the Euphrates*. Oxford: Oxford University Press.

Morgan, R.P.C. 1995. *Soil Erosion and Conservation*. Harlow: Longman.

Morren, G.E.B. 1986. *The Miyanmin: Human Ecology of a Papua New Guinea Society.* Ann Arbor: UMI Research Press.

Mountain, M-J. 1991a. *Highland New Guinea hunter-gatherers: the evidence of Nombe Rockshelter, Simbu, with emphasis on the Pleistocene.* Unpublished PhD thesis, Australian National University.

Mountain, M-J. 1991b. Landscape use and environmental management of tropical rainforest by pre-agricultural hunter-gatherers in northern Sahulland. *Bulletin of the Indo-Pacific Prehistory Association* 11: 54–68.

Moy, C.M., G.O. Seltzer, D.T. Rodbell and D.M. Anderson 2002. Variability of El Niño/Southern oscillation activity at millennial timescales during the Holocene epoch. *Nature* 420: 162–165.

Muke, J. 1984. *The Huon discoveries: a preliminary report on the stone artefacts and a comparative analysis of the distribution of waisted tools in Greater New Guinea.* Unpublished B.Litt thesis, Department of Prehistory and Anthropology, Australian National University.

Muke, J., T.P. Denham and V. Genorupa 2007. Nominating and managing a World Heritage Site in the highlands of Papua New Guinea. *World Archaeology* 39: 324–338.

Muke, J. and H. Mandui 2003. In the shadows of Kuk: evidence for prehistoric agriculture at Kana, Wahgi Valley, Papua New Guinea. *Archaeology in Oceania* 38: 177–185.

Murdock, G.P. 1959. *Africa: Its Peoples and Their Culture History.* New York: McGraw-Hill.

Neumann, K. 2003. New Guinea: a cradle of agriculture. *Science* 301: 180–181.

O'Connell, J.F. and J. Allen 2007. Pre-LGM Sahul (Australia-New Guinea) and the archaeology of early modern humans. In P. Mellars, K. Boyle, O. Bar-Yosef and C. Stringer (eds.), *Rethinking the Human Revolution*, pp. 395–410. Cambridge: McDonald Institute for Archaeological Research.

O'Connor, S., A. Barham, K. Aplin, K. Dobney, A. Fairbairn and M. Richards 2011. The power of paradigms: examining the evidential basis for early to mid-Holocene pigs and pottery in Melanesia. *Journal of Pacific Archaeology* 2: 1–25.

O'Connor, S., R. Ono and C. Clarkson 2011. Pelagic fishing at 42,000 years before present and the maritime skills of modern humans. *Science* 334: 1117–1121.

O'Garra, A. 1981. *A palaeoecological study of recent environmental change in the Highlands of Papua New Guinea.* Unpublished PhD thesis, University of Liverpool.

Oldfield, F. 1988. Magnetic and elemental analysis of recent lake sediments from the Highlands of Papua New Guinea. *Journal of Biogeography* 15: 529–553.

Oldfield, F., P.G. Appleby and R. Thompson 1980. Palaeoecological studies of lakes in the Highlands of Papua New Guinea I. The chronology of sedimentation. *Journal of Ecology* 68: 457–477.

Oldfield, F., A.T. Worsley and A.F. Baron 1985. Lake sediments and evidence for agricultural intensification: a case study from the highlands of Papua New Guinea. In I. Farrington (ed.), *Prehistoric Intensive Agriculture in the Tropics*, pp. 385–391. Oxford: British Archaeological Reports, International Series 232, Part I.

Pain, C.F. and R.J. Blong 1976. Late Quaternary tephras around Mount Hagen and Mount Giluwe, Papua New Guinea. In R.W. Johnson (ed.), *Volcanism in Australasia*, pp. 239–251. Elsevier: Amsterdam.

Pain, C.F., C.J. Pigram, R.J. Blong and G.O. Arnold 1987. Cainozoic geology and geomorphology of the Wahgi Valley, central highlands of Papua New Guinea. *BMR Journal of Australian Geology and Geophysics* 10: 267–275.

Pasveer, J.M. 2004. *The Djief Hunters: 26,000 Years of Rainforest Exploitation on the Bird's Head of Papua, Indonesia.* Modern Quaternary Research in Southeast Asia 17. Leiden: A.A. Balkema Publishers.

Pauketat, T.R. 2001. Practice and theory in archaeology: an emerging paradigm. *Anthropological Theory* 1: 73–98.

Paull, R.E., C-S. Tang, K. Gross and G. Uruu 1999. The nature of the taro acridity factor. *Postharvest Biology and Technology* 16: 71–78.

Pearsall, D.M. 2007. Modelling prehistoric agriculture through the palaeoenvironmental record: theoretical and methodological issues. In T.P. Denham, J. Iriarte and L. Vrydaghs (eds.), *Rethinking Agriculture: Archaeological and Ethnoarchaeological Perspectives*, pp. 210–230. Walnut Creek: Left Coast.

Perrier, X., E. De Langhe, M. Donohue, C. Lentfer, L. Vrydaghs, F. Bakry, F. Carreel, I. Hippolyte, J-P. Horry, C. Jenny, V. Lebot, A-M. Risterucci, K. Tomekpe, H. Doutrelepont, T. Ball, J. Manwaring, P. de Maret and T.P. Denham 2011. Multidisciplinary perspectives on banana (*Musa* spp.) domestication. *Proceedings of the National Academy of Sciences (USA)* 108: 11311–11318.

Perry, L. and K.V. Flannery 2007. Precolumbian use of chilli peppers in the valley of Oaxaca, Mexico. *Proceedings of the National Academy of Sciences (USA)* 104: 11905–11909.

Piper, P.J., H. Hung, F.Z. Campos, P. Bellwood and R. Santiago 2009. A 4000 year-old introduction of domestic pigs into the Philippine Archipelago: implications for understanding routes of human migration through Island Southeast Asia and Wallacea. *Antiquity* 83: 687–695.

Piperno, D.R. 1989. Non-affluent foragers: resource availability, seasonal shortages, and the emergence of agriculture in Panamanian tropical forests. In D.R. Harris and G.C. Hillman (eds.), *Foraging and Farming: The Evolution of Plant Exploitation*, pp. 538–554. London: Unwin Hyman.

Piperno, D.R. 1998. Paleoethnobotany in the Neotropics from microfossils: new insights into ancient plant use and agricultural origins in the tropical forest. *Journal of World Prehistory* 12: 393–449.

Piperno, D.R. 2006. *Phytoliths: A Comprehensive Guide for Archaeologists and Paleoecologists.* Lanham: AltaMira.

Piperno, D.R. and I. Holst 1998. The presence of starch grains on prehistoric stone tools from the humid neotropics: indications of early tuber use and agriculture in Panama. *Journal of Archaeological Science* 25: 765–777.

Piperno, D.R. and D.M. Pearsall 1998. *The Origins of Agriculture in the Lowland Neotropics.* San Diego: Academic Press.

Piperno, D.R., J.A. Ranere, I. Holst and P. Hansell 2000. Starch grains reveal early root crop horticulture in the Panamanian tropical forest. *Nature* 407: 894–897.

Plarre,W. 1995. Evolution and variability of special cultivated crops in the highlands of West New Guinea (Irian Jaya) under present Neolithic conditions. *Plant Genetic Resources Newsletter* 103: 1–13.

Powell, J.M. 1970a. *The impact of man on the vegetation of the Mount Hagen region, New Guinea*. Unpublished PhD thesis, Australian National University.

Powell, J.M. 1970b. The history of agriculture in the New Guinea Highlands. *Search* 1: 199–200.

Powell, J.M. 1974. A note on wooden gardening implements of the Mount Hagen region, New Guinea. *Records of the PNG Museum* 4: 21–28.

Powell, J.M. 1976. Vegetation. In K. Paijmans (ed.), *New Guinea Vegetation*, pp. 23–105. Canberra: CSIRO and ANU Press.

Powell, J.M. 1980. Studies of New Guinea vegetation history. *Proceedings of the IVth International Palynology Conference, Lucknow (1976–7)* 3: 11–20.

Powell, J.M. 1981. The origins of agriculture in New Guinea. *Proceedings of the IVth International Palynology Conference, Lucknow (1976–7)* 3: 295–310.

Powell, J.M. 1982a. Plant resources and palaeobotanical evidence for plant use in the Papua New Guinea Highlands. *Archaeology in Oceania* 17: 28–37.

Powell, J.M. 1982b. The history of plant use and man's impact on the vegetation. In J.L. Gressitt (ed.), *Biogeography and Ecology of New Guinea. Volume 1*, pp. 207–227. The Hague: Junk.

Powell, J.M. 1982c. *A preliminary report on Lake Ambra Vegetation History Site, Mount Hagen*. Unpublished manuscript, Kuk archive, ANH, RSPAS, ANU.

Powell, J.M. 1984. *Ecological and palaeoecological studies at Kuk I: below the grey clay*. Unpublished manuscript, Kuk archive, ANH, RSPAS, ANU.

Powell, J.M. and S. Harrison 1982. *Haiyapugua: Aspects of Huli Subsistence and Swamp Cultivation*. Occasional Paper No. 1 [New Series]. Port Moresby: Department of Geography, UPNG.

Powell, J.M., A. Kulunga, R. Moge, C. Pono, F. Zimike and J. Golson 1975. *Agricultural Traditions in the Mount Hagen Area*. Occasional Paper No. 12. Port Moresby: Department of Geography, UPNG.

Pretty, G.L. 1965. Two stone pestles from western Papua and their relationship to prehistoric pestles and mortars from New Guinea. *Records of the South Australian Museum* 15: 119–130.

Ramanatha, R.V., P.J. Matthews, P.B. Eyzaguirre and D. Hunters (eds.) 2010. *The Global Diversity of Taro: Ethnobotany and Conservation*. Rome: Bioversity International.

Rappaport, R. 1968. *Pigs for the Ancestors: Ritual in the Ecology of a New Guinea People*. New Haven: Yale University Press.

Read, J., G.S. Hope and R. Hill 1990. The dynamics of some Nothofagus-dominated rainforests in Papua New Guinea. *Journal of Biogeography* 17: 185–204.

Reimer, P.J., M.G.L. Baillie, E. Bard, A. Bayliss, J.W. Beck, P.G. Blackwell, C. Bronk Ramsey, C.E. Buck, G.S. Burr, R.L. Edwards, M. Friedrich, P.M. Grootes, T.P. Guilderson, I. Hajdas, T.J. Heaton, A.G. Hogg, K.A. Hughen, K.F. Kaiser, K.F. Kromer, F.G. McCormac, S.W. Manning, R.W. Reimer, D.A. Richards, J.R. Southon, S. Talamo, C.S.M. Turney, J. van der Plicht and C.E. Weyhenmeyer 2009. IntCal09 and Marine09 radiocarbon age calibration curves, 0–50,000 years cal BP. *Radiocarbon* 51: 1111–1150.

Renfrew, C. 1973. *Before Civilization: The Radiocarbon Revolution and Prehistoric Europe*. Harmondsworth: Penguin.

Renfrew, C. 2002. "The emerging synthesis": The archaeogenetics of farming/language dispersals and other spread zones. In P. Bellwood and C. Renfrew (eds.), *Examining the Farming/Language Dispersal Hypothesis*, pp. 3–16. Cambridge: McDonald Institute for Archaeological Research.

Richerson, P.J., R. Boyd and R.L. Bettinger 2001. Was agriculture impossible during the Pleistocene but mandatory during the Holocene? A climate change hypothesis. *American Antiquity* 66: 387–411.

Riebe, I. 1974. '…*And then we killed': an attempt to understand the fighting history of the upper Kaironk Valley Kalam from 1914–1962*. Unpublished MA thesis. Sydney: Department of Anthropology, University of Sydney.

Roberts, R.G., T.F. Flannery, L.K. Ayliffe, H. Yoshida, J.M. Olley, G.J. Prideaux, G.M. Laslett, A. Baynes, M.A. Smith, R. Jones and B.L. Smith 2001. New ages for the last Australian megafauna: continent-wide extinction about 46,000 years ago. *Science* 292: 1888–1892.

Roscoe, P. 2002. The hunters and gatherers of New Guinea. *Current Anthropology* 43: 153–162.

Roullier, C., L. Benoit, D.B. McKey and V. Lebot 2013. Historical collections reveal patterns of diffusion of sweet potato in Oceania obscured by modern plant movements and recombination. *Proceedings of the National Academy of Sciences (USA)* 110: 2205–2210.

Ruan, J., F. Kherbouche, D. Genty, D. Blamart, H. Cheng, E. Dewilde, S. Hachi, R.L. Edwards, E. Regnier and J-L. Michelot 2016. Evidence of a prolonged drought ca. 4200 yr BP correlated with prehistoric settlement abandonment from the Gueldaman GLD1 Cave, Northern Algeria. *Climate of the Past* 12: 1–14.

Ruddiman W.F., E.C. Ellis J.O. Kaplan and D.Q. Fuller 2015. Defining the epoch we live in. *Science* 348: 38–39.

Sahlins, M. 1967. Poor man, rich man, big-man, chief: political types in Melanesia and Polynesia. *Comparative Studies in Society and History* 5: 285–303.

Salisbury, R.F. 1962. *From Stone to Steel: Economic Consequences of a Technological Change in New Guinea*. Melbourne: Melbourne University Press.

Sauer, C.O. 1952. *Agricultural Origins and Their Dispersals*. New York: American Geographical Society.

Sayer, A. 1984. *Method in Social Science: A Realist Approach*. London: Hutchinson.

Scarre, C. 1988. *Past Worlds: The Times Atlas of Archaeology*. London: The Times.

Schiffer, M.B. 1987. *Formation Processes of the Archaeological Record*. Albuquerque: University of Mexico Press.

Serpenti, L.M. 1965. *Cultivators in the Swamps: Social Structure and Horticulture in a New Guinea Society*. Assen, The Netherlands: Van Gorcum.

Shuji, Y. and Matthews, P.J. (eds.) 2002. *Proceedings of the International Area Studies Conference VII: Vegeculture in Eastern Asia and Oceania*. JCAS Symposium Series No. 16. Osaka: National Museum of Ethnology.

Sillitoe, P. 1983. *Roots of the Earth*. Kensington: UNSW Press.

Sillitoe, P. 1996. *A Place Against Time: Land and Environment in the Papua New Guinea Highlands*. Amsterdam: Harwood Academic Publishers.

Sillitoe, P. 2002. Always been farmer-foragers? Hunting and gathering in the Papua New Guinea highlands. *Anthropological Forum* 12: 45–76.

Simmonds, N.W. 1976a. Sugarcanes: Saccharum (Gramineae-Andropogoneae). In N.W. Simmonds (ed.), *Evolution of Crop Plants*, pp. 104–108. London: Longman.

Simmonds, N.W. 1976b. Bananas: Musa (Musaceae). In N.W. Simmonds (ed.), *Evolution of Crop Plants*, pp. 211–215. London: Longman.

Smith, B.D. 1992. *Rivers of Change: Essays on Early Agriculture in the Eastern North America*. Washington, DC: Smithsonian Books.

Smith, B.D. 1998a. Between foraging and farming. *Science* 279: 1651–1652.

Smith, B.D. 1998b. *The Emergence of Agriculture*. New York: Scientific American Library.

Smith, B.D. 2001. Low-level food production. *Journal of Archaeological Research* 9: 1–43.

Smith, B.D. 2012. A cultural niche construction theory of initial domestication. *Biological Theory* 6: 260–271.

Smith, M.A., M.J. Bird, C.S.M. Turney, L.K. Fifield, G.M. Santos, P.A. Hausladen and M.L. di Tada 2001. New ABOX AMS-14C ages remove dating anomalies at Puritjarra rock shelter. *Australian Archaeology* 53: 45–47.

Sniderman, J.M.K., J. Finn and T.P. Denham 2009. A late Holocene palaeoecological record from Ambra Crater in the highlands of Papua New Guinea and implications for agricultural history. *The Holocene* 19: 449–458.

Souter, G. 1963. *New Guinea: The Last Unknown*. Sydney: Angus and Robertson.

Specht, J. 2003. On New Guinea hunters and gatherers. *Current Anthropology* 44: 209.

Specht, J., T.P. Denham, J. Goff and J.E. Terrell 2014. Deconstructing the Lapita cultural complex in the Bismarck archipelago. *Journal of Archaeological Research* 22: 89–140.

Spriggs, M. 1996. Early agriculture and what went before in Island Melanesia: continuity or intrusion? In D.R. Harris (ed.), *The Origins and Spread of Agriculture and Pastoralism in Eurasia*, pp. 524–537. London: University College London Press.

Spriggs, M. 1997. *The Island Melanesians*. Oxford: Blackwell.

Staubwasser, M., F. Sirocko, P. Grootes and M. Segl 2003. Climate change at the 4.2 ka BP termination of the Indus valley civilization and Holocene south Asian monsoon variability. *Geophysical Research Letters* 30: 1–4.

Steensberg, A. 1980. *New Guinea Gardens: A Study of Husbandry With Parallels in Prehistoric Europe*. London: Academic Press.

Stone, B.C. 1982a. New Guinea Pandanaceae: first approach to ecology and biogeography. In J.L. Gressitt (ed.), *Biogeography and Ecology of New Guinea*, pp. 401–436. The Hague: W. Junk.

Stone, B.C. 1982b. *Agronomic study of Karuka*. Unpublished report prepared for the Department of Primary Industry, Government of Papua New Guinea.

Stone, B.C. 1984. Pandanus from Ok Tedi Region, Papua New Guinea, collected by Debra Donoghue. *Economic Botany* 38: 304–313.

Stoops, G. 1989. Relict properties in soils of humid tropical regions with special reference to Central Africa. In A. Bronger and J.A. Catt (eds.), *Paleopedology: Nature and Application of Palaeosols*, Catena Supplement 16: 95–106.

Stover, R.H. and N.W. Simmonds (eds.) 1987. *Bananas*, 3rd edition. Tropical Agriculture Series. Harlow: Longman.

Strathern, A. 1969. Why is the pueraria a sweet potato? *Ethnology* 8: 189–198.

Strathern, A. 1971. *The Rope of Moka: Big-Men and Ceremonial Exchange in Mount Hagen, New Guinea*. Cambridge: Cambridge University Press.

Strathern, A. 1972. *One Father, One Blood: Descent and Group Structure Among the Melpa People*. Canberra: Australian National University Press.

Stringer, C. 2000. Coasting out of Africa. *Nature* 327: 293–297.

Stuiver, M. and P.J. Reimer 1993. Extended 14C data base and revised CALIB 3.0 14 C calibration program. *Radiocarbon* 35: 215–230.

Stuiver, M., P. Stuiver, E. Bard, J.W. Beck, G.S. Burr, K.A. Hughen, G. Kromer, G. McCormac, J. van der Plicht and M. Spurk 1998. INTCAL98 radiocarbon age calibration, 24,000–0 cal BP. *Radiocarbon* 40: 1041–1083.

Sullivan, M.E. and P.J. Hughes 1986. The geomorphic setting of prehistoric garden terraces in the Eastern Highlands of Papua New Guinea. In V. Gardiner (ed.), *International Geomorphology Part II*, pp. 569–582. Chichester: John Wiley and Sons.

Sullivan, M.E. and P.J. Hughes 1991. Patterned micro-relief grasslands in cold upland valleys in Papua New Guinea. *Zeitschrift für Geomorphologie Suppl.* 83: 83–94.

Sullivan, M.E., P.J. Hughes and J. Golson 1986. Prehistoric engineers of the Arona Valley. *Science in New Guinea* 12: 27–41.

Sullivan, M.E., P.J. Hughes and J. Golson 1987. Prehistoric garden terraces in the Eastern Highlands of Papua New Guinea. *Tools and Tillage* V(4): 199–213, 260.

Summerhayes, G. 2003. Modelling differences between Lapita obsidian and pottery distribution patterns in the Bismarck Archipelago. In C. Sand (ed.), *Pacific Archaeology: Assessments and Prospects*, pp. 139–149. Nouméa: Le Cahiers de l'Archéologie en Nouvelle-Calédonie 15.

Summerhayes, G. and J. Allen 1993. The transport of Mopir obsidian of late Pleistocene New Ireland. *Archaeology in Oceania* 28: 144–148.

Summerhayes, G.R., M. Leavesley and A. Fairbairn 2009. Impact of human colonization on the landscape: A view from the western Pacific. *Pacific Science* 63: 725–745.

Summerhayes, G.R., M. Leavesley, A.S. Fairbairn, H. Mandui, J. Field, A. Ford and R. Fullagar 2010. Human adaptation and plant use in highland New Guinea 49,000 to 44,000 years ago. *Science* 330: 78–81.

Sutton, A., M-J. Mountain, K. Aplin, S. Bulmer and T.P. Denham 2009. Archaeozoological records for the highlands of New Guinea: a review of current evidence. *Australian Archaeology* 69: 41–58.

Swadling, P. 1973. *The Human Settlement of the Arona Valley, Eastern Highlands Province, Papua New Guinea*. Port Moresby: University of Papua New Guinea (for the Electricity Commission).

Swadling, P. 1983. *How long have people been in the ok Tedi impact region?* National Museum Record No. 8. Boroko: PNG National Museum and Art Gallery.

Swadling, P., N. Araho and B. Ivuyo 1991. Settlements associated with the inland Sepik-Ramu Sea. *Bulletin of the Indo-Pacific Prehistory Association* 11: 92–110.

Swadling, P., J. Chappell, G. Francis, N. Araho and B. Ivuyo 1989. A Late Quaternary inland sea and early pottery in Papua New Guinea. *Archaeology in Oceania* 24: 106–109.

Swadling, P. and R. Hide 2005. Changing landscapes and social interaction: looking at agricultural history from a Sepik – Ramu perspective. In A. Pawley, R. Attenborough, J. Golson and R. Hide (eds.), *Papuan Pasts: Cultural, Linguistic and Biological Histories of Papuan-Speaking Peoples*, pp. 289–328. Pacific Linguistics 572. Canberra: RSPAS, ANU.

Swadling, P. and G. Hope 1992. Environmental change in New Guinea since human settlement. In J. Dodson (ed.), *The Naive Lands: Prehistory and Environmental Change in Australia and the Southwest Pacific*, pp. 13–42. Melbourne: Longman Cheshire.

Symon, D.E. 1985. The Solanaceae of New Guinea. *Journal of the Adelaide Botanic Gardens* 8: 1–171.

Szabó, K. and S. O'Connor 2004. Migration and complexity in Holocene Island Southeast Asia. *World Archaeology* 36: 621–628.

Tanno, K.-I. and G. Willcox 2006. How fast was wild wheat domesticated? *Science* 311: 1886.

Telban, B. 1988. Firemaking and stone mortar use in the Upper Yuat, Papua New Guinea. *Canberra Anthropology* 11: 31–43.

Terrell, J.E. 2002. Tropical agroforestry, coastal lagoons and Holocene prehistory in Greater Near Oceania. In Y. Shuji and P.J. Matthews (eds.), *Proceedings of the International Area Studies Conference VII: Vegeculture in Eastern Asia and Oceania*, pp. 195–216. Osaka: National Museum of Ethnology.

Terrell, J.E. 2004. Introduction: 'Austronesia' and the great Austronesian migration. *World Archaeology* 36: 586–590.

Terrell, J.E., J.P. Hart, S. Barut, N. Cellinese, A. Curet, T.P. Denham, H. Haines, C.M. Kusimba, K. Latinis, R. Oka, J. Palka, M.E.D. Pohl, K.O. Pope, J.E. Staller and P.R. Williams 2003. Domesticated landscapes: The subsistence ecology of plant and animal domestication. *Journal of Archaeological Method and Theory* 10: 323–368.

Tharp, T.M. 1984. Sediment characteristics and stream competence in ephemeral and intermittent streams, Fairborn, Ohio. In A.P. Schick (ed.), *Channel Processes – Water, Sediment, Catchment Controls*, pp. 121–136. CATENA Supplement 5. Brannschweig: Catena-Verlag.

Thomas, J. 1996a. The cultural context of the first use of domesticates in continental Central and Northwest Europe. In D.R. Harris (ed.), *The Origins and Spread of Agriculture and Pastoralism in Eurasia*, pp. 310–322. London: University College London Press.

Thomas, J. 1996b. *Time, Culture and Identity: An Interpretive Archaeology*. London and New York: Routledge.

Thomas, J. 1999. *Understanding the Neolithic*. London: Routledge.

Thompson, R. and F. Oldfield 1986. *Environmental Magnetism*. London: Allen and Unwin.

Thomson, V.A., O. Lebrasseur, J.J. Austin, T. Hunt, D. Burney, T.P. Denham, N.J. Rawlence, J.R. Wood, J. Gongora, L.G. Flink, A. Linderholm, K. Dobney, G. Larson and A. Cooper 2014. Using ancient DNA to study the origins and dispersal of ancestral Polynesian chickens and human migration across the Pacific. *Proceedings of the National Academy of Sciences (USA)* 111: 4826–4831.

Torrence, R. and H. Barton (eds.) 2006. *Ancient Starch Research*. Walnut Creek: Left Coast Press.

Torrence, R. and P. Swadling 2008. Social networks and the spread of Lapita. *Antiquity* 82: 600–616.

Trafford, B.D. 1975. Drainage design. In A.J. Thomasson (ed.), *Soils and Field Drainage*, pp. 5–17. Soil Survey Technical Monograph No. 7. Harpendon: Soil Survey of England and Wales.

Trigger, B.G. 2007. *A History of Archaeological Thought*, 2nd edition. Cambridge: Cambridge University Press.

Tudhope, A.W., C.P. Chilcott, M.T. McCulloch, E.R. Cook, J. Chappell, R.M. Ellam, D.W. Lea, J.M. Lough and G.B. Shimmield 2001. Variability in the El Niño – Southern Oscillation through a glacial-interglacial cycle. *Science* 291: 1511–1517.

Turney, C.S.M., M.I. Bird, L.K. Fifield, R.G. Roberts, M. Smith, C.E. Dortch, R. Grün, E. Lawson, L.K. Ayliffe, G.H. Miller, J. Dortch and R.G. Cresswell 2001. Early human occupation at Devil's Lair, Southwestern Australia 50,000 years ago. *Quaternary Research* 55: 3–13.

Turney, C.S.M., P. Kershaw, S. Clements, N. Branch, P. Moss and L.K. Fifield 2004. Millennial and orbital variations of El Niño/Southern Oscillation and high-latitude climate in the last glacial period. *Nature* 428: 306–310.

Tworek-Matuszkiewicz, B. 2001. The Ambun Stone. *Artonview* (winter): 59.

Vavilov, N.I. 1992. *Origin and Geography of Cultivated Plants*. D. Löve, trans. and ed. Cambridge: Cambridge University Press.

Veth, P. 1993. *Islands in the Interior: The Dynamics of Prehistoric Adaptations Within the Arid Zone of Australia*. Archaeological Series 3. Ann Arbor, MI: International Monographs in Prehistory.

Vigilante, T. and D.M.J.S. Bowman 2004. Effects of individual fire events on the flower production of fruit-bearing tree species, with reference to Aboriginal people's management and use, at Kalumburu, North Kimberley, Australia. *Australian Journal of Botany* 52: 405–415.

Waddell, E. 1972. *The Mound Builders: Agricultural Practices, Environment and Society in the Central Highlands of New Guinea*. Seattle and London: University of Washington Press.

Waddell, E. 1973. Raiapu Enga adaptive strategies: structure and implications. In H.C. Brookfield (ed.), *The Pacific in Transition*, pp. 25–54. London: Edward Arnold.

Walker, D. and J.R. Flenley 1979. Late Quaternary vegetational history of the Enga Province of upland Papua New Guinea. *Philosophical Transactions of the Royal Society of London* 286: 265–344.

Watson, J.B. 1964. A previously unreported root crop from the New Guinea highlands. *Ethnology* 3: 1–5.

Watson, J.B. 1965. From hunting to horticulture in the New Guinea highlands. *Ethnology* 4: 295–309.

Watson, J.B. 1968. Pueraria: names and traditions of a lesser crop of the central highlands, New Guinea. *Ethnology* 7: 268–279.

Watson, V.D. and J.D. Cole 1977. *Prehistory of the Eastern Highlands of New Guinea*. Seattle: University of Washington Press.

Weiss, E., M.E. Kislev and A. Hartmann 2006. Autonomous cultivation before domestication. *Science* 312: 1608–1610.

Weissner, P. 2005. Social, symbolic, and ritual roles of the sweet potato in Enga, from its introduction until first contact. In C. Ballard, P. Brown, R.M. Bourke and T. Harwood (eds.), *The Sweet Potato in the Pacific: A Reappraisal*, pp. 121–130. Ethnology Monographs 19 and Oceania Monograph 56. Pittsburgh and Camperdown: University of Pittsburgh and University of Sydney.

Whistler, W.A. 1990. The other Polynesian gourd. *Pacific Science* 44: 115–122.

White, J.P. 1967. *Taim bilong bipo: investigations towards a prehistory of the Papua New Guinea Highlands*. Unpublished PhD thesis, Australian National University.

White, J.P. 1972. *Ol Tumbuna: Archaeological Excavations in the Eastern Central Highlands, Papua New Guinea*. Terra Australis 2. Canberra: RSPAS, ANU.

White, J.P., K.A.W. Crook and B.P. Ruxton 1970, Kosipe: a late Pleistocene site in the Papua Highlands. *Proceedings of the Prehistoric Society* 36: 152–170.

Wickler, S. 2001. *The Prehistory of Buka: A Stepping Stone Island in the Northern Solomons*. Terra Australis 16. Canberra: RSPAS, ANU.

Wilde, W.J.J.O. and B.E.E. Duyfjes 2010. Cucurbitaceae. *Flora Malesiana* 19.

Willis, K.J., L. Gillson and T.M. Brncic 2004. How 'virgin' is virgin rainforest? *Science* 304: 402–403.

Wilson, S.M. 1985. Phytolith evidence from Kuk, an early agricultural site in New Guinea. *Archaeology in Oceania* 20: 90–97.

Wittfogel, K. 1957. *Oriental Despotism: A Comparative Study of Total Power*. New Haven: Yale University Press.

Wong, S., R. Kiew, G. Argent, O. Set, S.K. Lee and Y.Y. Gan 2002. Assessment of the validity of the sections in *Musa* (Musaceae) using ALFP. *Annals of Botany* 90: 231–238.

Wood, A.W. 1987. The humic brown soils of the Papua New Guinea Highlands: a reinterpretation. *Mountain Research and Development* 7: 145–156.

Worsley, A.T. and F. Oldfield 1988. Palaeoecological studies of three lakes in the highlands of Papua New Guinea II. Vegetational history over the last 1600 years. *Journal of Ecology* 76: 1–18.

Wright, D., T.P. Denham, D. Shine and M. Donohue 2013. An archaeological review of western New Guinea. *Journal of World Prehistory* 26: 25–73.

Wroe, S., J. Field, M. Archer, D.K. Grayson, G.J. Price, J. Louys, J.T. Faith, G.E. Webb, I. Davidson and S.D. Mooney 2013. Climate change frames debates over extinction of megafauna in Sahul (Pleistocene Australia New Guinea). *Proceedings of the National Academy of Sciences (USA)* 110: 8777–8781.

Yang, X., H. Barton, Z. Wan, Q. Li, Z. Ma, M. Li, D. Zhang and J. Wei 2013. Sago-type palms were an important plant food prior to rice in southern tropical China. *PLoS One*: e63148. doi: 10.1371/journal.pone.0063148. Print 2013.

Yen, D.E. 1973. The origins of Oceanic agriculture. *Archaeology and Physical Anthropology in Oceania* 8: 68–85.

Yen, D.E. 1974. *The Sweet Potato and Oceania: An Essay in Ethnobotany*. Honolulu: Bishop Museum Bulletin 236.

Yen, D.E. 1982. The history of cultivated plants. In R.J. May and H. Nelson (eds.), *Melanesia: Beyond Diversity*, pp. 281–296. Canberra: Research School of Pacific Studies, Australian National University.

Yen, D.E. 1985. Wild plants and domestication in Pacific islands. In V.N. Misra and P. Bellwood (eds.), *Recent Advances in Indo-Pacific Prehistory*, pp. 315–326. New Delhi: Oxford and IBH Publishing.

Yen, D.E. 1989. The domestication of environment. In D.R. Harris and G.C. Hillman (eds.), *Foraging and Farming: The Evolution of Plant Exploitation*, pp. 55–75. London: Unwin Hyman.

Yen, D.E. 1990. Environment, agriculture and the colonisation of the Pacific. In D.E. Yen and J.M.J. Mummery (eds.), *Pacific Production Systems: Approaches to Economic Prehistory*, pp. 258–277. Canberra: RSPAS, ANU.

Yen, D.E. 1991. Domestication: the lessons from New Guinea. In A. Pawley (ed.), *Man and a Half: Essays in Pacific Anthropology and Ethnobiology in Honour of Ralph Bulmer*, pp. 558–569. Auckland: The Polynesian Society.

Yen, D.E. 1995. The development of Sahul agriculture with Australia as bystander. *Antiquity* 69, Special Number 265: 831–847.

Yen, D.E. 1996. Melanesian aboriculture: historical perspectives with emphasis on the genus Canarium. In B.R. Evans, R.M. Bourke and P. Ferrar (eds.), *South Pacific Indigenous Nuts*, pp. 36–44. Canberra: ACIAR.

Yen, D.E. 1998. Subsistence to commerce in Pacific agriculture: some four thousand years of plant exchange. In H.D.V. Pendergast, N.L. Etkin, D.R. Harris and P.J. Houghton (eds.), *Plants for Food and Medicine*, pp. 161–183. Kew: Royal Botanic Gardens.

Yen, D.E. 2003. The nature of domestication in agricultural formation. Paper presented at *World Archaeological Congress 5*, Washington, DC.

Zeder, M., D. Bradley, E. Emschwiller and B.D. Smith (eds.) 2006. *Documenting Domestication*. Berkeley: University of California Press.

Zerega, N.J.C., D. Ragone and T.J. Motley 2004. Complex origins of breadfruit (*Artocarpus altilis*, Moraceae): implications for human migrations in the Pacific. *American Journal of Botany* 91: 760–766.

Zerega, N.J.C., D. Ragone and T.J. Motley 2006. Breadfruit origins, diversity, and human-facilitated distribution. In T.J. Motley, N. Zerega and H. Cross (eds.), *Darwin's Harvest: New Approaches to the Origins, Evolution, and Conservation of Crops*, pp. 213–238. New York: Columbia University Press.

Zhao, Z. 2011. New archaeobotanic data for the study of the origins of agriculture in China. *Current Anthropology* 52: S295–306.

Zhukovsky, P.M. 1962. *Cultivated Plants and Their Wild Relatives*. P.S. Hudson, Abridged trans. Farnham Royal: Commonwealth Agricultural Bureaux.

Zohary, D., M. Hopf and E. Weiss 2012. *Domestication of Plants in the Old World: The Origin and Spread of Domesticated Plants in Southwest Asia, Europe and the Mediterranean Basin*. Oxford: Oxford University Press.

Zvelebil, M. 1986. *Hunters in Transition: Mesolithic Societies of Temperate Eurasia and the Transition to Farming*. Cambridge: Cambridge University Press.

Zvelebil, M. 1996. The agricultural frontier and the transition to farming in the circum-Baltic region. In D.R. Harris (ed.), *The Origins and Spread of Agriculture and Pastoralism in Eurasia*, pp. 323–345. London: UCL Press.

Index

Note: Page numbers in italics indicate figures and bold indicates tables on the corresponding pages.